Mountain Biodiversity

A Global Assessment

Mountain Biodiversity

A Global Assessment

Edited by Ch. Körner and E.M. Spehn

The Parthenon Publishing Group
International Publishers in Medicine, Science & Technology

A CRC PRESS COMPANY
BOCA RATON   LONDON   NEW YORK   WASHINGTON, D.C.

Library of Congress Cataloging-in-Publication Data
   Mountain biodiversity : a global assessment / [edited by] Christian Körner and Eva Spehn.
     p. cm.
   Includes bibliographical references (p.).
   ISBN 1-84214-091-4 (alk. paper)
   1. Mountain ecology. 2. Biological diversity.
I. Körner, Christian II. Spehn, Eva.
III Title.
QH541.5.M65 M72 2002
577.5′3--dc21
                 2002019191

British Library Cataloguing in Publication Data
   Mountain biodiversity : a global assessment
   1. Mountain ecology 2. Biological diversity
   3. Mountain plants 4. Mountain animals
   I. Korner, Christian II. Spehn, Eva
   578.7′53

   ISBN 1842140914

Published in the USA by
The Parthenon Publishing Group
345 Park Avenue South, 10th Floor
New York, NY 10010, USA

Published in the UK and Europe by
The Parthenon Publishing Group
23–25 Blades Court
Deodar Road
London SW15 2NU, UK

Copyright © 2002 The Parthenon Publishing Group

Typeset by Siva Math Setters, Chennai, India
Printed and bound by Bookcraft (Bath) Ltd., Midsomer Norton, UK

Contents

Preface

What is a mountain? Even in the context of this assessment, views may differ widely on this. For some, it is the 8000-m tops of the Himalayas, for others a 100-m hill in a coastal area. For some, a mountain starts where the cable car starts, for others it is where rivers shed their sediments into the ocean. With the latter concept, the world would consist only of mountains, not an appropriate definition for a new research agenda. An obvious need to focus on a coherent, practical convention led a consortium of scientists during a 1999 planning meeting in Glion, Switzerland to define a high elevation focus of the Global Mountain Biodiversity Assessment (GMBA), as described below.

For this assessment NOT to become a general 'slope' research network, it was agreed that the principal interest of GMBA is threefold:

(1) The alpine life zone above the climatic treeline,
(2) The treeline ecotone itself (the transition from the forest zone to the alpine zone), and
(3) The uppermost mountain forests (close to the natural treeline) or their man-made substitutes (highland pastures).

The terminology used here follows that of the classical literature, and differentiates between the naturally forested montane zone, the treeline ecotone, i.e. the upper end of the closed montane forest (should it still be in place) up to the tree limit, and the treeless alpine life zone. The word 'alpine' has pre-Indogermanic roots, probably coming from Asia, with 'alpo' referring to steep slopes in the Basque language today. The word 'alpine' is applied here strictly as a 'terminus technicus' for the life zone above the treeline worldwide (with no special reference to the European Alps). This also contrasts with the common language where 'alpine' is often used to indicate mountainous regions in general (Körner, 1999).

With this pragmatic focus on the highlands, the consortium did not overlook the enormous ecological problems at lower elevations; the heavily used and overpopulated lower montane zone in particular. It was felt, however, that the impact of land use, deforestation and intense agriculture at mid to lower elevations already receives a lot of attention (though never enough!), whereas the life zone above never became a focus of international concern. There was no other reason than this for this book and the GMBA network focus on the highest elevation biota. Lower elevations were included only as part of elevational gradients in comparative approaches. The high elevation climatic treeline (not its regionally anthropogenically depressed position) has been selected as a bioclimatological reference line because it occurs worldwide at similar temperatures, irrespective of latitude (Körner, 1998). Therefore, the Mexican altiplano at ca. 2000 m elevation is far below the GMBA core region, but the Scandinavian alpine tundra above the 600 m treeline is well within it.

In other words, this volume deals with the biological richness, its function and change at the cool, high elevation end of the biosphere. This life zone covers nearly 5% of the global terrestrial land area (or twice as much if cold and hot deserts are disregarded). The treeless alpine life zone alone accounts for ca. 3% of the land area. Often considered barren, it hosts a vast biological richness, exceeding that of many low elevation biota (Körner, 1995).

The chapters of this volume have been selected in a peer reviewing process from the presentations offered during the first international conference on mountain biodiversity, its function and change, held at the Rigi-Kaltbad

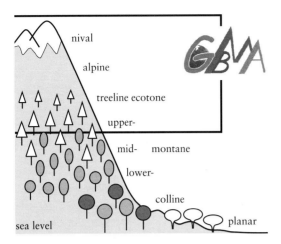

A schematic representation of the elevational life zones and the GMBA focal range of altitudes

mountain resort in central Switzerland, between 7 and 10 September 2000. Within sight of ice covered peaks of 4000 m elevation, roofing the traditionally managed mountain pastures of the Alps, 120 scientists from 34 different nations presented their views on how much richness in their concerned biota is present for selected groups of organisms; how they think this richness developed; and what future changes might be expected. There was a strong representation from developing countries, illustrating their difficulties in sustainably handling their uplands in view of poverty and land demand. The human dimension was present as a cross-cutting issue in almost all presentations, highlighting land use as the most important element of global change in the mountain world, exceeding the significance of atmospheric changes and climate warming.

The Rigi conference of GMBA was a start, illuminating the status of our knowledge. In fulfilment of Chapter 13 of Agenda 21, the international network activity of GMBA should now work at both widening and deepening the understanding of upland biota in the fields of many disciplines. It appears that a greater representation of animal sciences would be desirable, given how much detailed wisdom may be latent within the community, and how little crosscutting analysis surfaced from literature surveys. Land use will certainly remain a focal point in this cooperative project.

The organisers wish to express their sincere thanks to the Swiss Academy of Sciences (SANW, Bern) and DIVERSITAS (Paris) who initiated this network and supported it with so much sympathy. We gratefully acknowledge the financial contributions of the Swiss Academy of Sciences, the Swiss Agency for the Environment, Forests and Landscape, the Swiss Federal Office for Agriculture, the Swiss Agency for Development and Cooperation, the Swiss Science Foundation, the Austrian Federal Ministry of Education, Science and Culture (Vienna), and UNU, the United Nations University (Tokyo). In person, we wish to thank Mdm Anne-Christine Clottu (SANW), without her, and Prof. Bruno Messerli's continuous encouragement and support, this project would not have happened. We hope that the GMBA initiative will flourish and will contribute to respect and sustainable use of the wild and tamed nature of the world's high mountains.

Christian Körner and Eva Spehn
Basel, September 2001

www.unibas.ch/gmba

Contributors

Okmir Agakhanjanz
Department of Geography
University of Minsk
Gamarnik str. 9-1-31
220090 Minsk
Bielo-Russia

Wilhelm Barthlott
Botanisches Institut
Universität Bonn
Meckenheimer Allee 170
D-53115 Bonn
Germany
barthlott@uni-bonn.de

Cynthia M. Beall
Department of Anthropology
238 Mather Memorial Building
Case Western Reserve University
Cleveland OH 44106-7125
USA
cmb2@po.cwru.edu

Urs Bloesch
Mittelstrasse 26
CH-2502 Biel
Switzerland
bloesch@swissonline.ch

Andreas Bosshard
Planning & Research for Ecology &
   Landscape
Litzibuch
CH-8966 Oberwil-Lieli, Switzerland
andreas.bosshard@litzibuch.ch

William D. Bowman
Mountain Research Station
Institute of Arctic and Alpine Research
   and Department of Environmental,
   Population, and Organismic Biology

University of Colorado
Boulder CO 80309-0334
USA
bowman@spot.colorado.edu

Gerald Braun
Space Management - Earth Observation
German Aerospace Center (DLR)
Königswinterer Str. 522-524
D-53227 Bonn-Oberkassel
Germany
Gerald.braun@dlr.de

Siegmar-W. Breckle
Department of Ecology
University of Bielefeld
Universitätstr. 25
D-33619 Bielefeld
Germany
sbreckle@biologie.uni-bielefeld.de

Manab Chakraborty
Kadoorie Farm and Botanic Garden
Lam Kam Road, Tai Po, NT, Hong Kong
China
mc1956@hotmail.com

Mary Damm
Mountain Research Station
Institute of Arctic and Alpine Research
   and Department of Environmental,
   Population, and Organismic Biology
University of Colorado
Boulder CO 80309-0334
USA
Mary.Damm@colorado.edu

Bernhard Wolf Dickoré
Albrecht-von-Haller-Institut für
   Pflanzenwissenschaften
Universität Göttingen

Untere Karspüle 2
D-37073 Göttingen
Germany
wdickor@gwdg.de

Thomas Dirnböck
Institute for Ecology and Conservation Biology
University of Vienna
Althanstrasse 14
A-1091 Vienna
Austria
dirn@pflaphy.pph.univie.ac.at

Stefan Dullinger
Institute for Ecology and Conservation
    Biology
University of Vienna
Althanstrasse 14
A-1091 Vienna
Austria
dull@pflaphy.pph.univie.ac.at

Daniel B. Fagre
US Geological Survey
Northern Rocky Mountain
    Science Center
Glacier National Park
West Glacier MT 59936
USA
Dan_fagre@usgs.gov

Myriam Gaudeul
Laboratoire de Biologie des Populations
    d'Altitude
CNRS UMR 5553
Université Joseph Fourier
F-38041 Grenoble cedex 9
France
myriam.gaudeul@ujf-grenoble.fr

Michael Gottfried
Institute for Ecology and Conservation
    Biology
University of Vienna
Althanstrasse 14
A-1091 Wien,
Austria
gottf@pflaphy.pph.univie.ac.at

Georg Grabherr
Institute for Ecology and Conservation
    Biology
University of Vienna
Althanstrasse 14
A-1091 Vienna
Austria
grab@pflaphy.pph.univie.ac.at

Ken Green
National Parks and Wildlife Service
Snowy Mountains
PO Box 2228
Jindabyne NSW 2627
Australia
ken.green@npws.nsw.gov.au

Stephan R.P. Halloy
New Zealand Institute for Crop &
    Food Research, Ltd.
PB 50034
Mosgiel
New Zealand
halloys@crop.cri.nz

Lawrence S. Hamilton
World Commission on Protected
    Areas
342 Bittersweet Lane
Charlotte, Vermont 05445
USA
Hamiltonx2@mindspring.com

Stuart A. Harris
Department of Geography
University of Calgary
Calgary
Alberta T2N 1N4
Canada
harriss@ucalgary.ca

Jianquan Liu
Northwest Plateau Institute
    of Biology
Chinese Academy of Science
59 XiGuan Street
810001 Xining, Qinghai
China
Ljqdxy@public.xn.qh.cn

Rüdiger Kaufmann
Dept. of Zoology and Limnology
University of Innsbruck
Technikerstrasse 25
A-6020 Innsbruck
Austria
ruediger.kaufmann@uibk.ac.at

Michael Kessler
Albrecht-von-Haller-Institut für
    Pflanzenwissenschaften
Untere Karspüle 2
D-37073 Göttingen
Germany
106606.464@compuserve.com

James Barrie Kirkpatrick
School of Geography and
    Environmental Studies
University of Tasmania
Box 252-78, GPO
Hobart
Tasmania 7001
Australia
J.Kirkpatrick@utas.edu.au

Frank Klötzli
Department of Environmental Science/
    Geobotanical Institute
Swiss Federal Institute for
    Technology (SFIT)
Zürichbergstr. 38
CH-8044 Zürich
Switzerland
rentsch@geobot.umnw.ethz.ch

Christian Körner
Institute of Botany
University of Basel
Schoenbeinstrasse 6
CH-4056 Basel
Switzerland
Ch.Koerner@unibas.ch

David Jury McDonald
Botanical Society of
    South Africa
Private Bag X 10
Claremont 7735

South Africa
davemcd@mweb.co.za

Bruno Messerli
Institute of Geography
University of Bern
Hallerstr. 12
3012 Bern
Switzerland
messerli@giub.unibe.ch

G.F. Midgley
Climate Change Research Group
National Botanical Institute
Private Bag X 7
Claremont 7735
South Africa

Georg Miehe
Fachbereich Geographie
Universität Marburg
Deutschhausstr. 10
D-35032 Marburg
Germany
miehe@mailer.uni-marburg.de

Mohamed Aliyar Mohamed-Saleem
(Formerly staff of the International
    Livestock Research Institute,
    Ethiopia)
17A, Boswell Place,
Colombo-6,
Sri-Lanka
saleem_mohamed@hotmail.com

Maximina Monasterio
Instituto de Ciencias Ambientales y
    Ecológicas (ICAE)
Facultad de Ciencias
Universidad de los Andes
Mérida 5101
Venezuela
maximina@cantv.net

Jens Mutke
Botanisches Institut
Universität Bonn
Meckenheimer Allee 170
D-53115 Bonn

Germany
Jens.Mutke@uni-bonn.de

Harald Pauli
Institute for Ecology and Conservation
    Biology
University of Vienna
Althanstrasse 14
A-1091 Vienna
Austria
pauli@pflaphy.pph.univie.ac.at

David L. Peterson
US Geological Survey
Forest and Rangeland Ecosystem
    Science Center
University of Washington
Box 352100
Seattle WA 98195
USA
Wild@u.washington.edu

Catherine M. Pickering
School of Environmental and
    Applied Sciences
Griffith University
PMB 50 Gold Coast Mail
    Centre
QLD 9726
Australia
C.Pickering@mailbox.gu.edu.au

Les Powrie
Climate Change Research
    Group
National Botanical Institute
Private Bag X 7
Claremont 7735
South Africa
Powrie@nbict.nbi.ac.za

Aditya N. Purohit
High Altitude Plant Physiology Research
    Centre
H.N.B. Garhwal University
Srinagar
Garhwal – 246174
India
purohit_aditya@hotmail.com

Hanta Rabetaliana Schachenmann
Association des Population de Montagnes
    du Monde (APMM)
Villa Magali, Ivory-Nord
301 Fianarantsoa
Madagascar
hrabetaliana@vitelcom.mg

Corinna Raffl
Department of Botany
University of Innsbruck
Sternwartestrasse 15
A-6020 Innsbruck
Austria
corinna.e.raffl@uibk.ac.at

Martin G. Raphael
USDA Forest Service
Pacific Northwest Research
    Station
3625 93rd Avenue
Olympia WA 98512-9193
USA
mraphael@fs.fed.us

Andreas Reder
German Aerospace
    Center (DLR)
Porz-Wahnheide, Linder Höhe
D-51147 Köln
Germany

Karl Reiter
Institute for Ecology and Conservation
    Biology
University of Vienna
Althanstrasse 14
A-1091 Wien
Austria
reiter@pflaphy.pph.univie.ac.at

Lina Sarmiento
Instituto de Ciencias Ambientales y
    Ecológicas (ICAE)
Facultad de Ciencias
Universidad de los Andes
Mérida 5101
Venezuela
lsarmien@ciens.ula.ve

Peter Schachenmann
African and Madagascar Mountain
    Association (AMMA)
Villa Magali, Ivory-Nord
301 Fianarantsoa
Madagascar
pschachenmann@vitelcom.mg

Julia K. Smith
Instituto de Ciencias Ambientales y
    Ecológicas (ICAE)
Facultad de Ciencias
Universidad de los Andes
Mérida 5101
Venezuela
julia@ciens.ula.ve

Eva M. Spehn
GMBA office
Institute of Botany
University of Basel
Schönbeinstrasse 6
CH-4056 Basel
Switzerland
gmba@ubaclu.unibas.ch
http://www.unibas.ch/gmba

Markus Staudinger
Institute for Ecology and Conservation
    Biology
University of Vienna
Althanstrasse 14
A-1091 Vienna
Austria
stand@pflaphy.pph.univie.ac.at

Irène Till-Bottraud
Laboratoire de Biologie des Populations
    d'Altitude
CNRS UMR 5553
Université Joseph Fourier
F-38041 Grenoble cedex 9
France
irene.till@ujf-grenoble.fr

Risto Virtanen
Department of Biology
University of Oulu
PO Box 3000
University of Oulu
FIN-90014
Finland
rvirtane@cc.oulu.fi

Barbara C. Wales
USDA Forest Service
Pacific Northwest Research Station
1401 Gekeler Lane
La Grande OR 97850
USA

Qiji Wang
Northwest Plateau Institute of Biology
Chinese Academy of Sciences
59 XiGuan Street
81000 Xining, Qinghai
China
wqj@mail.nwipb.ac.cn

Douglas Williamson
Forestry Department
Food and Agriculture Organisation of
    the United Nations
Viale delle Terme di Caracalla
I-00100 Roma
Italy
douglas.williamson@fao.org

Michael J. Wisdom
USDA Forest Service
Pacific Northwest Research Station
1401 Gekeler Lane
La Grande OR 97850
USA
Mwisdom@fs.fed.us

Thomas Wohlgemuth
Swiss Federal Institute for Forest, Snow
    and Landscape Research
Zürcherstrasse 111
CH-8903 Birmensdorf
Switzerland
wohlgemuth@wsl.ch

Zerihun Woldu
Department of Biology
Addis Ababa University
P.O. Box 3434

Addis Ababa
Ethiopia
zerihun.herbarium@telecom.net.et,
    zerihunw@hotmail.com

Xinquan Zhao
Northwest Plateau Institute of Biology
Chinese Academy of Sciences

59 XiGuan Street
81000 Xining, Qinghai
China
xqzhao@public.xn.qh.cn

Part I

Introduction

Mountain Biodiversity, its Causes and Function: an Overview

1

Christian Körner

The high elevation biota of the world have been the subject of many synthetic approaches, including assessments of the variability and diversity of ecosystem properties and processes, ecophysiological traits and organismic functioning in general (e.g. Franz, 1979; Vuilleumier and Monasterio, 1986; Billings, 1988; Rundel *et al.*, 1994; Archibold, 1995; Wielgolaski, 1997; Körner, 1999). In this overview, questions relating to high elevation biodiversity will be specifically discussed (for previous attempts with a global perspective see for instance Chapin and Körner, 1995; Rahbek, 1995).

NAILS AND SCREWS AND THE LOWLAND–UPLAND CONTRACT OF SOCIETY

Mountainous terrain covers 24% of the global land area (Kapos *et al.*, 2000). The trivial fact that mountains have slopes causes them to directly or indirectly influence the life of half of the Earth's human population (Messerli and Ives, 1997). Slopes (and the peaks behind them) not only capture water, but guide it to the forelands and, via large river systems, feed the plains. Runoff and the sediments carried with it, both benefit (water supply and mineral nutrients) and disadvantage (floods and mud slides) downslope society. Slopes provide gravitational power to water, which can be converted to electric energy. Slopes guard and guide or constrain and endanger traffic routes. Slopes stop clouds (advection) or create them (convection), slopes exert mechanical force on any organism, installation or activity on them.

Unless made of solid rock, the only way in which loose substrates are secured to slopes is through vegetation. These living claws, screws and nails sustain slope benefits and prevent slope dangers. Slopes are only as stable and safe as the integrity and stability of their vegetation (Figure 1.1). Integrity of upslope vegetation is the centrepiece of downslope welfare in many parts of the world. There is, indeed, a too often overlooked link between far away mountains and densely populated lowlands. The recurring floods in China, for instance, which directly affected 200 million people's lives in 1999, the recent mud slides in the Philippines or the Andes, are impressive manifestations of uncontrolled upslope instability. It is obvious that there are many mutual upslope–downslope interdependencies. A substantial fraction of lowland productivity is required to cover the needs of upslope society and to maintain or enforce sustainable management in the highlands. This 'downslope–upslope contract', in essence, rests on the effectiveness of vegetation – natural or man-made – to control erosion.

BIODIVERSITY PROVIDES INSURANCE

The formula is simple: intact vegetation provides safety. Since the risks and threats on mountain slopes are manifold, they require a multitude of safeguards. The hammering of the ground by heavy raindrops or hail, the disruptive force of surface runoff, the mechanical sheer of loose ground, the exploration of deep substrate moisture, the resistance against

Figure 1.1 The integrity of steep alpine slopes such as this one, at 2500 m in the Swiss central Alps, is secured by plants, their belowground structures in particular. The cartoon to the right symbolises the need for a multitude of 'tools' to do the 'job'

trampling and grazing by large herbivores, and snow-gliding, are among many manifestations that require different mechanical solutions (see Figure 1.1). A multitude of plant structures can and do offer these services in a concerted manner. Yet, at times, any of these 'tools' may fail. Natural diseases, divergent life cycles, and varying sensitivity to stress and disturbance may eliminate different players, at least periodically. Hence, it requires each key tool to be present in various combinations, so that at least one functions in case of emergency. The benefits of diversity are manifold (Mooney *et al.*, 1995). In the simplest terms, Moffat, (as quoted by Stuart Pimm 1996) said that: 'There is a dance going on of compensatory changes. Something always benefits from a disaster, provided you have enough species'. Whether ecosystems with more species are less at risk of loss of integrity than systems with less species is not a generalizable issue, as was previously thought. It depends on the absolute number of species, the identity of the species and the circumstances under which the system is operating. Hence, the above discussion should be taken as a plausible trend, not a proven fact for any individual case.

According to the 'insurance hypothesis', one of the benefits of biological richness is that it insures against system failure. Functional redundancy may not play any role for long periods of time, but a single extreme event can cause it to become the life-preserver of fragile slopes that sustain ecosystem functioning. The richness of biota includes all levels of organismic diversity from genes to landscapes, it includes all groups of organisms from microbes to the largest herbivores.

By no means should ecosystem functioning be considered the only valid reason to conserve biological richness, but prioritising on this one function appears plausible, given that all other benefits and values depend on it. Other justifications are the ethical value (the right of existence), the aesthetic value (the beauty), the economic value (potential or instrumental economic uses). Each of these has value in its own right. In high mountains, several or all of these values apply together, as this book will illustrate.

Having said all this, the reader will find this introductory note somewhat plant-biased. The reason for this is, in part, my own profession. However, having joined many discussions with zoologists, there seems to be less synthetic knowledge available in high elevation animal sciences. Perhaps we are far from even simple generalisations which may hold for animal studies. As an example, plants become smaller with increasing elevation, if the same morphotypes are compared; not so animals: birds and rodents become larger, as do some groups of

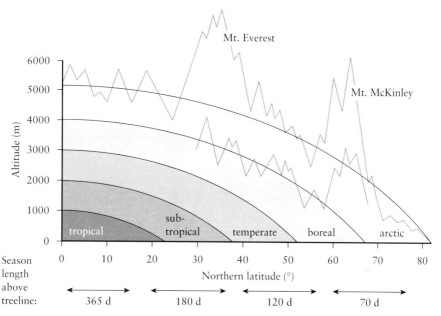

Figure 1.2 The correlation between altitudinal and latitudinal life zones. The elevational compression of biomes causes mountains to become hot spots of biological richness

insects, others do not significantly change in size and some also become smaller (such as fish). If they become smaller, are there more individuals needed to fulfil a particular ecosystem function (such as chopping down litter)? It seems, we do not know. Do developmental cycles become shorter? For grasshoppers, this does not seem to be the case. Thankfully a few animal scientists have joined this assessment and the doors are wide open for more to come in!

THE BIOLOGICAL RICHNESS OF HIGH ELEVATION BIOTA

As one ascents a humid, 5000-m high equatorial mountain, one passes through nearly all climatic zones of the world over a relatively short distance (Figure 1.2). On flat land, one would have to travel thousands of kilometres to pass through a similar series of thermal climates. Imagine a bear in the Rocky Mountains or the Carpathians, on a pleasant autumn afternoon it may be the berries in the alpine heathland which attract its attention, while the 'boreal' forest below provides shelter for it during the following cold night. Heavy snowfall

may lead the bear to spend the next morning in the deciduous forests of the temperate life zone a few kilometres downhill. Within a day, the bear has covered a 3000-km range of latitudinal climatic difference, while perhaps walking no more than 8 km. This compression of life zones explains why, on a 100 km grid scale, no landscape can beat the biological richness of mountains (Barthlott *et al.*, 1996). Nowhere else is it possible to protect and conserve so much biological diversity within a relatively restricted region, than in mountains and in the tropical mountains in particular (Klötzli, 1997). Tropical mountains may span the range from the humid lowland forest to glaciated mountain tops, and it is therefore no surprise that these regions become hot spots of biodiversity.

The richness of organismic taxa declines with elevation (at least in humid areas), and biologists have long recognised that elevational and latitudinal species richness gradients mirror each other (Figure 1.2; Rahbek, 1995 and references therein). Once the glacial region is reached, it is unlikely that more than 5% of all species of a given mountain region and its

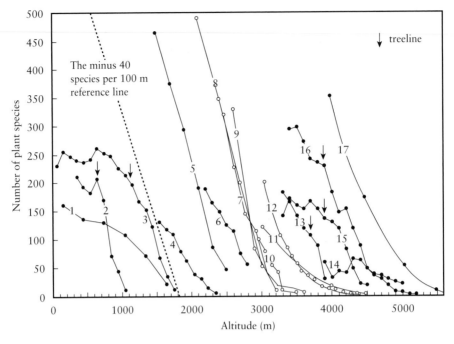

Figure 1.3 Examples of the elevational reduction of plant species diversity in different mountain ranges. (1) E, NE Greenland (Schwarzenbach, 2000); (2) Glen Clova, Scotland (Willis and Burkill in Raunkiaer, 1908); (3) Aurland, S Norway (Odland and Birks, 1999); (4) Jotunheimen, S-Norway (Jorgensen, 1932 in Odland and Birks, 1999); (5) Tatra Alps (Kotula in Raunkiaer, 1908); (6) Olympus, Greece (Polunin, 1980); (7) Swiss Alps (Heer in Raunkiaer, 1908); (8) Bernina, Swiss Alps (Brockmann-Jerosch in Raunkiaer, 1908); (9) West Alps (Thompson in Raunkiaer, 1908); (10) Bernina, Swiss Alps (Rübel, 1911); (11) Oetztaler Alps, Tyrol (Reisigl and Pitschmann, 1958); (12) Montafon, Alps (Grabherr et al., 1995); (13) Oytagh; (14) K2 North; (15) Batura; (16) Nanga Parbat, Karakorum (Dickoré and Miehe, Chapter 10); (17) Hindukush (Breckle in Walter and Breckle, 1991)

immediate lowland surroundings can thrive. Within the alpine belt, the total plant species diversity of a given region commonly declines by about 40 species of vascular plants per 100 m of elevation (Figure 1.3). Surprisingly, this rate of decline does not differ greatly across a wide range of mountains (see Nagy et al., in prep.). This decrease in species richness follows the decrease in mean temperature (Figure 1.4), but often accelerates in the upper part of such transects.

Species richness in animals declines with elevation in a similar way as in plants. Rahbek (1995) reviewed this field most recently and, from his analysis of 97 publications with 163 examples, it is quite obvious that altitude and latitude operate in the same direction. However, Rahbek noted that this decline is not necessarily monotonic (Figure 1.5), and Meyer and Thaler (1995) showed that trends strongly depend on the taxonomic group considered. The reasons for this elevational decline of species richness are not fully resolved. Since similar trends are observed in plant and animal diversity, we are either dealing with a general evolutionary bottleneck phenomenon (for whatever reason), affecting all organismic groups similarly but independently, or the reduced plant diversity induces the decline in diversity of all other organismic categories. Cryptogams (mosses and lichens) may deviate from this pattern over certain ranges of elevation and increase in abundance as the climate becomes cooler and more humid (Virtanen et al., 2002), but ultimately the number of cryptogam species also drops drastically as one

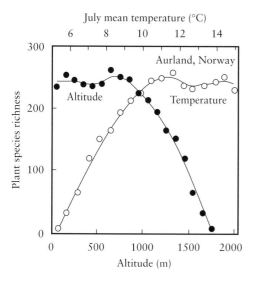

Figure 1.4 The reduction of species diversity and temperature with elevation in S Norway (Odland and Birks, 1999). Note the discontinuity at a certain elevation

reaches the highest peaks. Isolated individuals of higher plants have been found up to 5900 m in Tibet, 6300 m in the Himalayas, and up to 4450 m in the Alps (Rongfu and Miehe, 1988; Grabherr *et al.*, 1995; Grabherr, 1997; Körner, 2001), but these outposts often profit from local microclimatic peculiarities, with the majority of species finding a limit at elevations 1000 m lower.

Less steep declines are seen in genera and family numbers in plants (Figure 1.6), which means that certain families contribute more to high elevation biodiversity than others. Given that mountains are often functional islands (isolated archipelagos), endemism is high, i.e. taxa are abundant which are known from only a single mountain (Packer, 1974; Salgado-Labouriau, 1986; Steyermark, 1996; Dahl, 1990; Mark, 1995; Nakhutsrisvili and Gagnidze, 1999; Safford, 1999). However, the highest degree of endemism is commonly not found at extreme, but at moderate elevations (Breckle, 1974). Many species are restricted to high elevation environments and do not survive when brought to warmer, lower elevations ('exclusive' alpines). In plants, reasons for this may be related to the small plant size (plants

become overgrown) or evolutionary adaptation of metabolism to low temperatures, with some of the highest-reaching plant species operating at a Q_{10} of 3–5 (Larigauderie and Körner, 1995). Similar metabolic adjustments have been reported for animals (Wieser, 1973).

The overall global plant species richness of the alpine life zone alone (i.e. above the tree-line) was estimated to be around 10,000 species, 4% of the global number of higher plant species which represents one third more species than would be expected from the 3% land coverage of the alpine terrain alone. No such estimates exist for animals but, based on flowering plants, high elevation biota are richer in species than might be expected from the land area they cover. The large species richness of high elevation biota does not correlate with their (often) low biomass per unit land area, as seems to be quite commonly found when natural or seminatural vegetation types are compared, rather than delimited experimental communities (Gough *et al.*, 1994).

CAUSES OF BIOLOGICAL RICHNESS AT HIGH ELEVATION

With increasing elevation, life conditions become more selective. Frost, the need to grow and reproduce at low average temperature, short growing seasons in extra-tropical mountains, and other climatic constraints, all impose selective barriers which exclude many species. As a consequence, one would expect much less biological richness than is actually found. However, once species are filtered by these environmental limits, life is not more 'stressful' than anywhere else. This hints at a widespread misconception with respect to limitation and stress (Körner, 1999): 'stress' (in human terms) has filtered out those species which cannot tolerate it. In fact, high elevation plants experience 'stress' when exposed to low elevation life conditions.

Hence, selection by stress may not be the only reason for the overall diminishing number of organisms at higher elevation. Unfortunately, the increasing climatic harshness coincides with a consistent **reduction in**

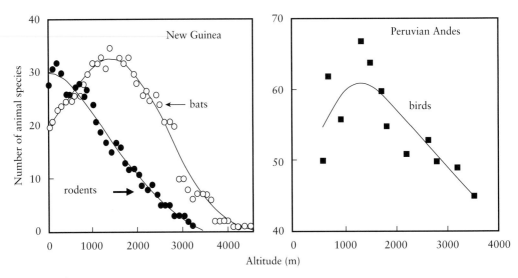

Figure 1.5 The reduction of species diversity of different animal groups with elevation (Rahbek, 1995). Examples are bats and rodents along an elevational transect in New Guinea and syntopic birds of an Amazonian slope of the Andes in Peru

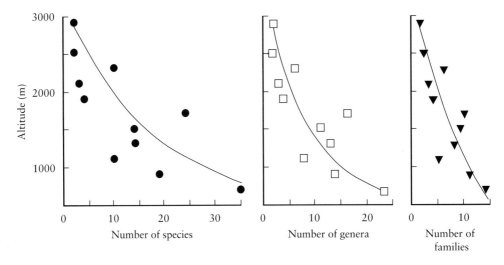

Figure 1.6 The reduction of species, genera and family number in trees in upper montane forests of Sichuan, China (Tang and Ohsawa, 1999)

land area with elevation (Figure 1.7). It may well be that biological richness and available space are linked, so that a reduction in space alone could cause a decline in species richness (Rahbek, 1995; Rosenzweig, 1995; Chown and Gaston, 2000; Körner, 2000b). This phenomenon is well known from island biogeography, where small islands are inhabited by less species (for reviews see Vitousek *et al.*, 1995; Pimm and Raven, 2000). The poor floristic diversity of tropical inselbergs seems to be in line with this concept (Barthlott *et al.*, 1993). However, alpine island size, distance to gene source, and latitude have all been found to only moderately correlate with species richness in the Rocky Mountains (Hadley, 1987),

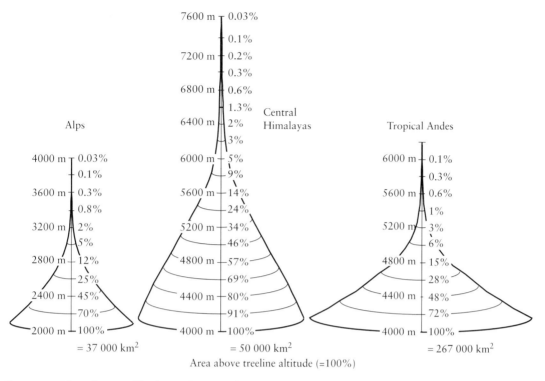

Figure 1.7 The reduction of land area above the alpine treeline in three different mountain regions (the total land area above the treeline elevation = 100%; modified from Körner, 2000b)

and species numbers in the Alps show only small increases once an alpine area of 20 km² is exceeded within a given mountain range (Wohlgemuth, 1993).

To some extent, the island nature of mountains may even have enhanced biodiversity, certainly the degree of endemism is increased (Salgado-Labouriau, 1986). The genetic and reproductive mechanisms which may facilitate high genetic diversity in alpine biota, despite spatial isolation, are **polyploidy** and a high degree of **self-incompatibility** (Körner, 1999; Till-Bottraud, Chapter 2). As Packer (1974) put it, 'it is not unreasonable to say that alpine species possess genetic systems directed towards the maintenance of genetically variable populations' – a prerequisite for life in rather unpredictable climates. To underline this statement, Packer refers to the fact that the autochthonous element reflected in percentage of endemic species in East Africa is 81%. According to Hedberg (1969, cited by Packer,

1974), the tree lobelias and the tree senecios provide beautiful examples of altitudinally vicarious taxa. High insect diversity and activity seems to well ensure pollination even in very high altitudes (Kalin-Arroyo et al., 1982; Körner, 1999). Unfortunately, very little comparative work on such questions has surfaced for animals.

The correlation between species richness and the actual land area available above a given elevation is illustrated in Figure 1.8. Although there is some variation, overall the correlation is not far from a 1:1 relationship. The land area reduction used here is derived from geographical information systems, and hence does not account for uninhabitable land, the fraction of which certainly increases with altitude. With this in mind, the high elevation species richness per unit available and inhabitable area increases, which is in strong contrast to the picture of depressing and very selective life conditions, permitting few species to thrive.

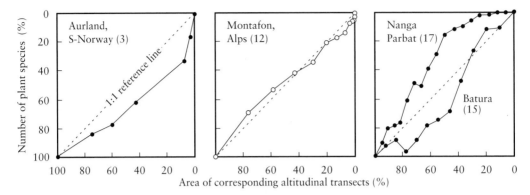

Figure 1.8 Plant species richness versus available land surface area with increasing elevation for three examples from Norway, the Alps and the Himalayas. The total area above a certain elevation is set equal to 100% so that certain elevational ranges became fractions of 100. In cases where mountain height exceeds the inhabitable range, I also defined upper limits for the 100% area. The actual ranges for 100% area are: > 1100 m in Norway, > 3050 m in the Alps, 3500–5200 m in the Nanga Parbat region and 3400–4500 m for the transect in Batura. Numbers refer to Figure 1.3

Instead of space, perhaps 'time' may be critical. For tropical forests it was shown that temporal patterns, especially seasonal patterns, increase the opportunity for regeneration (Iwasa *et al.*, 1993). When the whole life cycle is pressed into a single narrow window of time, species diversity tends to drop. This may well apply to the short season of temperate alpine environments, but would suggest a higher diversity in the tropical alpine zone per unit of land area, not what the literature seems to tell us so far. As short as the alpine season might be at higher latitudes, there is still a fair amount of differentiation in phenologies among species (Körner, 1999).

On top of habitat diversity, **plant size** itself may be a major cause of highland species diversity. On average, plants above the treeline reach only one tenth of the size of their closest lowland relatives (Körner, 1999). For this simple reason, a greater number of individuals and species can be nested in any given space, including the many microhabitats the highlands offer. A nice illustration of the enormous diversity that may be seen in the alpine environment is given in Figure 1.9 for an arctic–alpine site in northern Sweden.

Part of the explanation for the surprisingly high plant diversity at very high elevations may also lie in the increasing **isolation** of individuals at the patch scale, reducing competitive interactions. This is known to stimulate diversity in the alpine environment under moderate and regular disturbance (e.g. Fox, 1981, but see the removal experiments referred to below). Perhaps the strongest and simplest foothold toward understanding the comparatively high species richness at high elevations are slopes and the gravitational forces on them. These create a multitude of physical niches which, in turn, become **micro-habitats** of contrasting exposure, moisture, nutrients, snow cover, substrate structure, etc., each with its own rather distinctive assemblage of species. A recent, very detailed analysis of the flora of nival rock and scree fields strongly supports this idea (Pauli *et al.*, 1999). There is hardly any other environment which offers so many different microenvironments over such a short distance as found in the alpine belt.

Removal experiments seem to indicate that species interactions matter, because positive effects are seen when neighbours are taken away, in part, even when nutrients are applied to control plots (e.g. Onipchenko and Blinnikov, 1994; Aksenova *et al.*, 1998; Bowman and Damm, Chapter 3). The common observation that old ecosystems, such as some tropical forests, are very rich in species,

Figure 1.9 High species diversity per unit of area in an arctic-alpine fellfield at 1050 m elevation in N Sweden (Abisko, 68° N) is facilitated and explained in part by the tinyness of the individuals. The size of this plot is 18 × 28 cm. Of the total 34 species identified 12 are phanerogams and 22 are crytogams (disregarding algae). **Sedges** (2 species) 1 *Carex* sp., 2 *Carex* sp.; **Herbs** (2 species) 3 *Thalictrum alpinum*, 4 *Polygonum viviparum*; **Dwarf shrubs and cushion plants** (8 species) 5 *Cassiope tetragona*, 6 *Vaccinium vitis-idea*, 7 *Salix herbacea*, 8 *Cassiope hypnoides*, 9 *Loiseleuria procumbens*, 10 *Silene acaulis*, 11 *Empetrum nigrum*, 12 *Diapensia lapponica*; **Lichenes** (10 species) 13 *Cetraria islandica*, 14 *Cladina rangiferina*, 15 *Cladina* sp., 16 *Cladonia af. gracilis*, 17 *Stereocaulon* sp., **18–23** not shown: 6 other species, including *Cetraria nivalis* and *Thamnolia vermicularis*; **Bryophytes** (11 species) 24 *Pohlia* cf. *drummondii*, 25 *Ptilidium cilare*, 26 *Barbilophozia kunzeana*, 27 *Polytrichum sexangulare*, 28 cf. *Sanionia nivalis*, 29 *Conostomum tetragonum*, **30–34** not shown: *Lophozia sudetica*, *Cephaloziella* sp., *Gymnocolea inflata*, *Dicranum fuscescens*, *Cephalozia biscuspidata*; (bryophytes kindly identified by P. Geissler, Geneva)

whereas young ecosystems are dominated by a few vigorous pioneer species, does not seem to apply to the highlands, where rather the opposite appears to be true. The richest alpine communities are maintained in a relatively dynamic young stage, either by natural disturbance or by grazing. Old, late successional vegetation in alpine regions commonly hosts less taxa. This is partly due to the fact that such old communities are only found on undisturbed, mostly flat or slightly inclined ground. It is a general phenomenon, both in the arctic and alpine zones that slope dynamics enhance diversity (Chapin and Körner, 1995).

Figure 1.10 The destruction of a natural vegetation sward rich in clonal plants has ruined these upland sites. Left the fate of a ski slope construction in a montane forest of Tirol, Austria; right an overgrazed alpine meadow which became a victim of tourist horses in Tafi del Valle (NW Argentina)

In summary:

(1) Reduced space does not seem to truly restrict alpine species diversity, given that numbers often do not decline if based on the actually available land area.

(2) Reduced season length may, however, restrict alpine species diversity, given that all species show peak growth, flowering, feeding, and propagation at roughly the same time, with overlapping demands at least outside the tropics.

(3) Reduced competition at high elevations does seem to play a role for the coexistence of greater numbers of taxa.

(4) Habitat heterogeneity (fragmentation) also permits co-existence of many taxa in close proximity.

(5) Small size of individual plants adds to the likelihood of a diverse suite of taxa occurring in a small space.

(6) The genetic and breeding system of high elevation taxa facilitates development and maintenance of high genetic diversity, despite spatial isolation.

(7) Age of plant communities, often suggested as facilitating greater species richness, does not work in that direction in the highlands. It rather seems that late successional assemblages are poorer in species and dominated by a few persistent species.

(8) A great deal of biological richness in the highlands does emerge from disturbance, and the young or medium successional stages often bear the richest upland flora and, perhaps, fauna. It is exactly these areas which are at greatest risk of erosion, where the substrates are still loose. It is likely to be these earlier stages of vegetation development which are most fragile and most dependent on a rich suite of players to secure a site. These are not necessarily confined to pioneer scree-fields. Early successional stages may include dwarf shrub communities on former pastures or forests, with some fragile shrubs forming a transitory stage.

THE FUNCTIONAL SIGNIFICANCE OF HIGHLAND BIODIVERSITY

Apart from the theoretical advantages of biodiversity discussed in the first section of this Chapter (e.g. insurance), there is very little sound experimental or observational evidence on the functional significance of the richness of biota at high elevation. Among the hard facts, perhaps the lack of success of almost all re-vegetation attempts on high elevation building sites, using seeds sown on raw soils, is the most obvious. The presence of clonal species seems imperative in rapidly securing loose slopes before erosion can carry them away. Operative clones cannot easily be sown at high elevations, and take a very long time to establish the

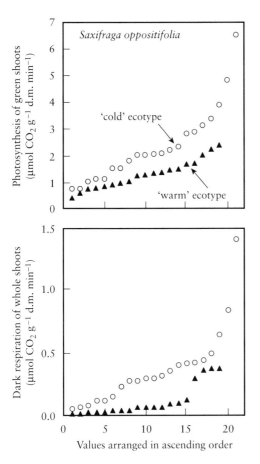

Figure 1.11 A demonstration of genetic differentiation of metabolic responses to temperature across small topographical gradients (from Crawford and Chapman, 1995). The diagram shows ranked rates of assimilation and respiration for individuals collected at a warm and a cold arctic microhabitat. Similar responses are to be expected at high elevations

required strong vegetation felt. A multitude of such slow growing clonal plants is certainly essential to prevent the situations shown in Figure 1.10.

More subtle examples of this largely unexplored field are situations where a keystone species becomes damaged, and subdominants suddenly become the safeguards for a site. For example, when the fragile carpets of creeping ericoids such as *Loiseleuria procumbens* in the Alps are trampled, cut by ski edges, 'burned' by winter desiccation or moulded under late-laying wet snow, the dark brown, very acid humus

becomes exposed and may heat to 70 °C in the midsummer sun, substantially reducing the chance of re-vegetation (Chapin and Körner, 1995). In such situations the scattered presence of clones of graminoids such as *Juncus trifidus* may become essential in re-vegetation. Fast germinating, obligatory seeders also need to be present in early spring to exploit such gaps. In montane forests, a single species forest containing species such as *Betula* or *Larix* may become heavily affected by outbreaks of certain moths, so it is best if these forests are intermingled with evergreen pines. One-species stands are always at risk of becoming the victim of a single damaging agent. How many species of graminoids would it require to hold a 30° slope in place for 50 years on calcareous gravel in the temperate zone at an altitude of 2300 m? How would the presence of a specific rodent or the stocking rate of sheep affect successful regeneration? Certainly this is a field requiring far more research.

In addition to taxonomic diversity, diversity of ecotypes of species adapted to very specific microenvironmental conditions has great importance. Crawford and Chapman (1995) demonstrated that arctic–alpine species may be different with respect to the thermal acclimation of their basic metabolism across spatial distances of a few metres (Figure 1.11). They also found well differentiated snow duration ecotypes at such small scales (Crawford *et al.*, 1995). It has long been known that trees of the montane forest are ecotypically (evolutionary) very different from lower elevation provenances (Engler, 1913; Holzer and Nather, 1974; Müller, 1981; Tranquillini and Havranek, 1985; Sulkinoja and Valanne, 1987). The functional implications of such genetic diversification are obvious.

UPLAND LAND USE AND BIODIVERSITY

'*Sola dosis facit venenum*' (only the dose causes something to become a poison) was the motto of Theophrastus Bombastus von Hohenheim (also known as Paracelsus), the famous medieval plague doctor and alchemist. Both montane and alpine regions have always

Figure 1.12 Traditional, sustainable utilisation of the highlands created man-made ecosystems of high cultural, economic and biological value. This montane hayfield replaces the uppermost forest in the central Alps and bears rare orchids and world ranking medical plants. It may have been in use for many hundreds or even thousands of years

been exposed to many forms of threats and disturbances, and grazing in particular. In most parts of the world, human land use has become a substitute and extension to what was the natural disturbance regime. The history of human land use in the uplands dates back thousands of years (e.g. Beug and Miehe, 1998), so upland pasturing cannot be considered a new influence on this vegetation. Problems only arise when the tolerable dose is exceeded. There are many examples of this excess but also cases where pastures improved with grazing (e.g. Schiechtl and Neuwinger, 1980; Pignatti and Wikus, 1987; Mahaney and Linyuan, 1991; Ram, 1992; Rikhari et al., 1992; Wilson, 1994; Bock et al., 1995; Hofstede et al., 1995; Trimble and Mendel, 1995; Quaranta et al., 1996; Pucheta et al., 1998; Körner, 2000a). But who knows the point of no return? Who knows what is sustainable (sustainable in the sense that the system remains intact)? Traditional shepherds knew, of course, but this knowledge has been lost on the whole, or was never incorporated into some sort of management plan for the uppermost grazed biota concerned. Lowland wisdom is hardly applicable to such situations.

The Alps, the Carpathians, the Caucasus, the Hindukush and Himalayas, the Mongolian mountains and others have been praised by travellers for their colourful highland mats – all of which are man-made ecosystems. These pastures, with their wooden fences and stone walls, dwellings and shrines, drainage and irrigation systems, specific soil dynamics and very special flora also represent a unique cultural heritage (Figure 1.12). This traditional, and thus commonly sustainable, management only exceptionally leads to erosion. These forms of land use are currently disappearing due to population growth, increasingly intense agriculture, or extensivation for marginal economic gain. Such changes induce substantial alterations in ecosystem functioning and biodiversity, with fire management exerting particularly dramatic effects (e.g. Hedberg, 1964; Billings, 1969; Cernusca, 1978; Beck et al., 1986; Deshmukh, 1986; De Benedetti and Parsons, 1979; Mark, 1992; Hofstede et al., 1995; Ramsay and Oxley, 1996; Tappeiner and Cernusca, 1998; Miehe and Miehe, 2000). For the following four reasons, measures to secure sustainable agriculture in traditional alpine pasture land near the treeline are highly recommended for many parts of the world:

(1) To conserve biologically highly diverse, stable and also attractive plant communities;
(2) To maintain a healthy, unpolluted food source for future generations;
(3) To retain a millennial cultural heritage; and
(4) To secure soils and water supply for downslope human communities.

In pastures near the treeline things can go wrong in three ways:

(1) Uncontrolled, non-traditional (i.e. patchy) grazing, causing spot-impact under otherwise low stocking rates;
(2) Stocking beyond carrying capacity or introduction of excessively heavy cattle; or
(3) Sudden abandonment of pastures.

All three of the above may affect soils and induce erosion or irreversible changes in ecosystem structure, such as invasion of shrubs into the grassland. Soil destabilisation after abandonment is related to sudden occlusion of man-made drainage systems (over-saturation of soils), and turf erosion caused by creeping late winter snow, frozen onto over-long grass. The perilous transition period, back to self-sustaining ground cover, may take more than half a century (Cernusca, 1978), but sensitivity to changed use varies with slope and vegetation type (Gigon, 1983). Following abandonment, shrub and tree invasion rapidly alter ecosystem properties that had slowly developed over millennia of manual work. The reversal of this succession will most likely be financially unaffordable, hence the loss of alpine pastures may be practically irreversible in many cases. Most treeline ecotones of the world are the result of substantial alterations by man (e.g. Lauer and Klaus, 1975; Klötzli, 1975) and are in a very dynamic state. For these areas to return to pastures will be dependent on fire or grazing pressure (e.g. Franklin et al., 1971).

With adequate control, grazing may have no negative effects on alpine vegetation. In places, it may even improve local ecosystem stability and favour biological richness (see the references in the first paragraph of this section). A recent study in the Swiss Alps revealed that six seasons of cattle *exclosure* (fence) from a 'natural' alpine *Carex curvula*-dominated grassland at 2500 m elevation (200–300 m above the treeline) had unexpected negative effects: Total (life) aboveground phanerogam biomass at the peak of the growing season (just before the first cattle visits) was reduced in the fenced exclusion area by one fifth, and there were clear indications of reduced abundance of rare forbs (Körner, 2000a). Hence, what is described in textbooks as one of the most typical types of natural alpine grassland, seems to benefit from sporadic cattle presence, both in terms of productivity and biodiversity. Dung deposition was found to create patch dynamics in this system, with a statistical rotation time of ca. 50 years. On average 2% of the plant cover is disturbed (largely killed) every year by dung deposits, but the surrounding vegetation gains in vigour. Overall, the patch dynamics induced by dung deposition stimulates biological diversity. A positive biomass response to traditional alpine grazing was also observed in the Garhwal Himalaya (Sundriyal, 1992). Preliminary work in the uplands of Ethiopia seems to support these observations (see Saleem and Woldu, Chapter 23). This is an important field that should be more widely explored within the GMBA programme in the future.

ATMOSPHERIC CHANGES AND MOUNTAIN BIODIVERSITY

Potential threats to high elevation biota from atmospheric changes such as warming, CO_2 enrichment, or enhanced soluble nitrogen deposition have been discussed in more detail elsewhere (see references below). In brief, these changes may be much less significant for biodiversity than changes in land use, but all of them are potentially important in affecting species richness.

Contrary to current assumptions, climatic warming is unlikely to induce a frontal movement of vegetation belts. Holocene vegetation history has taught us that responses to climatic change always operated at the species level (Ammann, 1995). Whole communities do not migrate, but single species or certain genotypes migrate and form new assemblages in target areas. What is too warm for one species may still be appropriate for another. We have genetic evidence that clones of alpine sedges have occupied the very same square metre of turf for the last few thousand years even though there have been quite dramatic changes in temperature during this time (Steinger et al., 1996). In contrast, Grabherr and Pauli (1994) and Gottfried et al. (1998 and references therein; Chapter 17) have demonstrated rapid shifts in summit communities over only a few decades in the Alps due to invasions of species from the communities below. Our general understanding of mountain plant biology (Körner, 1999) and the recent climatological evidence published by Pauli et al. (1999) suggest a local rearrangement of microhabitats, should there be climatic warming, rather than elevational isoline shifts (Körner, 2001). Should warming become associated with enhanced winter snow falls, we may even see adverse effects in some parts of the world (reduced season length because of snow pack).

The enrichment of the atmosphere with more CO_2 has been thought to provide a particular stimulation to growth at high elevations because the air is thinner, i.e. the partial pressure of all gases including CO_2, is reduced. However, a 4-year field trial in the Swiss Alps has not yielded such responses, irrespectively of whether more mineral nutrients were co-supplied with CO_2 (Körner et al., 1997). However, species tended to respond differently, with some gaining and others losing, in their biomass contribution to the system. In the long term, this may well lead to a change in community composition. It was noted that fast-growing species such as Poa alpina, responded more than the dominant slow-growing sedge. It was also shown that the tissue quality changed under elevated CO_2 so that proteins became depleted. This caused the dominant natural herbivores, grasshoppers, to enhance their food intake, with potential negative consequences on plants and for the fecundity of these animals. Hence, if anything, atmospheric CO_2 enrichment will – in subtle ways – affect mountain biodiversity, and this is likely to happen world-wide.

Among the other atmospheric changes, soluble nitrogen deposition seems to be most influential, but perhaps more restricted to certain polluted regions. A very low dose, corresponding to current central European lowland depositions, caused the above-mentioned high elevation sedge community to almost double its seasonal aboveground biomass within 4 years. Once more, the response was not symmetric across species. The faster-growing were stimulated more than the slow-growing species, i.e. it exerted a biodiversity effect (Schäppi and Körner, 1996). A recent high dose-nutrient experiment in pioneer vegetation in a glacier forefield also exhibited a massive effect on biodiversity of nutrient addition, causing the typically small stature, compact growth forms to be overgrown by the more vigorous grasses (Heer and Körner, 2002). Given that such 'fleshy'

grasses are much less resistant to any mechanical impact, the system's performance under such disturbance may become weakened. Biodiversity effects of nutrient addition in the alpine zone are well documented for a number of sites (e.g. Andrew and Johansen, 1978; Bowman et al., 1993; Theodose and Bowman, 1997). The effect may be seen even half a century after treatment, as documented by Hegg et al. (1992) in the Swiss Alps. Animals would always be co-affected, given the change in their food source.

CONCLUSION

Biological diversity in the world's uplands is richer than in most lowlands, if calculated on available land area. There seems to be a link between plant and animal diversity in uplands, both demonstrating a high degree of endemism and differentiation of ecotypes, but this has been explored much less for animals than for plants. The causes of high biological diversity are manifold, but microhabitat diversity, small size of organisms, isolation combined with effective reproductive systems, and moderate disturbance seem to contribute. Consequences for ecosystem integrity are largely related by slope stability, the centrepiece of any mountain ecology. It is suggested, but rock-hard data are still unavailable, that a highly structured, diverse ground cover is the best insurance for the maintenance of intact slopes under the many erosion forces. Of all global change impacts on mountain biodiversity, land use is the leading factor. It is encouraging that traditional upland grazing systems and land management have contributed to the establishment of rich biota in a sustainable way. This traditional knowledge is currently either overrun by population pressure and poverty or becomes lost as traditional land-use methods disappear. There is an urgent need to create broad awareness that what happens in these often remote uplands has immediate consequences for the lowlands. Mountain biodiversity is not only a scientific theme of high interest, it is perhaps the best indicator value of the integrity of mountain ecosystems. As a cultural heritage, diverse high-elevation grazing land also deserves our great respect and contains real and potential economic value.

This volume brings together a collection of regional assessments from around the world of how much biological richness there is in the uplands and how humans influence it. The basis on which to discuss the consequences or even the benefits of diversity as it currently exists and the potential drawbacks of change is thin. We hope that this attempt to bring together knowledge from around the world will assist in developing a deeper search for answers. The Global Mountain Biodiversity Assessment (GMBA) network was founded to guide this process into the future.

References

Ammann B (1995). Paleorecords of plant biodiversity in the Alps. In: Chapin FS III, Körner Ch (eds) *Arctic and alpine biodiversity: Patterns, Causes and Ecosystem Consequences*. Ecol Studies, 113:137–49, Springer, Berlin, Heidelberg, New York

Andrew CS and Johansen C (1978). Differences between pasture species in their requirements for nitrogen and phosphorus. In Wilson JR (ed.), *Plant relations in pastures*. CSIRO, Canberra, pp. 111–27

Aksenova AA, Onipchenko VG and Blinnikov MS (1998). Plant interactions in alpine tundra: 13 years of experimental removal of dominant species. *Ecoscience*, 5:258–70

Archibold OW (1995). *Ecology of World Vegetation*. Chapman & Hall, London, Glasgow, Weinheim

Barthlott W, Groger A and Porembski S (1993). Some remarks on the vegetation of tropical Inselbergs: diversity and ecological differentiation. *Biogeographica*, 69:105–24

Barthlott W, Lauer W and Placke A (1996). Global distribution of species diversity in vascular plants: Towards a world map of phytodiversity. *Erdkunde*, 50:317–27

Beck E, Scheibe R and Schulze ED (1986). Recovery from fire: Observations in the alpine vegetation of western Mt Kilimanjaro (Tanzania). *Phytocoenologia*, 14:55–77

Beug HJ and Miehe G (1998). Vegetationsgeschichtliche Untersuchungen in Hochasien-1. Anthropogene Vegetationsveränderungen im Langtang-Tal, Himalaya, Nepal. *Petermanns Geogr Mitt.*, 142:141–8

Billings WD (1969). Vegetational pattern near alpine timberline as affected by fire–snowdrift interactions. *Vegetatio*, 19:192–207

Billings WD (1988). Alpine vegetation. In Barbour MG and Billings WD (eds), *North American Terrestrial Vegetation*. Cambridge Univ Press, Cambridge, NY, Port Chester, pp. 392–420

Bock JH, Jolls CL and Lewis AC (1995). The effects of grazing on alpine vegetation: A comparison of the central Caucasus, Republic of Georgia, with the Colorado Rocky Mountains, USA. *Arctic and Alpine Research*, 27:130–6

Bowman WD, Theodose TA, Schardt JC and Conant RT (1993). Constraints of nutrient availability on primary production in two alpine tundra communities. *Ecology*, 74:2085–97

Breckle SW (1974). Notes on alpine and nival flora of the Hindu Kush, East Afghanistan. *Bot Notiser*, 127:278–4

Cernusca A (ed.) (1978). Ökologische Analysen von Almflächen im Gasteiner Tal. Veröff Oesterr MaB-Hochgebirgsprogramm Hohe Tauern Band 2, Universitätsverlag Wagner, Innsbruck

Chapin FS III and Körner Ch (1995). Patterns, changes, and consequences of biodiversity in arctic and alpine ecosystems. In Chapin FS III and Körner Ch (eds), *Arctic and Alpine Biodiversity: Patterns, Causes and Ecosystem Consequences*. Ecol Studies, 113:313–20, Springer, Berlin, Heidelberg, New York

Chown SL and Gaston KJ (2000), Areas, cradles and museums: the latitudinal gradient in species richness. *TREE*, 15:311–15

Crawford RMM and Chapman HM (1995). Climatic change and species polymorphism in the Arctic. *Ecosystems research report*, 10:115–21, European Commission, Brussels

Crawford RMM, Chapman HM and Smith LC (1995). Adaptation to variation in growing season length in arctic populations of *Saxifraga oppositifolia* L. *Botanical Journal of Scotland*, 41:177–92

Dahl E (1990). History of the Scandinavian alpine flora. In Gjaerevoll O (ed.), *Alpine Plants*. Tapir Publishers, Trondheim (Royal Norwegian Society of Science), pp. 16–21

De Benedetti SH and Parsons DJ (1979). Natural fire in subalpine meadows: A case description from the Sierra Nevada. *Journal of Forestry*, 77: 477–9

Deshmukh I (1986). *Ecology and Tropical Biology*. Blackwell Scientific, Palo Alto, Oxford, London

Engler A (1913). Einfluss der Provenienz des Samens auf die Eigenschaften der forstlichen Holzgewächse. *Mitteilungen Schweiz Centralanstalt für das forstliche Versuchswesen*, 10:190–386

Fox JF (1981). Intermediate levels of soil disturbance maximize alpine plant diversity. *Nature*, 293:564–5

Franklin JF, Moir WH, Douglas GW and Wiberg C (1971). Invasion of subalpine meadows by trees in the Cascade Range, Washington and Oregon. *Arctic and Alpine Research*, 3:215–24

Franz H (1979). *Hochgebirgsökologie*. Ulmer, Stuttgart

Gigon A (1983). Typology and principles of ecological stability and instability. *Mountain Research and Development*, 3:95–102

Gottfried M, Pauli H and Grabherr G (1998). Prediction of vegetation patterns at the limits of plant life: A new view of the alpine-nival ecotone. *Arctic and Alpine Research*, 30:207–21

Gough L, Grace JB and Taylor KL (1994). The relationship between species richness and community biomass: the importance of environmental variables. *Oikos*, 70:271–9

Grabherr G and Pauli MGH (1994). Climate effects on mountain plants. *Nature*, 369:448

Grabherr G (1997). The high mountain ecosystems of the alps. In Wielgolaski FE (ed.), *Ecosystems of the World 3, Polar and Alpine Tundra*. Elsevier, Amsterdam, Lausanne, New York, pp. 97–121

Grabherr G, Gottfried M, Gruber A and Pauli H (1995). Patterns and current changes in alpine plant diversity. In Chapin FS III and Körner Ch (eds), *Arctic and Alpine Biodiversity: Patterns, Causes and Ecosystem Consequences*. Ecological Studies, 113:167–81, Springer, Berlin

Hadley KS (1987). Vascular alpine plant distributions within the central and southern Rocky Mountains, USA. *Arctic and Alpine Research*, 19:242–51

Hedberg O (1964). Features of afroalpine plant ecology. *Acta Phytogeographica Suecica*, 49: 1–149

Heer C and Körner Ch (2002). High elevation pioneer plants are sensitive to mineral nutrient addition. Submitted to Basic and Applied Ecology, in press

Hegg O, Feller U, Dahler W and Scherrer C (1992). Long term influence of fertilization in a Nardetum. *Vegetatio*, 103:151–8

Hofstede RGM, Mondragon Castillo MX and Rocha Osorio CM (1995). Biomass of grazed, burned, and undisturbed Paramo grasslands, Colombia. I. Aboveground vegetation. *Arctic and Alpine Research*, 27:1–12

Holzer K and Nather J (1974). *Die Identifizierung von forstlichem Vermehrungsgut*. 100 Jahre Forstliche Bundesversuchsanstalt, Wien, pp. 15–42

Iwasa Y, Sato K, Kakita M and Kubo T (1993). Modelling biodiversity: Latitudinal gradient of forest species diversity. In Schulze E-D and Mooney HA (eds), *Biodiversity and Ecosystem Function*. Ecological Studies, 99:433–51, Springer, Berlin

Kalin Arroyo MT, Primack R and Armesto J (1982). Community studies in pollination ecology in the high temperate Andes of central Chile. I. Pollination mechanisms and altitudinal variation. *American Journal of Botany*, 69:82–97

Kapos V, Rhind J, Edwards M and Price MF (2000). Developing a map of the world's mountain forests. In Price MF and Butt N (eds), *Forests in Sustainable Mountain Development: A State-of-knowledge Report for 2000*. CAB International, Wallingford

Klötzli F (1975). Zur Waldfähigkeit der Gebirgssteppen Hoch-Semiens (Nordäthiopien). *Beitr naturk Forsch Südw-Dtl*, 34:131–47

Klötzli F (1997). Biodiversity and vegetation belts in tropical and subtropical mountains. In Messerli B and Ives JD (eds), *Mountains of the World: A global priority*. Parthenon Publishing Group, New York, London, pp. 232–5

Körner Ch (1995). Alpine plant diversity: a global survey and functional interpretations. In Chapin FS III and Körner Ch (eds), *Arctic and Alpine Biodiversity: Patterns, Causes and Ecosystem Consequences*. Ecological Studies, 113:45–62, Springer, Berlin

Körner Ch (1998). A re-assessment of high elevation treeline positions and their explanation. *Oecologia*, 115:445–59

Körner Ch (1999). *Alpine Plant Life*. Springer, Berlin

Körner C (2000a). The alpine life zone under global change. *Gayana Bot*, 57:1–17

Körner C (2000b). Why are there global gradients in species richness? Mountains might hold the answer. *Trends in Ecology and Evolution (TREE)*, 15:513–14

Körner C (2001). Alpine Ecosystems. *Encyclopedia of Biodiversity*. Academic Press, New York, pp. 133–44

Körner C, Diemer M, Schäppi B, Niklaus P and Arnone J (1997). The responses of alpine grassland to four seasons of CO_2 enrichment: a synthesis. *Acta Oecologica*, 18:165–75

Larigauderie A and Körner Ch (1995). Acclimation of leaf dark respiration to temperature in alpine and lowland plant species. *Annals of Botany*, 76:245–52

Lauer W and Klaus D (1975). Geoecological investigations on the timberline of Pico di Orziba, Mexico. *Arctic and Alpine Research*, 7: 315–30

Little EL Jr (1941). Alpine flora of San Francisco Mountain, Arizona. *Madrono*, 6:65–96

Mahaney WC and Linyuan Z (1991). Removal of local alpine vegetation and overgrazing in the Dalijia mountains, northwestern China. *Mountain Research and Development*, 11:165–7

Mark AF (1995). The New Zealand alpine flora and vegetation. *Quat Bull Alp Garden Soc*, 63: 245–59

Mark AF (1992). Indigenous grasslands of New Zealand. In Coupland RT (ed.), *Ecosystems of the World Vol. 8B. Natural Grasslands – Eastern Hemisphere*. Elsevier, Amsterdam, pp. 361–410

Messerli B and Ives JD (eds) (1997). *Mountains of the World: A global priority*. Parthenon Publishing Group, New York, London

Meyer E and Thaler K (1995). Animal diversity at high altitudes in the Austrian Central Alps. In Chapin FS III and Körner Ch (eds), *Arctic and Alpine Biodiversity: Patterns, Causes and Ecosystem Consequences*. Ecological Studies, 113:97–108, Springer, Berlin

Miehe G and Miehe S (2000). Comparative high mountain research on the treeline ecotone under human impact. *Erdkunde*, 54:34–50

Moffat AS (1996). Biodiversity is a boon to ecosystems, not species. *Science*, 271:1497

Mooney HA, Lubchenco J, Dirzo R and Sala OE (1995). Biodiversity and ecosystem functioning: Ecosystem analysis. In Heywood VH (ed.), *Global Biodiversity Assessment*. Cambridge University Press, Cambridge, pp. 328–452

Müller HN (1981). Messungen zur Beziehung Klimafaktoren – Jahrringwachstum von Nadelbaumarten verschiedener waldgrenznaher *Standorte. Mitt forstl. Bundesversuchsanstalt, Wien*, 142: 327–55

Nagy L, Grabherr G, Körner C and Thompson DBA (2002). *Alpine Biodiversity in Europe*. Submitted to Ecological Studies, Springer Verlag, Heidelberg

Nakhutsrishvili G and Gagnidze RI (1999). Die subnivale und nivale Hochgebirgsvegetation des Kaukasus. *Phytocoenosis*, 11:173–83

Odland A and Birks HJB (1999). The altitudinal gradient of vascular plant richness in Aurland, western Norway. *Ecography*, 22:548–66

Onipchenko VG and Blinnikov MS (eds) (1994). Experimental investigation of alpine plant communities in the northwestern Caucasus. *Veröff Geobot Inst ETH (Rübel)*, 115:3–118

Packer JG (1974). Differentiation and dispersal in alpine floras. *Arctic and Alpine Research*, 6:117–28

Pauli H, Gottfried M and Grabherr G (1999). Vascular plant distribution patterns at the low-temperature limits of plant life – the alpine–nival ecotone of Mount Schrankogel (Tyrol, Austria). *Phytocoenologia*, 29:297–325

Pignatti E and Wikus E (1987). Alpine grasslands and the effect of grazing. In Miyawaki A *et al.*, *Vegetation ecology and creation of new environments*. Tokai Univ Press, Tokyo, pp. 225–34

Pimm SL and Raven P (2000). Extinction by numbers. *Nature*, 403:843–5

Polunin O (1980). *Flowers of Greece and the Balkans: a field guide*. Oxford Univ Press, Oxford

Pucheta E, Cabido M, Diaz S and Funes G (1998). Floristic composition, biomass, and aboveground net plant production in grazed and protected sites in a mountain grassland of central Argentina. *Acta Oecologia*, 19:97–105

Quaranta A, Friz P, Tamanini C and Oberosler R (1996). Ethological observations of cattle on summer alpine pastures of Trentino Alto Adige (Italy). *Rivista di Biologia*, 89:221–32

Rahbek C (1995). The elevational gradient of species richness: a uniform pattern? *Ecography*, 18:200–05

Ram J (1992). Effects of clipping on aboveground plant biomass and total herbage yields in a grassland above treeline in central Himalaya, India. *Arctic and Alpine Research*, 24:78–81

Ramsay PM and Oxley ERB (1996). Fire temperatures and postfire plant community dynamics in Ecuadorian grass paramo. *Vegetatio*, 124:129–44

Raunkiaer C (1908). Livsformernes Statistik som Grundlag for biologisk Plantegeografi. *Botanisk Tidsskrift*, 29:42–83

Reisigl H and Pitschmann H (1958). Obere Grenzen von Flora und Vegetation in der Nivalstufe der zentralen Ötztaler Alpen (Tirol). *Vegetatio*, 8:93–129

Rikhari HC, Negi GCS, Pant GB, Rana BS and Singh SP (1992). Phytomass and primary productivity in several communities of a central Himalayan alpine meadow, India. *Arctic and Alpine Research*, 24:334–51

Rongfu H and Miehe G (1988). An annotated list of plants from Southern Tibet. *Willdenowia*, 18:81–112

Rosenzweig ML (1995). *Species Diversity in Space and Time*. Cambridge Univ Press, Cambridge

Rübel E (1911). *Pflanzengeographische Monographie des Berninagebietes*. Engler's Bot Jahrb 47, Leipzig

Rundel PW, Smith AP and Meinzer FC (eds) (1994). *Tropical Alpine Environments*. Cambridge University Press, Cambridge

Safford HD (1999). Brazilian Paramos I. An introduction to the physical environment and vegetation of the campos de altitude. *Journal of Biogeography*, 26:693–712

Salgado-Labouriau ML (1986). Late Quaternary paleoecology of Venezuelan high mountains. In Vuilleumier F and Monasterio M (eds), *High Altitude Tropical Biogeography*. Oxford Univ Press, New York, Oxford, pp. 202–17

Schäppi B and Körner Ch (1996). Growth responses of an alpine grassland to elevated CO_2. *Oecologia*, **105**:43–52

Schiechtl HM and Neuwinger I (1980). Regeneration von Vegetation und Boden nach Einstellung der Beweidung und Bodenstreunutzung in einem zentralalpinen Hochlagen-Aufforstungsgebiet. *Mittlg forstl Bundesversuchsanstalt, Wien*, **129**:63–80

Schwarzenbach FH (2000). *Altitude distribution of vascular plants in mountains of East and Northeast Greenland*. Bioscience 50, Danish Polar Centre

Steiger Th, Körner Ch and Schmid B (1996). Long-term persistence in a changing climate: DNA analysis suggests very old ages of clones of alpine *Carex curvula*. *Oecologia*, **105**:94–9

Steyermark JA (1986). Speciation and endemism in the flora of the Venezuelan tepuis. In Vuilleumier F and Monasterio M (eds), *High Altitude Tropical Biogeography*. Oxford Univ Press, New York, Oxford, pp. 317–73

Sulkinoja M and Valanne T (1987). Leafing and bud size in *Betula* provenances of different latitudes and altitudes. *Rep Kevo Subarctic Res Stat*, 20:27–33

Sundriyal RC (1992). Structure, productivity and energy flow in an alpine grassland in the Garhwal Himalaya. *Journal of Vegetation Science*, 3:15–20

Tang CQ and Ohsawa M (1999). Altitudinal distribution of evergreen broad-leaved trees and their leaf-size pattern on a humid subtropical mountain, Mt Emei, Sichuan, China. *Plant Ecology*, 145:221–33

Tappeiner U and Cernusca A (1998). Effects of land-use changes in the Alps on exchange processes (CO_2, H_2O) in grassland ecosystems. In Kovar K, Tappeiner U, Peters NE and Craig, RG (eds), *Hydrology, Water Resources and Ecology in Headwaters*. IAHS Publ 248:131–8

✳Theodose TA and Bowman WD (1997). Nutrient availability, plant abundance and species diversity in two alpine tundra communities. *Ecology*, **78**:1861–72

Tranquillini W and Havranek WM (1985). Influence of temperature on photosynthesis in spruce provenances from different altitudes. In Turner H and Tranquillini W (eds), *Establishment and tending of subalpine forest: Research and management*. Proc 3rd IUFRO Workshop, p. 1.07-00, 1984. Ber Eidg Anst forstl Versuchswes, 270:41–51

✳Trimble SW and Mendel AC (1995). The cow as a geomorphic agent – A critical review. *Geomorphology*, 13:233–53

Virtanen R, Dirnböck T, Dullinger S, Grabherr G, Pauli H, Staudinger M and Villar L (2002). Patterns in the Plant Species Richness of European High Mountain Vegetation. In Nagy L *et al.*, *Alpine Biodiversity in Europe*. Submitted to Ecological Studies Springer Verlag, Heidelberg

Vitousek PM, Loope LL and Adsersen H (1995). *Islands, Biological Diversity and Ecosystem Function*. Ecological Studies 115. Springer, Berlin

Vuilleumier F and Monasterio M (eds) (1986). *High Altitude Tropical Biogeography*. Oxford Univ Press, New York, Oxford, pp. 153–201

Walter H and Breckle SW (1991). *Ökologie der Erde*, vol 4. Fischer, Stuttgart

Wielgolaski FE (1997). *Polar and Alpine Tundra*. Ecosystems of the World 3 , Elsevier, Amsterdam

Wieser W (1973). *Effects of temperature on ectothermic organisms: ecological implications and mechanisms of compensation*. Springer, Berlin

Wilson SD (1994). The contribution of grazing to plant diversity in alpine grassland and heath. *Australian Journal of Ecology*, 19:137–40

Wohlgemuth T (1993). Der Verbreitungsatlas der Farn- und Blütenpflanzen der Schweiz (Welten und Sutter 1982) auf EDV: Die Artenzahlen und ihre Abhängigkeit von verschiedenen Faktoren. *Botanica Helvetica*, **103**:55–71

Part II

How much Mountain Biodiversity is there and Why?

Intraspecific Genetic Diversity in Alpine Plants

2

Irène Till-Bottraud and Myriam Gaudeul

INTRODUCTION

Biodiversity, defined as the diversity of the living world, has recently received much attention from both the scientific and political communities (see the Rio de Janeiro and follow-up international conferences on biodiversity). The awareness of a general decrease in biodiversity and of an increase in extinction rates in all taxa and all parts of the world is the main cause for this attention. However, the major problem, apart from understanding the functional meaning of the word, is still simply to correctly assess this biodiversity. On the one hand, we do not know the exact number of species living on this planet (and when it comes to bacteria, we are far from even having a correct estimate!). On the other hand, except for individuals from the same clone, we know that every individual of every species is genetically unique. Biodiversity measures should, ideally, include a sample from a large proportion of the genome of many individuals from most species in an ecosystem. This is unreasonable and unrealistic, as many species have not yet been identified. However, representative estimates should combine data at all levels. Biodiversity is indeed measured at different levels, but the measures have never been combined. One series of measures is based on the number of species in a community or a geographic area (or a combination of number and frequency), while another set measures the diversity of genotypes within a population or a species (genetic variability). In both types of measures the total diversity is sometimes partitioned into local or broader scale diversity (α, β or γ for species diversity; within or among populations for genetic diversity). The measures based on the number of species are essentially used by ecologists and stem from an early taxonomic idea that species are homogeneous entities and can be well represented by a single type specimen. According to this view, estimating species diversity should give a good estimate of total biodiversity. However, species are not monotypic and, as shown below, within species diversity is important for the persistence of each species. Biodiversity therefore represents a continuum that goes from high taxonomic levels (orders, families, species) all the way to among-individuals diversity, and all taxonomic levels have to be taken into account to obtain an accurate estimate of biodiversity.

Biodiversity, both in terms of species and within-species diversity, has proved important in the dynamics of ecological systems. The number of species in an ecosystem has a strong influence on its stability (the 'insurance policy' hypothesis) and on its productivity (Hector *et al.*, 2000; but this is still controversial, see Huston *et al.*, 2000). On the other hand, the amount of genetic variability of a species will determine its short-term (adaptive capacity) and long-term (evolutionary potential) ability to respond to environmental change. Persistence of populations has been shown to be positively linked to genetic variability (Saccheri *et al.*, 1998; Frankham and Ralls, 1998). Although Lande (1988) argued that demographic factors were more important than genetic factors in determining the short-term fate of a population, it is now accepted that they often act synergistically. Genetic variability may interact

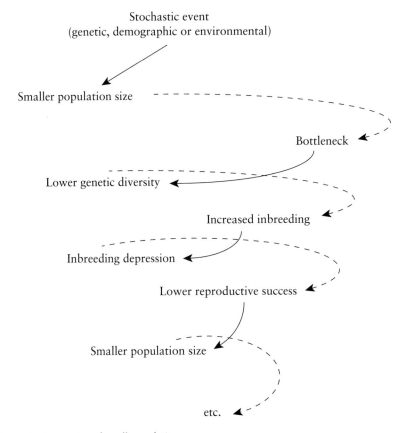

Figure 2.1 The extinction vortex of small populations

with demographic effects to produce the 'extinction vortex' of small populations (Gilpin and Soulé, 1986; Figure 2.1): if the size of a population is substantially reduced by a stochastic event (either environmental, demographic or genetic), genetic consequences such as inbreeding depression and/or decreased variability may be triggered. This, in turn, will affect the overall fitness and adaptive potential of the population and the effective population size will steadily decline, further accentuating the initial cause of the vicious spiral, ultimately leading to extinction. These phenomena are valid for all species and all types of environments. They are, however, more crucial in species with a small geographical range or with a narrow ecological niche because environmental changes will have a stronger impact on these.

Alpine species will be particularly vulnerable in the case of global warming because fragmentation of the landscape is high, their limited range and location on top of mountains (they will not be able to move higher; see later Chapters). It is therefore crucial to evaluate their genetic diversity in order to estimate their adaptive potential. Another question concerns determination of the kind of within-species diversity patterns that can be expected in alpine environments due to ecological specificity. Many hypotheses can be formulated and five mutually non-exclusive hypotheses are addressed. The first three concern within-population diversity, the fourth concerns genetic differentiation among populations, and the last is a null hypothesis.

(1) Because alpine environments are harsh, plants are exposed to strong directional selection. Moreover, selection is probably stronger at higher altitudes. The action of selection is to retain the fittest individuals that generally share some adaptive characteristics and to eliminate the others, reducing diversity. We thus expect lower genetic diversity in alpine compared to plain-dwelling plants, and reduced within-population diversity with increasing altitude.

(2) Alpine habitats are characterised by high environmental heterogeneity over small distances and high temporal (climatic) variation. These conditions should lead to the opposite effect of (1), i.e. cause high within-population diversity or select for high plasticity (i.e. the same genotype produces different – adapted – phenotypes in different environments).

(3) During the last Ice Age, that ended some 10 000 years ago, most mountains were bare of vegetation and recolonisation is recent. For the species that have an arctico–alpine distribution, we expect different levels of variability in the Arctic compared to the Alps: diversity should be higher in the Alps for species with southern refugia, whereas diversity should be lower in the Alps for species with Arctic refugia. Indeed, recolonisation of an area requires the foundation of new populations. This generally occurs via the migration of a very low number of individuals into new areas (founder effects). Therefore, new populations usually have lower diversity than the populations of origin. As a result, if recolonisation is a stepping stone process, the most recent populations will have the lowest diversity.

(4) Alpine habitats are very fragmented, with little connection between areas of similar ecology and have sharp boundaries. Migration between populations should thus be rare and gene exchange between populations will be low. Therefore, populations will diverge at random (genetic drift). This should lead to stronger among-population differentiation in alpine habitats than at lower altitudes.

(5) If the biology of the species (life form, reproductive system, etc.) is an important factor driving its diversity, as reviewed by Hamrick and Godt (1989), we globally expect the same variability levels in alpine plants as in lowland species, unless life form, reproductive system, or other characteristics are different in alpine compared to lowland species.

MATERIAL AND METHODS

Literature data

Studies that specifically address the genetic diversity of alpine plants (*sensu stricto*, i.e. above-treeline vegetation, Körner, 1995) are scarce. Therefore, in order to work on a large enough sample of studies, we extended our search to montane species, supposing that our hypotheses also hold for these species. Moreover, as we consider hypotheses addressing altitude variations, this will enable comparison of alpine and montane species. We also included phylogeographic studies, despite their low within-population sample sizes that can lead to underestimation of within-population diversity (Ellstrand and Roose, 1987). Several alpine or montane species were studied because they are rare or endangered, or because they exhibit peculiar biological characteristics (such as clonality, which is quite frequent at high altitudes). This again could lead to a biased estimate of diversity and will be kept in mind when analysing the data. We thus were left with 24 studies representing 19 species.

How is genetic variability estimated?

To measure genetic variability, one must obviously identify some genetically inherited traits. These can be directly observable morphological characters such as flower colour (or the colour of other plant parts), leaf shape, size, flowering time, etc. Flower colour is often a simple character (it has only a few modalities) and has a simple inheritance (one or two genes). However, the other traits are quantitative measures that generally show complex inheritance (polygenic) and are, moreover, both genetically and environmentally determined. Thus, the first step is to extract the genetic part of the variability (genetic variance, V_G) from total variability (genetic + environmental). This can only be done if the plants are cultivated in controlled conditions to ensure that all plants are submitted to the same environment. Plasticity (i.e. the same genotype produces different – adapted – phenotypes in different environments) is measured as the genetic x environmental variance and can be estimated if the same genotypes are cultivated in at least two different controlled conditions. The genetic variance can be further decomposed into additive variance (V_A, which is directly inherited and subject to selection) and non-additive variance (such as dominance or more complex interactions among genes). This partitioning of variance is done through a standard analysis of variance (Falconer, 1960). Genetic differentiation among populations can be similarly estimated through an analysis of variance.

Less directly observable types of traits for the estimation of genetic variability are molecular markers (Karp *et al.*, 1996). They represent variability at the protein (isozyme) or DNA level (AFLP, RFLP, microsatellites, RAPD, sequence, etc.). They are more commonly studied than morphological markers because they have a simple inheritance (one gene) and are either co-dominant (heterozygotes are different from homozygotes) or dominant (heterozygotes and one of the homozygote genotypes are identical). DNA variation can also be studied using chloroplastic or mitochondrial DNA. In this case, the marker is generally maternally inherited and is haploid. From these sets of markers, population geneticists use different types of parameters to estimate (i) genetic diversity within a species or a population and (ii) genetic differentiation among populations, i.e. to partition the diversity at different levels: within populations, among populations, among geographic regions, etc.

If the markers are diploid and codominant, the most frequently used parameters of global diversity are the proportion of polymorphic loci (P), the mean number of alleles per locus (A) and the expected heterozygosity (H_e, a measure combining the number of alleles (n) and their frequency (p_i); $H_e = 1/n\Sigma p_i(1 - p_i)$). H_e ranges from 0 (no variability) to 1. The fixation index (Fst; Wright, 1951) is often used to estimate the genetic structure of the populations. Fst measures the proportion of the variance in allele frequencies that is attributable to genetic differentiation among populations. For a locus with two alleles of average frequencies p and q in the whole sample, Fst = var(p)/2pq; var(p) being the variance in allele frequency. Fst varies from 0 (identical populations) to 1 (no shared allele in the different populations). An analogue of Fst has been derived for polygenic characters (Qst; Spitze, 1993).

For haploid markers, the global diversity is usually estimated through a similarity index (e.g. Dice, 1945; $S_{ij} = 2n_{ij}/n$; with n_{ij} the number of bands shared by populations i and j and n the total number of bands) or the Shannon index ($H = -(1/n)\Sigma p_i \log_2 p_i$; with p_i the frequency of band i and n the total number of bands; Lewontin, 1972). An equivalent of the analysis of variance, the AMOVA (Analysis of MOlecular Variance; Excoffier *et al.*, 1992), gives the proportion of the total diversity attributable to within and among populations (or additional subdivisions if needed). For AMOVA, the values obtained are relative values but not absolute estimates of diversity.

Diploid, dominant markers can be analysed as haploid or diploid co-dominant markers, but this requires an assumption about the genotypic structure of the populations (e.g. Hardy–Weinberg proportions) in order to estimate the allele frequencies.

Contrary to morphological variations that can lead to drastic differences in the adaptation of a plant (difference in fitness), most variations at the molecular level have very little influence on the fitness of individuals: they are not subject to selection and are called 'neutral markers'. Genetic variability is then almost entirely due to the combined action of mutation (that generates this variability), migration (gene exchange), and random genetic drift, all of which depend on population size.

Different diversity estimators are not directly comparable, and the different types of markers have different intrinsic levels of variability (e.g. mitochondrial DNA has a slower mutation rate than microsatellite DNA, leading to a much lower within-population diversity). Moreover, in most cases, invariant or ambiguous loci are discarded and only variable bands are studied, thus artificially increasing the levels of variability and rendering comparisons meaningless. Comparisons will thus be made only between studies using the same estimators, and differences among markers will be kept in mind.

RESULTS AND DISCUSSION

Does diversity decrease with altitude?

Selection

The hypothesis that diversity decreases as selection pressures increase is obviously true for the selected characters (only the 'best' morphology is maintained), but is also expected for neutral characters if selection is very strong because genes are linked together on the chromosomes. We found only two studies on morphological variation within species (*Dryas octopetala* in the Alaskan mountains, McGraw, 1985; *Polemonium viscosum* in the Rockies, Galen *et al.*, 1991). The first addresses ecotypic differentiation, with two ecotypes in different environments or microenvironments. Although this type of study is important and shows morphological (and adaptive) variation across very short distances, the results are not adequate for testing our hypothesis as we are looking for morphological diversity within a given environment. The second studied variation in morphological characters in two populations 500 m apart in elevation, but no estimate of the genetic variance within each sub-population was made.

Only one study addresses the question of neutral markers directly. Gugerli *et al.* (1999) studied five pairs of populations of *Saxifraga oppositifolia* from three areas in the Swiss Alps with 84 polymorphic RAPD markers. The pairs of populations were 400 m apart in elevation. The data were analysed with AMOVA and no significant effect of altitude could be detected, although the diversity was slightly lower in high elevation populations. It could be argued here that the altitudinal difference was too small (only 400 m) to observe any significant differences.

No general pattern emerges for all the species for which the expected heterozygosity (He,) or other parameter was measured within populations (Table 2.1). Expected heterozygosity is very variable among species, ranging from 0 in *Saxifraga cernua* in the Alps and *Bensoniella oregona* in northern America (but this may be expected as they are both clonal; Bauert *et al.*, 1998; Soltis, 1992), to 0.29 in *Populus tremuloides* in Canada (more surprisingly as it is a clonal species; Jelinski and Cheliak, 1992). A similar range was found in 468 lowland species assessed for isozymes: He values ranged from: 0.35 to 0 (Hamrick *et al.*, 1991), with an average of 0.113 (Hamrick and Godt, 1989). Note how the expected heterozygosity value is dependent on what loci are considered: in *Haplostachys haplostatchya* (Morden and Loefler, 1999), He = 0.166 when all loci are taken into account, whereas it increases to 0.357 when only the polymorphic loci are considered. The other parameters that were estimated, especially the Shannon index values, were not all computed similarly and are thus impossible to compare. This emphasises the care that should be taken when such comparisons are made. When corrections to the Shannon index are made to render the data comparable, the values are not greater than 0.04 for the two endemic species, and the maximum value was found in *Eryngium alpinum* (0.34; Gaudeul *et al.*, 2000), whereas the values can reach 0.23 in common lowland species (see Gaudeul *et al.*, 2000). There is, therefore, no strong support for our hypothesis.

Environmental heterogeneity

The hypothesis of selection for plasticity cannot be tested as neither of the two studies on morphological variation estimated the plasticity of the plants (McGraw, 1985;

Table 2.1 Within-population genetic diversity. General survey

Species	Author	Pop/ind[a]	Marker	No[b]	Diversity[c]	a/m[d]	Biology
Populus tremuloides	Jelinski & Cheliak, 1992	6/26	Isozymes	14	$H_e = 0.29$	a	clonal
Lloydia serotina	Jones et al., 1998	16/25	Isozymes	10	$H_e = 0.064$	a	clonal
Silene acaulis	Abbott et al., 1995	9/20	Isozymes	8	$H_e = 0.20$	a	outcross
Saxifraga cernua	Bauert et al., 1998	7/10	RAPD	30	$H_e = 0$	a	clonal
Primula maguirei	Wolf & Sinclair, 1997	8/25	Isozymes	13	$H_e = 0.11$	m	outcross, endemic
Pinus sylvestris	Sinclair et al., 1999	7/30	mtDNA		$H_e = 0.586$	m	outcross
Prunus africana	Dawson & Powell, 1999	10/7	RAPD	48	$H_e = 0.075$	m	outcross
Haplostachys haplostachya	Morden & Loefler, 1999	3/15	RAPD	54 122	$H_e = 0.357$ $H_e^* = 0.166$	m	outcross, endemic
Bensoniella oregona	Soltis et al., 1992	6/6–45	Isozymes	24	$H_e = 0$	m	clonal
Veratrum album	D. Kleijn T. Steinger (2002)	4 pastures/50 4 hayfields/50	RAPD	19	G/N = 0.91 G/N = 0.54	m	clonal outcross
Saxifraga cernua	Gabrielsen & Brochmann, 1998	2/46	RAPD	38	G/N = 0.33	a	clonal
Fitzroya cupressoides	Allnutt et al., 1999	12/5	RAPD	54	Sh = 0.34–0.64	m	outcross
Erodium paularense	Martin et al., 1999	6/10–20	RAPD	171	Sh = 0.99–3	m	outcross, endemic
Eryngium alpinum	Gaudeul et al., 2000	14/24	AFLP	62	Sh = 0.19–0.34 $H_e = 0.13–0.23$	m	outcross
Astragallus cremnophylax var. cremnophylax	Travis et al., 1996	3/47	AFLP	352	P = 0.38 $H_e = 0.13–0.02$	m	outcross, endemic
Carex curvula	Steinger et al., 1996	1/116	RAPD	95	$S_{ij} = 0.65$	a	clonal

a: number of populations/individuals per population studied

b: total number of loci

c: mean population genetic diversity estimates;

H_e: expected heterozygosity,

H_e^*: H_e estimated from polymorphic loci (monomorphic loci were excluded), in the other cases, there is no mention of the type of loci used; G/N: number of genotypes/number of samples; Sh: Shannon index; S_{ij}: similarity index; P: proportion of polymorphic loci

d: alpine/montane species

Table 2.2 Among-population genetic diversity. Comparison of high and low elevation populations for *Saxifraga oppositifolia* and *Saxifraga cernua*. *S. oppositifolia* in the Alps (Gugerli *et al.*, 1999) vs. *S. oppositifolia* in the Arctic (different regions; Gabrielsen *et al.*, 1997); *S. cernua* in the Alps (Bauert *et al.*, 1998) vs. *S. cernua* in the Arctic (Svalbard; Gabrielsen and Brochmann, 1998)

AMOVA	Pop[a]	Ind[b]	Vwp[c]	Vbp(r)[d]	Vbr[e]	No. of RAPD markers
S. oppositifolia						
Alps	10	20	95%	5%		84
Arctic	18	5	64%	28%	9%	35
S. cernua						
Alps	7	7–12	0	0	100%	30
Svalbard	2	46	59%	41%		38

[a]: number of populations; [b]: number of individuals per population; [c]: variance within populations; [d]: variance between populations within regions; [e]: variance between regions

Galen *et al.*, 1991). However, in both studies, there is substantial variation across short distances indicating that habitat heterogeneity over short distances can potentially maintain genetic variability. The data for neutral markers (see above) do not indicate any general trend. The hypothesis of increased within-population diversity with elevation is thus not confirmed.

Post-glacial recolonisation

This question is actually tautological: the refuge areas, or the present populations closest to former refugia, are supposed to be the most diverse because they required less migration, thus fewer founder events are characterised by a loss of genetic diversity. To correctly address this hypothesis would require independent information about the location of refugia on the one hand, and genetic diversity on the other hand. However, all studies we found are based only on genetic diversity to identify refuge areas of a species (see for example Abbott *et al.*, 1995 and 2000 and Gabrielsen *et al.*, 1997 for *Saxifraga oppositifolia*; Sinclair *et al.*, 1999 for *Pinus sylvestris*).

Is there stronger among-population differentiation at higher elevations?

In two species of *Saxifraga* (*S. cernua* and *S. oppositifolia*) studies have been performed both at high elevation in the Alps (above 2000 m; Bauert *et al.*, 1998; Gugerli *et al.*, 1999) and at low elevation in the arctic (Gabrielsen *et al.*, 1997, Gabrielsen and Brochmann, 1998; Table 2.2). All studies used RAPD markers (although not necessarily the same markers and not the same number) and partitioned the variance with AMOVA. In *S. cernua* a unique genotype is found in each of the three regions of the Swiss Alps that were studied, leading to a 100% variance among regions, whereas in Svalbard some variation was found within populations and the variance among populations is only 41%. It should be emphasised that these values are relative and that in the Alps the global diversity is extremely low (only three genotypes!). The 100% variance among populations, although technically true, does not mean much in terms of a partition of diversity. *S. cernua* reproduces by bulbils, but is probably able to set seeds in the arctic while remaining purely clonal in the Alps (Gabrielsen and Brochmann, 1998). In *S. oppositifolia* the variance between populations or regions is lower in the Alps than in the arctic, but only five individuals were sampled per population in the arctic study, thus probably underestimating the within-population part of the variance and overestimating the among-population diversity (again because these values are relative values). Moreover, as the high and low elevation areas are very far apart,

Table 2.3 Among-population genetic diversity. Species studied with the fixation index (Fst)

Species	Author	Pop/ind[a]	Marker	No[b]	Fst	a/m[c]	Biology
Lloydia serotina	Jones *et al.*, 1998	16/25	Isozymes	10	0–0.10	a	clonal
Polemonium viscosum	Galen *et al.*, 1991	2/20	Isozymes	3	0.09	a	outcross
Eryngium alpinum	Gaudeul *et al.*, 2000	14/24	AFLP	62	0.17–0.63	m	outcross
Populus tremuloïdes	Jelinski & Cheliak, 1992	6/26	Isozymes	14	0.03	m	clonal

[a]: number of populations/individuals per population studied; [b]: total number of loci [c]: alpine/montane species

refugia and recolonisation routes probably have a strong impact on the diversity, and the effect of these various mechanisms cannot be distinguished. Therefore, no conclusions can be drawn from these studies.

Other alpine or montane species have been studied. Four species were studied using Fst (Table 2.3). The Fst values are not greater in alpine than in montane species. Here the range in which the species has been studied, or its habitat, should be considered. We expect a stronger divergence among populations as geographic distance increases (because migration across populations is expected to be lower), and some habitats are less prone to gene exchange. In *Polemonium viscosum* two populations, 1.5 km apart, were studied and the Fst value was low, although significantly different from zero. On the contrary, two *Eryngium alpinum* populations were studied across the French Alps and both have a high global Fst, and high pairwise Fst (most are above 0.2) even for populations less than 10 km apart. This species lives in open habitats within the forested elevation range in the Alps. This characteristic at least partly explains the strong differentiation observed (Gaudeul *et al.*, 2000).

The AMOVA was used in ten species (Table 2.4). There is a lower proportion of the variance among populations in three (out of four) alpine species compared to montane species, however, all three are outcrossers. The fourth alpine species is selfing and has, as expected, a high proportion of the variance among populations. Four of the six montane

species have a relatively high proportion of the variance among populations. One is highly selfing (*Gentianella germanica*). In another, *Eryngium alpinum*, Fst values (Table 2.3) are very high, indicating strong differentiation despite lower variance among populations. This is again because AMOVA gives relative values: the within population diversity being very high (Table 2.1), the among-population variance, although high in absolute values, appears relatively low. The variability of *Pinus sylvestris* was studied using mitochondrial markers. These markers have a slow mutation rate and are therefore not as variable within populations, but exhibit high variation levels among populations (which is why they are commonly used to study variation over the entire range of widespread species). It is, therefore, not surprising that most of the variance observed is found among populations and is probably not due to the fact that it is a montane species. Here again, no clear trend appears with elevation. However, as we have seen, the type of marker or estimator used, or the biological characteristics of the species appear to be of importance.

How important is the biology of the species in determining the levels of variability?

Hamrick and Godt (1989) and Hamrick *et al.* (1991) have surveyed isozyme diversity in plant species characterised by various regional distributions and reproductive modes. Their

Table 2.4 Among-population genetic diversity. Species studied with Analysis of MOlecular Variance (AMOVA)

Species	Author	Pop/ind[a]	Marker	Vup[b]	Vbp(r)[c]	Vbr[d]	a/m[e]	Biol
Silene acaulis	Abbott et al., 1995	9/20	Isozymes	87%	13%		a	outcross
Saxifraga oppositifolia	Gugerli et al., 1999	10/20	RAPD	95%	5%		a	outcross
Draba aizoides	Widmer & Baltisberger, 1999	6/5	cpDNA	72%	28%		a	outcross
Saxifraga cespitosa	Tollefsrund et al., 1998	30/5	RAPD	26%	42%	32%	a	selfing
Gentianella germanica	Fischer & Matthies, 1998	11/5	RAPD	37%	14%	49%	m	70% selfing
Erodium paularense	Martin et al., 1999	6/15	RAPD	68%	16%	17%	m	outcross, endemic
Primula maguirei	Wolf & Sinclair, 1997	8/25	Isozymes	57%	43%		m	outcross, endemic
Pinus sylvestris (Spain)	Sinclair et al., 1999	7/30	mtDNA	20%	80%		m	outcross
Prunus africana	Dawson & Powell, 1999	10/7	RAPD	27%	7%	66%	m	outcross
Eryngium alpinum	Gaudeul et al., 2000	14/24	AFLP	38%	62%		m	outcross

[a]: number of populations/individuals per population studied; [b]: variance within populations; [c]: variance between populations within regions; [d]: variance between regions; [e]: alpine/montane species

31

general conclusions follow. 'Boreal–temperate' species have a lower within- and among-population diversity than 'temperate and tropical' species (but this difference accounted for only a small proportion of the observed variance). Conversely, the breeding system has a relatively high impact on both diversity levels. Outcrossing species generally have a very high level of within population diversity and low differentiation among populations, whereas selfing species have low within population diversity and high differentiation among populations. Last, endemics have lower global diversity than widespread species but partition it in a similar fashion.

The within-population values of diversity obtained in the studies we have surveyed are not particularly low, but are extremely variable (Table 2.1). Similarly, the degree of differentiation among populations is quite variable (Tables 2.3 and 2.4). No particular trend emerges with elevation. However, the reproductive system of the species could explain a lot of this diversity. The two selfing species (*Saxifraga caespitosa* and *Gentianella germanica*) have low within- and high among-population variance compared to the outcrossers. The four studies on endemic species showed average levels of expected heterozygosity.

Clonality is a common trait of the alpine flora and, as such, could influence the levels of variability of alpine plants. A common idea is that clonal plants exhibit little within-population variation because of reduced sexual reproduction. This is, however, not the case. According to Hamrick and Godt (1989; data from 66 species), populations of clonal species can be either totally monomorphic, but this is more the exception than the rule, or exhibit a normal diversity pattern. Similarly, Ellstrand and Roose (1987; data from 27 species) found that clonal species often exhibit considerable variation both within and among populations, and that most species are multiclonal. In our survey (Table 2.1), some clonal species do show very low within population diversity (H = 0 for *Saxifraga cernua* in the Alps and *Bensoniella oregana*; H = 0.06 for *Lloydia serotina*), while others have average levels of diversity (*Populus*

tremuloides, *Carex curvula*, *Veratrum album*). Similarly, 32 different genotypes were found in a 10 m × 20 m area of a *Rhododendron ferrugineum* population in the French Alps, assessed with AFLP markers (Escaravage *et al.*, 1998). In some cases, such as in *Veratrum album* (D. Kleijn and T. Steinger pers. com.), the level of diversity depends on the type of site (hayfield or pasture).

CONCLUSION

Although the number of studies we could use is relatively low (24, corresponding to 19 species), and the data often are not adequate (limited sample sizes leading to an under-estimation of within-population diversity; endangered or clonal species), the levels of within- and among-population diversity found in alpine and montane plants are comparable to those found at lower elevations. There does not seem to be any significant reduction in diversity at higher elevations, and alpine species are thus not immediately at risk because of their habitat, from the genetic viewpoint, in the event of environmental change, assuming that neutral variation correlates with adaptive potential. This does not, however, mean that this risk is irrelevant in the future, as any decrease in population size might lead some populations into the extinction vortex.

All the results point to no general trend either for within-population diversity, or for genetic structure. This might be due to the heterogeneity in sample size (number of populations, number of individuals per population), surveyed area (continent vs. mountain range), marker type (low vs. high mutation rates markers have different intrinsic diversity levels) and type of analysis (measuring absolute or relative levels of diversity) from one study to the other that renders comparisons difficult. However, diversity and differentiation patterns seem related to the reproductive characteristics of the species (selfing vs. outcrossing) rather than to elevation. Clonality, on the other hand, does not appear to be an important factor acting on diversity.

The most parsimonious conclusion at present is, therefore, that the reproductive biology of the species seems to override any specific effect of the alpine environment. However, we found very few suitable studies on alpine and montane species and general trends are thus difficult to observe, especially as the different studies used different markers and estimators. Moreover, we found only one study directly addressing one of our precise hypotheses. Indeed, most studies cannot distinguish the effect of various biotic or abiotic mechanisms in shaping the genetic diversity, and more specific studies are needed before firm conclusions can be drawn.

ACKNOWLEDGEMENTS

We are most indebted to Christian Körner who suggested this review, and to Gordon Luikart and an anonymous referee for comments on the manuscript.

References

Abbott RJ, Chapman HM, Crawford RMM and Forbes DG (1995). Molecular diversity and derivations of populations of *Silene acaulis* and *Saxifraga oppositifolia* from the high Arctic and more southern latitudes. *Molecular Ecology*, 4:199–207

Abbott RJ, Smith LC, Milne RI, Crawford RMM, Wolff K and Balfour J (2000). Molecular analysis of plant migration and refugia in the arctic. *Science*, 289:1343–6

Allnutt TR, Newton AC, Lara A, Premoli A, Armesto JJ and Vergara R (1999). Genetic variation in *Fitzroya cupressoides* (alerce), a threatened South American conifer. *Molecular Ecology*, 8:975–87

Bauert MR, Kälin M, Baltisberger M and Edwards PJ (1998). No genetic variation detected within isolated relict populations of *Saxifraga cernua* in the Alps using RAPD markers. *Molecular Ecology*, 7:1519–27

Dawson IK and Powell W (1999). Genetic variation in the Afromontane tree *Prunus africana*, an endangered medicinal species. *Molecular Ecology*, 8:151–6

Dice LR (1945). Measures of the amount of ecological association between species. *Ecology*, 26: 297–302

Ellstrand NC and Roose ML (1987). Patterns of genotypic diversity in clonal plant species. *American Journal of Botany*, 74:123–31

Escaravage N, Questiau S, Pornon A, Doche B and Taberlet P (1998). Clonal diversity in a *Rhododendron ferrugineum* L. (Ericaceae) population inferred from AFLP markers. *Molecular Ecology*, 7:975–82

Excoffier L, Smouse PE and Quattro JM (1992). Analysis of molecular variance inferred from metric distances among DNA haplotypes: application to human mitochondrial DNA restriction data. *Genetics*, 131:479–91

Falconer DS (1989). *Introduction to Quantitative, Genetics*, 3rd edition. Longman Sci. and Tech., Harlow, UK

Fischer M and Matthies D (1998). RAPD variation in relation to population size and plant fitness in the rare *Gentianella germanica* (Gentianaceae). *American Journal of Botany*, 85:811–19

Frankham R and Ralls K (1998). Inbreeding leads to extinction. *Nature*, 392:441–2

Gabrielsen TM, Bachmann K, Jakobsen KS and Brochmann C (1997). Glacial survival does not matter: RAPD phylogeography of Nordic *Saxifraga oppositifolia*. *Molecular Ecology*, 6:831–42

Gabrielsen TM and Brochmann C (1998). Sex after all: high levels of diversity detected in the arctic clonal plant *Saxifraga cernua* using RAPD markers. *Molecular Ecology*, 7:1701–08

Galen C, Shore JS and Deyoe H (1991). Ecotypic divergence in alpine *Polemonium viscosum*: genetic structure, quantitative variation, and local adaptation. *Evolution*, 45:1218–28

Gaudeul M, Taberlet P and Till-Bottraud I (2000). Genetic diversity in an endangered alpine plant, *Eryngium alpinum* L. (Apiaceae), inferred from amplified fragment length polymorphism markers. *Molecular Ecology*, 9:1625–39

Gilpin ME and Soulé ME (1986). Minimum viable populations: processes of species extinction. In Soulé ME (ed.), *Conservation Biology, the Science of Scarcity and Diversity*. Sinauer, Sunderland, MA, pp 19–34

Gugerli F, Eichenberger K and Schneller JJ (1999). Promiscuity in populations of the cushion plant *Saxifraga oppositifolia* in the Swiss Alps as inferred from random amplified polymorphic DNA (RAPD). *Molecular Ecology*, 8:453–61

Hamrick JL and Godt MJW (1989). Allozyme diversity in plant species. In Brown A, Clegg MT, Kahler AL and Weir BS (eds), *Plant population genetics, breeding and genetic resources.* Sinauer Sunderland, MA, pp 43–63

Hamrick JL, Godt MJW, Murawski DA and Loveless MD (1991). Correlations between species traits and allozyme diversity: implications for conservation biology. In Falk DA and Holsinger KE (eds), *Genetics and Conservation of Rare Plants.* Oxford University Press, New York, pp 75–86

Hector A, Schmid B, Beierkuhnlein C, Caldeira MC, Diemer M, Dimitrakopoulos PG, Finn JA, Freitas H, Giller PS, Good JRH, Högberg P, Huss-Danell K, Joshi J, Jumpponen A, Körner C, Leadley PW, Loreau M, Minns A, Mulder CPH, O'Donovan G, Otway SJ, Pereira JS, Prinz A, Read DJ, Scherer-Lorenzen M, Schulze ED, Spehn EM, Terry AC, Troumbis AY, Woodward FI, Yachi S and Lawton JH (2000). Plant diversity and productivity experiments in European grasslands. *Science,* **286**:1123–7

Huston MA, Aarsen LW, Austin MP, Cade BS, Fridley JD, Garnier E, Grime JP, Hodgson J, Lauenroth WK, Thompson K, Vandermeer JH and Wardle DA (2000). No consistent effect of plant diversity on productivity. *Science,* **289**:1255

Jelinski DE and Cheliak W (1992). Genetic diversity and spatial subdivision of *Populus tremuloides* (Salicaceae) in a heterogeneous landscape. *American Journal of Botany,* **79**:728–36

Jones B and Gliddon C (1999). Reproductive biology and genetic structure in *Lloydia serotina.* *Plant Ecology,* **141**:151–61

Karp AA, Seberg O and Buiatti M (1996). Molecular techniques in the assessment of botanical diversity. *Annals of Botany,* **78**:143–9

Kleijn D and Steinger T (2002). Contrasting effects of grazing and hay cutting on the spatial and genetic population structure of *Veratrum album,* and unpalatable, long-lived, clonal plant species. *Journal of Ecology,* **in press**

Körner C (1995). Alpine plant diversity: a global survey and functional interpretations. In Chapin SL and Körner C (eds), *Arctic and Alpine Biodiversity.* Springer Verlag, Berlin Heidelberg, pp 45–62

Lande R (1988). Genetics and demography in biological conservation. *Science,* **241**:215–44

Lewontin RC (1972). The apportionment of human diversity. *Evolutionary Biology,* **6**:381–94

Martin C, Gonzales-Benito ME and Iriondo JM (1999). The use of genetic markers in the identification and characterisation of three recently discovered populations of a threatened plant species. *Molecular Ecology,* **8**:S31–S40

McGraw JB (1985). Experimental ecology of *Dryas octopetala* ecotypes. III Environmental factors and plant growth. *Arctic and Alpine Research,* **17**:229–39

Morden CW and Loefler W (1999). Fragmentation and genetic differentiation among subpopulations of the endangered Hawaiian mint *Haplostachys haplostachya* (Lamiaceae). *Molecular Ecology,* **8**:617–25

Saccheri I, Kuussaari M, Kankare M, Vikman P, Fortelius W and Hanski I (1998). Inbreeding and extinction in a butterfly metapopulation. *Nature,* **392**:491–4

Sinclair WT, Morman JD and Ennos RA (1999). The post-glacial history of Scots pine (*Pinus sylvestris* L.) in western Europe: evidence from mitochondrial DNA variation. *Molecular Ecology,* **8**:83–8

Soltis PS, Soltis DE, Tucker TL and Lang FA (1992). Allozyme variability is absent in the narrow endemic *Bensoniella oregona* (Saxifragaceae). *Conservation Biology,* **6**:131–4

Spitze K (1993). Population structure of *Daphnia obtusa*: quantitative genetic and allozymic variation. *Genetics,* **135**:367–74

Steinger T, Körner C and Schmid B (1996). Long-term persistence in a changing climate: DNA analysis suggests very old ages of clones of alpine *Carex curvula.* *Oecologia,* **105**:94–9

Tollefsrud MM, Bachmann K, Jakobsen KS and Brochmann C (1998). Glacial survival does not matter – II: RAPD phylogeography of Nordic *Saxifraga cespitosa.* *Molecular Ecology,* **7**: 1217–32

Travis SE, Maschinski J and Keim P (1996). An analysis of genetic variation in *Astragallus cremnophylax* var. *cremnophylax*, a critically endangered plant, using AFLP markers. *Molecular Ecology,* **5**:735–45

Widmer A and Baltisberger M (1999). Extensive intraspecific chloroplast DNA 'cpDNA' variation in the alpine *Draba aizoides* (L. Brassicaceae): haplotypes relationships and population structure. *Molecular Ecology,* **8**:1405–15

Wolf PG and Sinclair RB (1997). Highly differentiated populations of the narrow endemic plant Maguire Primrose (*Primula maguirei*). *Conservation Biology,* **11**:375–81

Wright S (1951). The genetical structure of populations. *Annals of Eugenism,* **15**:323–54

Causes and Consequences of Alpine Vascular Plant Diversity in the Rocky Mountains

3

William D. Bowman and Mary Damm

INTRODUCTION

The alpine zone of the Rocky Mountains extends for ca. 5100 km in a series of nearly continuous mountain highlands, from the Brooks Range of northern Alaska to the Sangre de Cristos of northern New Mexico (Figure 3.1; Billings, 1988). The mountain ranges vary in the amount of land above the treeline, their proximity to the Arctic circle and other mountain ranges, and the spatial and temporal extent of Pleistocene glaciation. These and other geographic features, such as soil parent material and climatic regime, have influenced the current regional patterns of alpine vascular plant diversity. Variation in plant diversity and community composition at the landscape scale (100s–1000s m^2) correlates with differential winter snow distribution (Billings, 1973; Walker *et al.*, 1993) which influences diversity through its effects on rates of soil resource supply and disturbance regimes. Finally, variation in microclimate, plant species effects on soil biogeochemistry and disturbance can influence fine-scale (< 10 m^2) variation in plant diversity.

The goals of this Chapter are: (1) to examine the geographic patterns of alpine vascular plant diversity in the Rockies, as derived from literature sources, in order to suggest some biogeographic factors which influence latitudinal variation in diversity; (2) to examine the influence of soil nutrient supply and other factors, such as disturbance, on landscape and local patterns of alpine plant diversity; and (3) to

suggest mechanisms by which plant diversity influences alpine ecosystem function. The focus of this review is primarily on the Central and Southern Rocky Mountain provinces, reflecting the greater availability of information from this part of the Rockies. Although our review is limited to vascular plants, we recognise the important, yet poorly understood, role of cryptogamic species as components in the alpine zone. We hope this review will point to the areas where additional research is needed to further our understanding of the factors controlling variation in plant diversity and its role in ecosystem functioning.

BIOGEOGRAPHIC PATTERNS OF ROCKY MOUNTAIN ALPINE PLANT DIVERSITY

Biogeographically, the Rockies can be divided into four main provinces (Brouillet and Whetstone, 1993, Figure 3.1): (1) the Brooks Range Province, extending from western Alaska to the eastern Northwest Territories; (2) the Northern Rockies, a relatively narrow system of north–south trending mountain ranges running from the Yukon to central Idaho and Montana; (3) the Central or Middle Rockies, with ranges of diverse orientation running from southern Montana to the intermountain basins in central Wyoming; and (4) the Southern Rockies which extend from central Wyoming and northern Utah to central New Mexico.

Pleistocene glaciation, which reached a last maximum ca. 18 000 years BP, is an important

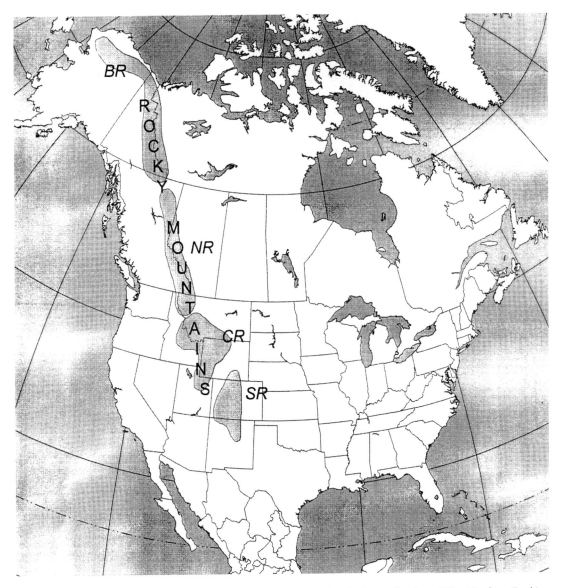

Figure 3.1 Biogeographic provinces of the Rocky Mountains; BR = Brooks Range Province, NR = Northern Rockies, CR = Central Rockies, and SR = Southern Rockies (from Brouillet and Whetstone, 1993)

factor potentially influencing latitudinal patterns of plant diversity. The Northern Rockies were covered or surrounded by the Cordilleran Ice Sheet, while the Central and Southern Rockies were subjected to local ice sheets and glaciers, usually of lesser coverage. This differential extent of glaciation may have resulted in lower species diversity in the Northern Rockies relative to more southerly regions (Takhtajan, 1986). Since the end of the major Pleistocene glaciation (12 000 BP), climatic fluctuations have resulted in periodic increases and decreases in the spatial extent of alpine areas, creating disjunctions where a continuous alpine zone formerly occurred (Figure 3.2). This shrinking and swelling of alpine areas, and changes in

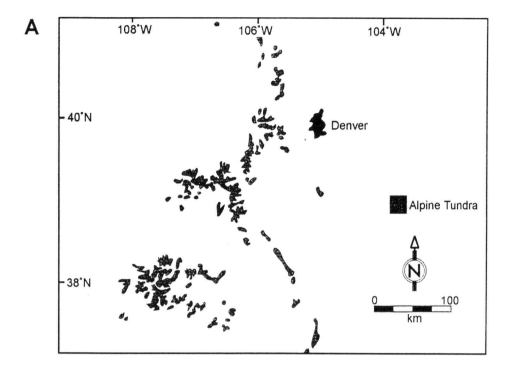

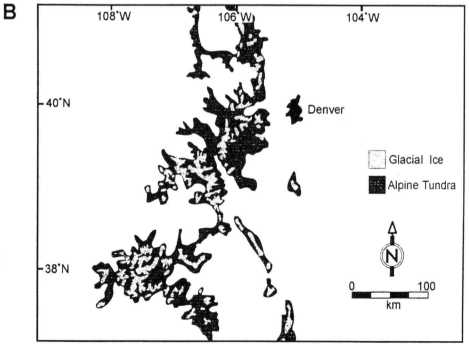

Figure 3.2 Map of Colorado showing current spatial distribution of the alpine zone (a, shaded regions), and distribution of glacial ice and alpine vegetation in the late Pinedale (ca. 15 000 BP). Figure from Elias (2001), with permission from Oxford University Press

37

their connection with the arctic flora during the Holocene (latest interglacial period), have probably influenced the current biogeographic patterns of alpine plant diversity (Richmond, 1962; Billings, 1978; Hadley, 1987). The potential contribution of local lowland flora to the alpine zone may be greatest in the southern part of the Rockies due to the lower spatial extent of glaciation.

We examined the latitudinal patterns of species richness, using available floristic studies (Table 3.1), to determine if species richness decreases from south to north, consistent with greater glaciation in the northern part of the Rockies, and greater isolation and contribution from local floras in the south (Weber, 1965; Takhtajan, 1986; Hadley, 1987). This analysis should be viewed with some caution as it lacks sufficient information from the most northern part of the Rockies. Although there is a trend towards decreasing richness with increasing latitude, it is statistically non-significant (Figure 3.3a). Most alpine floras of the Rockies have between 150–300 species, a range similar to that noted in a global survey of alpine plant diversity by Körner (1995).

Evidence to support greater speciation and a larger contribution of lowland species to the floras of the Southern Rockies, due to greater distance from the arctic species pool and less influence of glaciation, is mixed. There is a smaller contribution of arctic species in the floras of mountain ranges to the south, indicating a larger contribution of more narrowly distributed species in the more southern ranges (Table 3.1, Figure 3.3b). The proportion of endemic species is significantly higher in the south, but is low throughout the alpine floras of the Rockies (2–7%, Figure 3.3c) relative to those of the Sierra Nevada (17%; Major and Bamberg, 1968) and the European Alps (Wohlgemuth, Chapter 8). The low degree of endemism in the Rockies indicates the biogeographic separation among the mountain ranges of the Rockies has not been sufficient to promote substantial local speciation. This suggestion is supported by a high degree of similarity among the alpine floras of the Southern, Central, and Northern Rockies, which share at least 50% of their species, except

where major geographic barriers exist, such as the Wyoming desert and Colorado Plateau (Komárková, 1979; Hadley, 1987). Hadley (1987) analysed data for the alpine floras of the Central and Southern Rocky Mountain alpine zones, using an island biogeography perspective, and suggested that variation in richness among mountain ranges in the Rockies was related to the area above the treeline (saturating at ca. 1200 km^2, smaller than the area of the studies used in our analysis), distance to nearest alpine highland, and latitude (higher diversity to the south).

Throughout the Rocky Mountain Cordillera, arctic species tend to grow in moist or wet sites, indicative of their low tolerance to drought (Chabot and Billings, 1972; Brunsfeld, 1981; Baker, 1983). In contrast, species with narrower geographic distributions are more common on dry, rocky sites. In east-central Idaho, Brunsfeld (1981) found a five-fold difference between the number of arctic species in moist meadow versus dry meadow sites, while more narrowly distributed and endemic species were more common in drier sites. In Colorado, the majority of endemic species occur on dry, rocky sites, often on scree slopes or in fellfield habitats (Weber and Wittmann, 1996a, b).

VARIATION IN ALPINE COMMUNITY TYPES

Alpine plant communities in the Rockies are spatially distributed in a regular fashion in association with topographic relief. While the composition of species in these communities may vary, there are physiognomic similarities among the communities found throughout the Rockies (Billings, 1988). Billings (1973, 1988) and Walker *et al.* (1993) described the repeated assemblages of plants as being controlled primarily by differential distribution of snow, which accumulates in leeward areas but is thin or absent in windward areas due to strong prevailing westerly winds. There is no evidence to suggest that elevation is an important determinant of community composition in the Rocky Mountain alpine zone, although the proportion of different communities changes

Table 3.1 Species richness and geographical ranges of alpine flora of ranges of the Rocky Mountains. The biogeographical classification is based on Komárková (1979), and is intended to serve as an indicator of the spatial distribution of the flora, and not indicate species origins. This classification combines several traditional Holarctic categories used by Hultén (1962) into a general circumpolar category. The biogeographical ranges are listed in order of decreasing spatial extent. Species with a circumpolar distribution are widespread in the Holarctic, occurring in at least two of the three northern hemisphere continents. North American species occur from the Rocky Mountain Cordillera to the interior of the continent, as far east as at least the Great Lakes region. Western North American species are found only in the Rocky Mountain Cordillera. Rocky Mountain species occur throughout a large portion of the Rocky Mountains. Regional species have a narrower distribution along the Rocky Mountain Cordillera, often limited to within one province (e.g. Northern, Central, or Southern Rockies). Endemic species are found predominantly in the State in which the study took place, but may occur just across the border in an adjacent State. Species geographic ranges were verified using Harrington (1954), Hitchcock et al. (1955–1969), Hulten (1968), Cronquist et al. (1972–1997), Moss (1983), Dorn (1984, 1992), Great Plains Flora Association (1986), Hickman (1993), Welsh et al. (1993), Weber and Wittmann (1996a,b), Nelson and Hartman (1997), and Hartman and Nelson (2001)

	Banff National Park, Alberta[1]	Lewis Mountains, Montana[2]	Big Snowy Mountains, Montana[2]	Flint Ck Mountains, Montana[2]	Beartooth Plateau, Wyoming-Montana[3]	Teton Range, Wyoming[4]	Front Range, Colorado[5]	Gore Range, Colorado[6]	Mosquito Range, Colorado[7]	Sawatch Range, Colorado[8]	West Elk Mountains, Colorado[9]	San Juan Mountains, Colorado[10]	Needle Mountains, Colorado[11]	Sangre de Cristo Mountains, New Mexico[12]	Sierra Nevada, California[13]
Latitude	51° 42'	49° 48'	46° 50'	46° 10'	45° 00'	43° 45'	40° 05'	39° 40'	39° 20'	39° 05'	38° 56'	37° 50'	37° 38'	36° 45'	
Species richness	120	181	120	158	237	216	253	204	166	289	220	190	204	163	615
Geographical range (species %)															
Circumpolar	47	38	33	35	41	31	38	28	33	27	28	35	30	–[14]	11
North America	14	12	18	10	11	10	9	8	9	14	12	10	6	–[14]	4
Western North America	32	30	28	33	27	40	29	36	30	32	32	26	32	28	41
Rocky Mountains	6	13	17	15	16	13	12	16	13	13	14	14	14	17	16
Regional	2	6	2	6	4	6	10	11	11	12	13	12	14	18	11
Endemic	0	1	2	1	1	0	2	1	4	2	1	3	4	5	17

[1]Broad (1973), [2]Bamberg and Major (1968), [3]Johnson and Billings (1962), [4]Spence and Shaw (1981), [5]Komárková (1979), [6]Hogan (1992), [7]Hartman and Rottman (1985a, [8]1988, [9]1987, [10]1985b), [11]Michener-Foote and Hogan (1999), [12]Baker (1980), [13]Major and Bamberg (1968), [14]separation of circumpolar and North American geographical ranges was not possible using the information provided

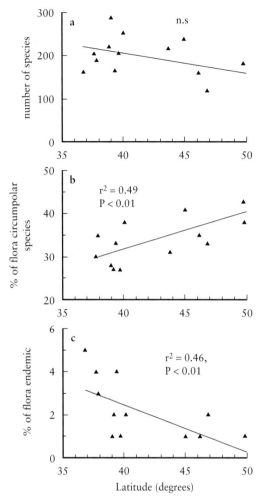

Figure 3.3 Relationship between latitude and species richness (a), percent of the flora containing circumpolar species (b), and percent of the flora which is endemic to the region (c), for floras of individual mountain ranges listed in Table 3.1. Lines fit using least squares linear regression

with changing elevation as a result of greater exposure to wind and subsequent loss of snow at high altitude.

Snow interacts with vegetation to influence the length of the growing season, exposure to low soil and air temperatures during the winter, and the supply of soil resources (Fisk *et al.*, 1998; Walker *et al.*, 2001). Oberbauer and Billings (1981) found an association between plant water relations and the topographic

position of the plant communities, indicating that aridity during the growing season is a factor determining the composition of species within the communities. Nutrient supply and subsequent competitive interactions also influence plant species composition, as determined by experimental addition of nutrients and neighbour removals (Bowman *et al.*, 1993, 1995; Theodose and Bowman, 1997a,b).

The strong physical and biotic gradients result in relatively sharp boundaries between communities and associated plant dominance across Rocky Mountain alpine landscapes. As a result, the turnover of species (similar to β diversity of Whittaker (1975)) is relatively high. The similarity of species composition among communities may be as low as 25% (Table 3.2), although the rapid change in dominance gives the perception of even greater dissimilarity in species composition among communities.

LANDSCAPE VARIATION IN PLANT DIVERSITY: RESOURCE GRADIENTS

As indicated above, there is a strong microclimatic gradient across alpine landscapes in the Rocky Mountains, associated with differential snow distribution and subsequent spatial variation in soil resource availability (Oberbauer and Billings, 1981; Taylor and Seastedt, 1994; Fisk *et al.*, 1998). Soil moisture, soil organic matter and rates of nutrient cycling are all influenced at the landscape scale by the amount and duration of snow cover. Experimental manipulation of soil resources in the alpine zone has indicated the supply of nitrogen and phosphorus are important to the establishment and abundance of alpine species (Fox, 1992; Bowman *et al.*, 1993, 1995; Theodose and Bowman, 1997a). Thus, it is reasonable to hypothesise that landscape patterns of alpine plant diversity should be related to the supply of N and P, as controlled by landscape patterns of snow accumulation.

A unimodal relationship between diversity and soil resource availability has been proposed (Grime, 1973; Huston, 1994; Gross *et al.*, 2000).

Table 3.2 Floristic similarity among alpine plant communities (noda) from Niwot Ridge, Colorado, using data from May and Webber (1982). Numbers are percent similarity of the flora, calculated using the Sorenson index $(2C/(N_1 + N_2))$, where C is the number of species in common between the communities, and N_1 and N_2 are species richness of the two separate communities

	Fellfield	*Dry meadow*	*Moist meadow*	*Wet meadow*	*Snowbed*
Fellfield	–				
Dry meadow	73	–			
Moist meadow	48	66	–		
Wet meadow	25	44	67	–	
Snowbed	25	45	77	67	–

In the absence of disturbance, we hypothesise that diversity in the alpine zone follows a unimodal distribution, related to stress tolerance and competition for belowground resources at low soil resource supply, competition for light at high soil resource supply, with maximum diversity at intermediate soil resource supply (Figure 3.4). We examined landscape patterns of species richness and diversity (using the Shannon index) in relation to rates of soil N and P supply, estimated using ion exchange resin bags (Binkley and Matson, 1983) and aboveground production, across Niwot Ridge, Colorado.

Diversity and richness of alpine plants tended to decrease with increasing soil resource supply and aboveground production in our analysis, although linear regressions gave relatively poor fits to the data, and were statistically non-significant (Figure 3.5). A unimodal distribution is indicated by the relationship between diversity and soil resource availability (Figure 3.5a), although the curve is relatively shallow and explains only 14% of the variation in observed diversity. Although this empirical support for a unimodal distribution is weak, fertilization experiments have provided evidence that the mechanisms suggested by Figure 3.4 may contribute to the control of nutrient supply on diversity. Fertilization of a relatively infertile, dry meadow community increased diversity, while fertilization of a more fertile, wet meadow decreased diversity (open triangles, Figure 3.5b; Bowman *et al.*, 1993; Theodose and Bowman, 1997a).

Factors other than soil nutrient supply, such as climatic stress and disturbance, including herbivory, are probable sources of variation in landscape diversity in the Rocky Mountain alpine zone. Cryoturbation, such as ice needle formation, occurs primarily in moist soils, which are usually of moderate to high fertility. Wind, and its associated mechanical and microclimatic influences on plants, varies considerably across the alpine landscape, with the highest velocities in dry meadows and fellfields (Greenland and Losleben, 2001). Biotic disturbances include pocket gopher digging and foraging and herbivory by voles. Pocket gophers are an important agent of soil disturbance in the alpine zone, influencing soil texture, nutrient supply and plant diversity for decades after the actual disturbance (Chambers, 1993; Davies, 1994; Litaor *et al.*, 1996; Cortinas and Seastedt, 1996; Sherrod, 1999). Preferential feeding by herbivores on graminoid species may help to increase plant diversity in the alpine zone (Dearing, 2001). The frequency and intensity of disturbances are important determinants of plant diversity in the alpine zone. Fox (1981, 1983) provided evidence that intermediate levels of soil disturbance are associated with the highest level of plant diversity in the alpine zone, consistent with the intermediate disturbance hypothesis (Connell, 1978).

PLANT DIVERSITY AND ALPINE ECOSYSTEM FUNCTION AND RESILIENCE

The role of plant diversity in ecosystem functioning and resilience is currently the subject of debate (Grime, 1997). Higher diversity of alpine vascular plant species increases the variation of

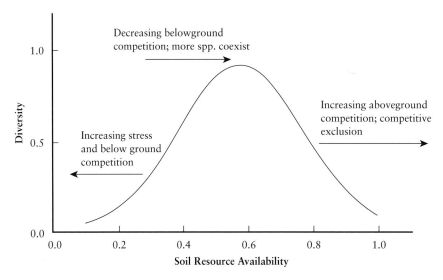

Figure 3.4 Hypothetical relationship between soil resource availability, sometimes approximated by production or biomass, and diversity

plant functional types, including different physiologies, morphologies, rooting depths and densities, and symbiotic relationships with soil microbes. Greater functional diversity may result in higher community resource use, greater potential net primary production (NPP), and greater capacity to withstand and recover from disturbances (Chapin *et al.*, 1998). In the Rocky Mountains alpine diversity may be important to soil stabilisation, maintenance of primary production in the face of climatic disturbances, particularly drought, and minimising the effects of anthropogenic N deposition. Currently, there is no experimental evidence to support a causal relationship between plant diversity and ecosystem function and resilience in the alpine zone, but there is evidence that individual species or growth forms may have disproportionately large influences on temporal variation in NPP and spatial variation in N biogeochemistry.

Most alpine plants exhibit relatively low growth rates (Körner, 1999) as a result of vegetative developmental patterns that limit their growth capacity, including preformation of buds 2–4 years in advance of emergence (Bowman and Conant, 1994; Diggle, 1997; Aydelotte and Diggle, 1997) and reduced cell

numbers relative to lowland congeners (Körner and Pelaez Menendez-Riedl, 1989). However, some species, including grasses and some herbs, can produce vegetative shoots in the same growing season in which they emerge, increasing their responsiveness to variation in resource supply (Bowman *et al.*, 1995; Theodose and Bowman, 1997a). Thus, species composition in the alpine zone can have a strong impact on responsiveness of NPP to climatic variation. Walker *et al.* (1994) found that the previous growing season's precipitation explained variation in NPP of communities in dry microclimates better than precipitation during the current year, while communities from more moist microsites responded more to the current year's precipitation. Increases in the proportion of more responsive species in the alpine flora that may accompany climate change or increased N deposition may, therefore, lead to greater within-year association between climate and NPP, potentially increasing the amplitude of temporal variation in NPP.

As indicated above, nutrient supply, particularly N, is important to variation in NPP and plant species composition in the alpine zone. While spatial variation in N cycling at the

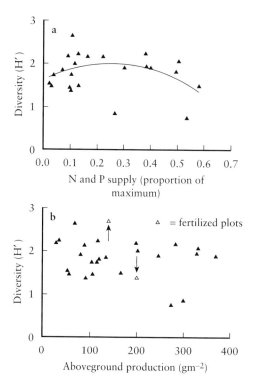

Figure 3.5 Relationship between resource availability and diversity for alpine vegetation of Niwot Ridge, Colorado. Diversity was estimated using the Shannon diversity index (H′), for plants sampled using the point intercept method in 23 1-m² plots across Niwot Ridge. Plots were selected to represent a broad range of microsites, encompassing all of the main community types, but excluding areas of obvious soil disturbance. Soil resource availability was estimated using both N and P supply (a) and aboveground production (b). N and P supply was estimated using ion exchange resins (Binkley and Matson, 1983), with four resin bags per plot sampled. Aboveground production was estimated using clip harvests at peak biomass. Two 0.04-m² subplots were clipped per plot. Linear regression analyses yielded non-significant results for all plots (data for species richness analyses not shown). The curve for b was fit using a second order polynomial ($r^2 = 0.14$, versus $r^2 = 0.03$ for a linear regression). The open triangles and arrows in b indicate the change in diversity found following N + P fertilization (Bowman et al., 1993, unpubl. data). The base of the arrows start at the level of diversity before fertilization, and show its change after fertilization

landscape scale is associated with variation in microclimate, particularly soil moisture (Fisk et al., 1998), variation within communities is associated with plant species influences as

much or more than microclimatic variation (Thomas and Bowman, 1998; Steltzer and Bowman, 1998; Steltzer, 1999). The influence of plant species on spatial variation in N cycling is associated primarily with symbiotic N_2 fixation and litter chemistry. *Trifolium* species are widespread in the Rocky Mountain alpine zone, obtaining much of their N via symbiotic N_2 fixation (Bowman et al., 1996), and augmenting soil pools of inorganic N. Thus, the presence of *Trifolium* can significantly alter small-scale variation in species composition and NPP (Thomas and Bowman, 1998). On Niwot Ridge, Colorado, *Trifolium* species make up 12% of the total plant cover, but they account for 25% of the total ecosystem N flux, which includes net N mineralization and N deposition (Bowman et al., 1996).

A less obvious example of plant control over N cycling occurs in moist meadow communities, which are found throughout the Rockies and are co-dominated by the herb *Acomastylis rossii* and the tillering grass *Deschampsia caespitosa*. Similar *Deschampsia–Geum* communities occur in the European alpine region. Rates of net N mineralization are 10-times higher and rates of net nitrification are four-times higher under canopies of *Deschampsia* than in soils under *Acomastylis* (Steltzer and Bowman, 1998). The differences in the rates of N cycling are associated with lower C:N ratios and faster fine root turnover in *Deschampsia*, but to a greater degree with very high concentrations of low molecular weight phenolics in *Acomastylis* (Steltzer and Bowman, 1998; Steltzer, 1999), which appear to stimulate higher microbial N immobilisation (Bowman and Steltzer, unpubl. data). Spatial analysis of the moist meadow alpine zone indicates these species have a significant influence on species composition in the areas adjacent to them (< 0.2 m), and are negatively correlated with spatial variation in species richness (Steltzer, 1999). This example illustrates the potential influence of plant secondary compounds on spatial variation in N cycling in the alpine zone (Hättenschwiler and Vitousek, 2000). Many slow-growing alpine herbs have relatively high concentrations of secondary compounds

(Dearing, 2001) which may influence both their susceptibility to herbivores as well as the biogeochemistry of the soils surrounding them.

The influence of plant species on spatial and temporal variation in ecosystem function suggests that the response of the alpine zone to environmental change may be mediated in large part by biotic changes (Bowman, 2000). In the Rocky Mountains most of the anthropogenic environmental threats are indirect, stemming from potential climate change and atmospheric pollution, particularly N deposition. Climate change scenarios for the Rockies are variable, although increases in precipitation have been noted at some sites (Williams *et al.*, 1996a; Bowman, 2000). Because most of the precipitation falls as snow, manipulation of snow cover has been a principal focus of climate change studies (Bock, 1976; Galen and Stanton, 1995, 1999; Walker *et al.*, 1994). Vegetation changes that accompany snow augmentation experiments include increases in both *Trifolium* species and *Acomastylis rossii*. Because these species have opposite influences on soil N cycling, the magnitude of spatial variation in N cycling in the alpine zone could increase with increasing winter precipitation. Additional work is needed to clarify the influence of changes in species composition on ecosystem function in relation to climate change in the alpine zone.

Atmospheric deposition of N is a regional problem, most apparent in the Front Range of the Southern Rockies (Sievering *et al.*, 1996; Williams *et al.*, 1996b). Significant biotic changes have been associated with elevated N deposition in forest and aquatic ecosystems of the Front Range (Baron *et al.*, 2000). Although significant changes in alpine plant species composition have been recorded during the past four decades (Korb and Ranker, 2000; Suding and Bowman, unpubl. data), the role of N deposition in these changes is uncertain. There is a significant correspondence between the species which have increased through time and those which increase in response to N fertilization. The responsive species are likely to have characters which promote higher rates of soil N cycling (e.g. *Deschampsia*, discussed above), resulting in a positive feedback to N deposition that may exacerbate the detrimental environmental impacts (Bowman and Steltzer, 1998).

ACKNOWLEDGEMENTS

Valuable comments on an earlier draft of this Chapter were provided by Julia Larson, Yan Linhart, Amy Miller, and Katherine Suding. Tim Hogan, University of Colorado Herbarium, and Ron Hartman, Rocky Mountain Herbarium, University of Wyoming, provided assistance with determination of the geographic distributions of species. Financial support for original research was provided by the Andrew W. Mellon Foundation and the Niwot Ridge Long Term Ecological Research Program.

References

Aydelotte AR and Diggle PK (1997). Analysis of developmental preformation in the alpine herb *Caltha leptosepala* (Ranunculacaea). *American Journal of Botany*, 84:1646–57

Baker WL (1980). Alpine vegetation of the Sangre de Cristo Mountains, New Mexico: Gradient analysis and classification. Unpublished thesis, University of North Carolina, Chapel Hill, NC

Baker WL (1983). Alpine vegetation of Wheeler Peak, New Mexico: Gradient analysis, classification and biogeography. MS thesis. *Arctic & Alpine Research*, 15:223–40

Bamberg SA and Major J (1968). Ecology of the vegetation and soils associated with calcareous parent materials in three alpine regions of Montana. *Ecological Monographs*, 38:127–67

Baron JS, Rueth HM, Wolfe AM, Nydick KR, Allstott EJ, Minear JT and Moraska B (2000). Ecosystem responses to nitrogen deposition in the Colorado Front Range. *Ecosystems*, 3:352–68

Billings WD (1973). Arctic and alpine vegetations: similarities, differences, and susceptibility to disturbance. *BioScience*, 23:697–704

Billings WD (1978). Alpine phytogeography across the Great Basin. *Great Basin Notational Memoirs*, 2:105–17

Billings WD (1988). Alpine Vegetation. In Barbour MG and Billings WD (eds), *North American Terrestrial Vegetation*. Cambridge University Press, Cambridge, pp 391–420

Binkley D and Matson PA (1983). Ion exchange resin bag method for assessing forest soil nitrogen availability. *Soil Science Society American Journal*, 47:1050–2

Bock JH (1976). The effects of increased snowpack on the phenology and seed germinability of selected alpine species. In Steinhoff HW and Ives JW (eds), *Ecological Impacts of Snowpack Augmentation in the San Juan Mountains of Colorado*. Final report to the Division of Atmospheric Water Resources Management, Bureau of Reclamation, Denver, Colorado, pp 265–80

Bowman WD (2000). Biotic controls over ecosystem response to environmental change in alpine tundra of the Rocky Mountains. *Ambio*, 29:396–400

Bowman WD and Conant RT (1994). Shoot growth dynamics and photosynthetic response to increased nitrogen availability in the alpine willow *Salix glauca*. *Oecologia*, 97:93–9

Bowman WD and Steltzer H (1998). Positive feedbacks to anthropogenic nitrogen deposition in Rocky Mountain alpine tundra. *Ambio*, 27:514–17

Bowman WD, Theodose TA, Schardt JC and Conant RT (1993). Constraints of nutrient availability on primary production in two alpine tundra communities. *Ecology*, 74:2085–97

Bowman WD, Theodose TA and Fisk MC (1995). Physiological and production responses of plant growth forms to increases in limiting resources in alpine tundra: Implications for differential community response to environmental change. *Oecologia*, 101:217–27

Bowman, WD, Schardt JC and Schmidt SK (1996). Symbiotic N_2 fixation in alpine tundra: Ecosystem input and variation in fixation rates among communities. *Oecologia*, 108:345–50

Broad J (1973). Ecology of alpine vegetation at Bow Summit, Banff National Park. MS thesis, University of Calgary, Calgary, Canada

Brouillet L and Whetstone RD (1993). Climate and physiography. In Flora of North America Editorial Committee (eds), *Flora of North America North of Mexico*. Oxford University Press, New York, pp 15–46

Brunsfeld SJ (1981). Alpine flora of east-central Idaho. MS thesis, University of Idaho, Moscow, Idaho, USA

Chabot, BF and Billings WD (1972). Origins and ecology of the Sierran alpine flora and vegetation. *Ecological Monographs*, 42:163–99

Chambers JC (1993). Seed and vegetation dynamics in an alpine herb field: effects of disturbance type. *Canadian Journal of Botany*, 71:471–85

Chapin FS, Sala OE, Burke IC, Grime JP, Hooper DU, Laurenroth WK, Lombard A, Mooney HA, Mosier AR, Naeem S, Pacala SW, Roy J, Steffen WL and Tilman D (1998). Ecosystem consequences of changing biodiversity. *BioScience*, 48:45–52

Connell JH (1978). Diversity in tropical rain forests and coral reefs. *Science*, 199:1302–10

Cortinas MR and Seastedt TR (1996). Short- and long-term effects of gophers (*Thomomys talpoides*) on soil organic matter dynamics in alpine tundra. *Pedobiologia*, 40:162–70

Cronquist A, Holmgren AH, Holmgren NH, Reveal JL and Holmgren PK (1972–1997) *Intermountain Flora vols 1-6* The New York Botanical Garden, Bronx, NY, USA

Davies EF (1994). Disturbance in alpine tundra ecosystems: The effects of digging by northern pocket gopher (*Thomomys talpoides*). MS thesis, Utah State University, Logan, Utah, USA

Dearing MD (2001). Plant herbivore interactions, In Bowman WD and Seastedt TR (eds), *Structure and Function of an Alpine Ecosystem: Niwot Ridge, Colorado*. Oxford University Press, New York

Diggle PK (1997). Extreme preformation in alpine *Polygonum viviparum*: An architectural and developmental analysis. *American Journal of Botany*, 84:154–69

Dorn RD (1984). *Vascular Plants of Montana*. Mountain West Publishing, Cheyenne, USA

Dorn RD (1992). *Vascular Plants of Wyoming*. 2nd Edition. Mountain West Publishing, Cheyenne, USA

Elias SA (2001). Paleoecology and late Quaternary environments of the Colorado Rockies, In Bowman WD and Seastedt TR (eds), *Structure and Function of an Alpine Ecosystem: Niwot Ridge, Colorado*. Oxford University Press, New York, pp 285–303

Fisk MC, Schmidt SK and Seastedt TR (1998). Topographic patterns of above- and below ground production and N cycling in alpine tundra. *Ecology*, 79:2253–66

Fox JF (1981). Intermediate levels of soil disturbance maximize alpine plant diversity. *Nature*, 293:564–5

Fox JF (1985). Plant diversity in relation to plant production and disturbance by voles in Alaskan tundra communities. *Arctic & Alpine Research*, 17:199–204

Fox JF (1992). Responses of diversity and growth-form dominance to fertility in Alaskan tundra fellfield communities. *Arctic & Alpine Research*, 24:233–7

Galen C and Stanton ML (1995). Responses of snowbed plant species to changes in growing-season length. *Ecology*, 76:1546–57

Galen C and Stanton ML (1999). Seedling establishment in alpine buttercups under experimental manipulations of growing-season length. *Ecology*, 80:2033–44

Great Plains Flora Association (1986). *Flora of the Great Plains*. University Press of Kansas, Lawrence, Kansas, USA

Greenland D and Losleben MV (2001). Climate. In Bowman WD and Seastedt TR (eds), *Structure and Function of an Alpine Ecosystem: Niwot Ridge, Colorado*. Oxford University Press, New York, USA

Grime JP (1973). Controls of species density in herbaceous vegetation. *Journal of Environmental Management*, 1:151–67

Grime JP (1997). Ecology – Biodiversity and ecosystem function: The debate deepens. *Science*, 277:1260–1

Gross KL, Willig MR, Gough L, Inouye R and Cox SB (2000). Patterns of species density and productivity at different spatial scales in herbaceous plant communities. *Oikos*, 89:417–27

Hadley KS (1987). Vascular alpine plant distributions within the central and southern Rocky Mountains, USA. *Arctic & Alpine Research*, 19: 242–51

Harrington HD (1954). *Manual of the Plants of Colorado*. Sage Books, Denver, Colorado

Hartman EL and Rottman ML (1985a). The alpine vascular flora of the Mt. Bross massif, Mosquito Range, Colorado. *Phytologia*, 57:133–51

Hartman EL and Rottman ML (1985b). The alpine vascular flora of three cirque basins in the San Juan Mountains, Colorado. *Madroño*, 32: 253–72

Hartman EL and Rottman ML (1987). Alpine vascular flora of the Ruby Range, West Elk Mountains, Colorado. *Great Basin Nat.*, 47:152–60

✳ Hartman EL and Rottman ML (1988). The vegetation and alpine flora of the Sawatch Range, Colorado. *Madroño*, 35:202–25

Hartman RL and Nelson BE (2001). *Checklist of the Vascular Plants of Colorado*. http://www.rmh.uwyo.edu

Hättenschwiler S and Vitousek PM (2000). The role of polyphenols in terrestrial ecosystem nutrient cycling. *Trends in Ecology & Evolution*, 15:238–43

Hickman JC (1993). *The Jepson Manual: Higher Plants of California*. University of California Press, Berkeley, CA, USA

Hitchcock CL, Cronquist A, Ownbey M and Thompson JW (1955–1969). *Vascular Plants of the Pacific Northwest*. vols 1–5 University of Washington Press, Seattle, WA, USA

✳ Hogan T (1992). A floristic survey of the Eagles Nest Wilderness Area in the southern Gore Range

of central Colorado. *Natural History Inventory of Colorado*, vol. 12, University of Colorado Museum, Boulder, CO, USA

Hultén E (1962). The Circumpolar Plants. *Kungl Svensk Vetenskapsakad Handl.*, ser 4, 8:5

Hultén E (1968). *Flora of Alaska and Neighboring Territories*. Stanford University Press, Stanford, USA

Huston MA (1994). *Biological diversity: the coexistence of species in changing landscapes*. Cambridge University Press, Cambridge, UK

Johnson PL and Billings WD (1962). The alpine vegetation of the Beartooth Plateau in relation to cryopedogenic process and patterns. *Ecological Monographs*, 32:105–35

Komárková V (1979). Alpine vegetation of the Indian Peaks area, Front Range, Colorado Rocky Mountains. *Flora et Vegetatio Mundi*, Bd VII, J Cramer, Vaduz

Korb JE and Ranker TA (2000). Changes in stand composition and structure between 1981 and 1996 in four Front Range plant communities in Colorado. *Plant Ecology*, 157:1–11

Körner, Ch (1995). Alpine plant diversity: A global survey and functional interpretations. In Chapin FS III and Körner Ch (eds), *Arctic and Alpine Biodiversity*. Ecological Studies, vol. 113, Springer-Verlag, Berlin, pp 45–62

Körner Ch (1999). *Alpine Plant Life – Functional plant ecology of high mountain ecosystems*. Springer, Berlin, 338 pp

Körner Ch and Pelaez Menendez-Riedl S (1989). The significance of developmental aspects in plant growth analysis. In Lambers H, Cambridge ML, Konings H and Pons TL (eds), *Variation in Growth Rate and Productivity of Higher Plant*. SPB Academic Publishing, The Hague pp 141–57

✳ Litaor MI, Mancinelli R and Halfpenny JC (1996). The influence of pocket gophers on the status of nutrients in alpine soils. *Geoderma*, 70:31–48

Major J and Bamberg SA (1968). Comparison of some North American and Eurasian alpine ecosystems. In Wright HE and Osburn WH (eds), *Arctic and Alpine Environments*. Indiana University Press, Bloomington, pp 89–118

May DE and Webber PJ (1982). Spatial and temporal variation of vegetation and its productivity on Niwot Ridge, Colorado. In: Halfpenny J (ed) *Ecological Studies in the Colorado alpine, a festschrift for John W. Marr*. Occasional paper no. 37, Institute of Arctic and Alpine Research, University of Colorado, Boulder, pp 35–62

Michener-Foote J and Hogan T (1999). The flora and vegetation of the Needle Mountains, San Juan Range, southwestern Colorado. *Natural History Inventory of Colorado*, vol. 18, University of Colorado Museum, Boulder

Moss EH (1983). *Flora of Alberta*. Second Edition, University of Toronto Press, Toronto

Nelson BE and Hartman RL (1997). *Checklist of the Vascular Plants of Wyoming.* http://www.rmh.uwyo.edu/species/index.htm

Oberbauer S and Billings WD (1981). Drought tolerance and water use by plants along an alpine topographic gradient. *Oecologia*, 50:325–31

Richmond GM (1962). *Quaternary stratigraphy of the La Sal Mountains, Utah.* US Geological Survey Paper 134

Sherrod SK (1999). A multiscale analysis of the northern pocket gopher (*Thomomys talpoides*) at the alpine site of Niwot Ridge, Colorado. PhD Thesis, University of Colorado, Boulder

Sievering H, Rusch D and Marquez L (1996). Nitric acid, particulate nitrate and ammonium in the continental free troposphere: Nitrogen deposition to an alpine tundra ecosystem. *Atmosphere & Environment*, 30:2527–37

Spence JR and Shaw RJ (1981). A checklist of the alpine vascular flora of the Teton Range, Wyoming, with notes on biology and habitat preferences. *Great Basin Naturalist*, 41:232–42

Steltzer H (1999). Plant species effects on spatial variation in nitrogen cycling in alpine tundra. PhD thesis, University of Colorado, Boulder

Steltzer H and Bowman WD (1998). Differential influence of plant species on soil N transformations within moist meadow alpine tundra. *Ecosystems*, 1:464–74

Takhtajan AL (1986). *Floristic Regions of the World.* University of California Press, Berkeley

Taylor RV and Seastedt TR (1994). Short- and long-term patterns of soil moisture in alpine tundra. *Arctic & Alpine Research*, 26:14–20

Theodose TA and Bowman WD (1997a). Nutrient availability, plant abundance, and species diversity in two alpine tundra communities. *Ecology*, 78:1861–72

Theodose TA and Bowman WD (1997b). The influence of interspecific competition on the distribution of an alpine graminoid: evidence for the importance of plant competition in an extreme environment. *Oikos*, 79:101–14

Thomas BD and Bowman WD (1998). Influence of a N$_2$-fixing *Trifolium* on plant species composition and biomass production in alpine tundra. *Oecologia*, 115:26–31

Walker DA, Halfpenny JC, Walker MD and Wessman CA (1993). Long-term studies of snow–vegetation interactions. *BioScience*, 43:287–301

Walker DA, Lewis BE, Krantz WB, Price ET and Tabler RD (1994). Hierarchic studies of snow–ecosystem interactions: A 100-year snow-alteration experiment. In Ferrik M (ed), *Proceedings of the Fiftieth Annual Eastern and Western Snow Conference*, Quebec City, Quebec, Canada, 8–10 June 1993, pp 407–14

Walker MD, Webber PJ, Arnold EA and Ebert-May D (1994). Effects of interannual climate variation on aboveground phytomass in alpine vegetation. *Ecology*, 75:393–408

Walker MD, Walker DA, Theodose TA and Webber PJ (2001). The vegetation: hierarchical species–environment relationships. In Bowman WD and Seastedt TR (eds), *Structure and Function of an Alpine Ecosystem: Niwot Ridge, Colorado.* Oxford University Press, New York

Weber WA (1965). Plant geography in the southern Rocky Mountains. In Wright HE and Frey DG (eds), *The Quarternary of the United States: a Review Volume for the VII Congress of the International Association for Quarternary Research.* Princeton University Press, Princeton, pp 453–68

Weber WA and Wittmann RC (1996a). *Colorado Flora: Eastern Slope.* University Press of Colorado, Niwot

Weber WA and Wittmann RC (1996b). *Colorado Flora: Western Slope.* University Press of Colorado, Niwot

Welsh SL, Atwood ND, Goodrich S and Higgins LC (eds) (1993). *A Utah Flora.* Second Edition, Revised. Brigham Young University, Provo

Whittaker RH (1975). *Communities and Ecosystems.* Macmillan, New York

Williams MW, Losleben M, Caine N and Greenland D (1996a). Changes in climate and hydrochemical responses in a high-elevation catchment in the Rocky Mountains, USA. *Limnology and Oceanography*, 41:939–46

Williams MW, Baron JS, Caine N, Sommerfeld R and Sanford R (1996b). Nitrogen saturation in the Rocky Mountains. *Environmental Science & Technology*, 30:640–6

Biodiversity of the Vascular Timberline Flora in the Rocky Mountains of Alberta, Canada

4

Stuart A. Harris

INTRODUCTION

The origin and distribution of arctic and alpine floras has always been a source of discussion since they obviously could not be directly evolved from the subtropical Tertiary floras that once lived in these regions. Darwin (1883) was the first to suggest that arctic species migrated south during glaciations when the life zones would have been displaced towards the equator and to lower altitudes. When the climate ameliorated, these floras were presumed to have found suitable habitats on the mountain tops. Hooker and Gray (1880) had previously recognised field evidence which supported this concept in the case of North America, and this has been generally accepted by most subsequent workers, e.g. Rydberg (1914) and Major and Bamberg (1968 a, b), although there is also evidence that the actual evolution and origin is much more complex (Tolmachev, 1959; Hultén, 1962).

In Eurasia, Hultén (1937; 1958; 1962; 1971) has been able to demonstrate the presence of a series of different geographical distributions of species. These include Circumarctic, Amphiatlantic, Eurasian (originating in central Asia), and Beringian floras, together with local endemics. However, the flora of North America presented special problems to him because of its large size, diverse climates, relative isolation from Eurasia, and limited detailed studies of its floras. Furthermore, there needed to be a rationalisation of the species terminology between North America and Eurasia (Weber, 1965). Accordingly, it was only possible to discuss the biogeography

of North America on the basis of qualitative studies until very recently (e.g. Weber, 1965; Major and Bamberg, 1968a). Subsequent work now makes it possible to attempt to quantify the plant distributions in selected areas.

In this Chapter, the distribution of the vegetation in the alpine and timberline ecotone of southwest Alberta is quantified for six specific locations along a 600-km section of the Rocky Mountains (Figures 4.1 and 4.2). It lies in the southern section of the key area that was almost completely glaciated during the Late Wisconsin glaciation, although it was not covered by ice during some of the earlier major cold events. Subsequently, it has been revegetated, and has also been under warmer conditions than now during the climatic optimum (Harris and Pip, 1973; Luckman and Kearney, 1986). This has resulted in an exceptionally varied, diverse flora with unique characteristics.

STUDY AREA

Southwest Alberta is bounded by the continental divide, which lies four ranges into the Rocky Mountains of the Eastern Cordillera of North America (Figure 4.1). The outer ranges consist of Palaeozoic limestones (Mountjoy, 1964), although Palaeozoic shales and conglomerates occur at Jasper (Price and Mountjoy, 1970). At Waterton National Park, PreCambrian limestones and shales underlie the Palaeozoic limestones.

Physiographically, these mountains form the main part of the Rocky Mountains that extends from southern Colorado north to Nahanni in the Northwest Territories. They

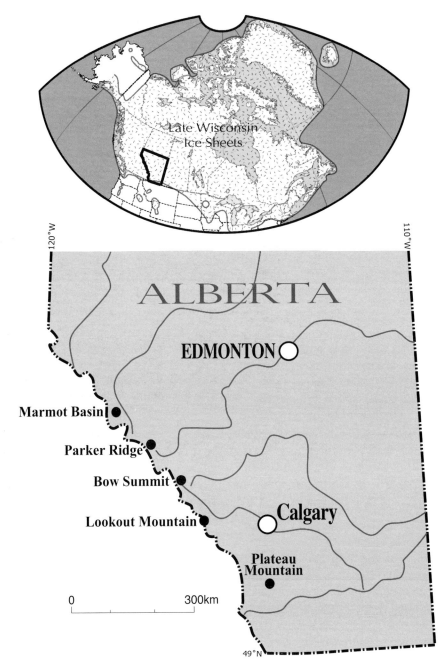

Figure 4.1 Location of the six study sites in southwest Alberta in relation to the area glaciated during the Late Wisconsin event

are continued northwards in Canada by the Mackenzie and Richardson ranges, that are composed of younger rocks. During the last (Late Wisconsin) glaciation (Figure 4.1), the Cordilleran and Laurentide ice sheets overlapped or partially coalesced along the mountain front (Alley, 1973; Roed, 1975). However, during at least some of the preceding 12 cold events (Harris, 1994; 2000), the ice cover was more limited in extent (Barendregt and Irving, 1998),

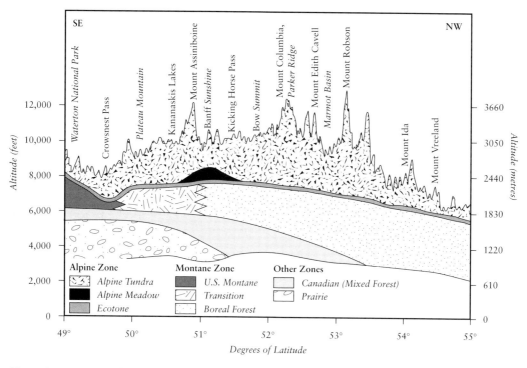

Figure 4.2 Location of the six study sites in relation to the topography and vegetation zones along the Rocky Mountains of southwest Alberta

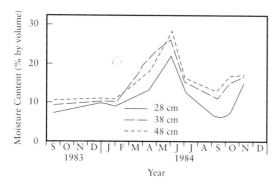

Figure 4.3 The annual variation in soil moisture at several depths at 2500 m on the top of Plateau Mountain (after Harris, 1998)

Howell and Harris (1978) discussed the effects of soil-forming factors and provided data on the effects of time on the end product.

The soil moisture regime is critical to plant growth. The soils have a limited period of moist conditions after snow melt at Plateau Mountain but are otherwise very dry (Figure 4.3). There is a much longer period of adequate moisture in the alpine soils north of Banff. Local snow banks alter these conditions in the alpine tundra everywhere (Ogilvie, 1969; Kuchar, 1975; Timoney, 1999), acting as a source of extra water in summer and by reducing the growing season due to late disappearance of the snow.

Table 4.1 provides the available data on the contemporary climate of the alpine and timberline ecotone at some of the study sites. The overall climate is continental and the eastern slopes of the mountains are subject to periodic winter Chinooks which increase in frequency southwards (Longley, 1972; Hare and Thomas, 1979). Mean annual precipitation (AES, no date) is

so that large areas of the landscape would have been subjected to alpine/arctic tundra conditions.

In the alpine zone, the soils are primarily Brunisols, Gleysols, Lithosols and Regosols (Baptie, 1968; Knapik, 1972; Knapik et al., 1973; Brewster, 1974; King et al., 1976).

Table 4.1 Climatic data from stations above treeline in southwest Alberta

Zone	Station	Latitude	Longitude	Elevation (m)	Mean annual air temperature (°C)	Mean winter snow cover (cm)	Years averaged
Alpine tundra	Marmot Basin #1	52° 47′ N	118° 07′ W	2286	−2.36	5	1979–1998
	Parker Ridge #3	52° 11′ N	117° 07′ W	2277	−3.95	16	1980–1999
	Sunshine A	51° 04′ N	115° 43′ W	2712	−6.98	9	1977–1979
	Plateau Mtn. #2	50° 13′ N	114° 31′ W	2500	−2.22	12	1975–1998
Timberline Ecotone	Sunshine C	51° 03′ N	115° 47′ W	2407	−3.45	73	1977–1979

400–500 mm/annum except at Sunshine–Columbia Icefields (up to 930 mm/annum) and at Waterton (1072 mm/annum).

The *timberline ecotone* is used in this paper for the zone above the treeline (upper limit of closed forest – Körner, 1998) and below the upper limit of isolated patches of forest (the *forestline* of Brockmann-Jerosch, 1919; Däniker, 1923; Hermes, 1955). The *alpine tundra* is normally found immediately above it, although an *alpine meadow* (>80% herbaceous cover) intervenes at Sunshine Meadows. The forest below treeline includes the *boreal forest* north of Banff, but exhibits a transition to the montane forest of the United States between there and Waterton National Park.

Table 4.2 The geographical ranges of the taxa at the six sites in southwest Alberta (actual numbers and percentages)

Origins	Alpine		Treeline ecotone	
Circumpolar	75	22%	78	19%
Beringian	31	9	35	8
Amphiatlantic	4	1	4	1
N. Cordilleran	21	6	14	3
M. Cordilleran	17	5	35	8
S. Cordilleran	59	17	68	16
Cordilleran	52	15	68	16
Total Cordilleran	149	43	185	43
Cordilleran–Prairie	22	6	27	6
N. American	4	13	80	19
E. American	0	0	0	0
Introduced	14	4	5	1
Distribution not known	0	0	2	0.4

METHODS

Figure 4.1 shows the location of the six sites chosen for this study. Detailed species lists are available for all these sites, i.e. Marmot Basin, Jasper (own data), Parker Ridge (Crack, 1977, and own data), Bow Summit (Broad, 1973, and own data), Sunshine Village (Knapik, 1973; Knapik *et al.*, 1973, and own data), Plateau Mountain (Bryant, 1968; Bryant and Scheinberg, 1970; Harris, 1991), and Waterton National Park (Kuchar, 1973; Kuijt, 1982). The methods used in these studies involved describing the species present and percentage cover for many randomly chosen 1 to 5 m plots representing the plant associations found at various elevations. About 40,000 herbarium sheets were examined at the University of Calgary Herbarium (UAC) to augment the published data, while the writer has been collecting data on the species distribution with altitude using random transects at these sites since 1974. Thus individual collecting biases should be minimal. Climatic data has been collected from most of the alpine areas since the 1970s.

The species nomenclature follows Moss (1983), updated by Cody (1996), and also the terminology used in the lists of threatened and endangered species. Obviously there are ongoing discussions as to the status of the subspecies and varieties of both the circumpolar species and the local endemics, so the exact number of taxa is a matter of opinion.

The assignment of geographic distribution follows Hultén (1937–1971), Hitchcock *et al.* (1969), Porsild and Cody (1979), Moss (1983),

Table 4.3 Species distribution at the treeline in southwestern Alberta. Both actual numbers of taxa and the percentage of the total taxa in that zone are listed for each site

Site	Alpine tundra		Zone Alpine meadow		Treeline ecotone	
Marmot Basin	101	28%	–	–	67	15%
Parker Ridge	66	18%	–	–	79	18%
Bow Summit	81	22%	–	–	132	30%
Sunshine	151	42%	142	100%	145	33%
Plateau Mountain	210	58%	–	–	150	35%
Waterton N. Park	125	35%	–	–	238	55%
Total number of taxa	361		142		434	

Table 4.4 Unique species at the study sites in southwestern Alberta. These are species that only occur at one of the six study sites (actual numbers and percentages for the site)

Location	Alpine tundra		Subalpine	
	n	%	n	%
Marmot Basin	21	21	12	9
Parker Ridge	7	11	8	6
Bow Summit	11	14	24	9
Sunshine	24	16	149	30
Plateau Mountain	60	29	39	15
Waterton National Park	38	30	115	35

and Cody (1996), checked with Peschkova (various dates). Cody lists the Northern Cordilleran endemics (limited to Alaska and the Yukon Territories), while those mainly restricted to Alberta and the adjacent portion of British Columbia are listed as Middle Cordilleran endemics.

RESULTS

Geographical ranges of species

Table 4.2 shows the distribution of geographic ranges of the taxa at the six study sites. Perhaps the most striking feature of the results is the number of species that only occur in North America. The number of Cordilleran species (43%) is about double that of the circumpolar species. About 20% of the flora is found throughout the North American alpine

zone. Four of the species are Amphiatlantic, and 14 have been introduced from Europe by man. Beringian species make up 9% of the flora. None of the Appalachian endemics have been collected in the alpine zone of southwest Alberta. Species composition changes markedly south of Banff. In general, the subalpine zone has twice the number of taxa of the adjacent alpine zone.

Species distribution in the zones

Table 4.3 summarises the species distribution at the six study sites. The alpine zone has a very varied flora, with no one site having over 60% of the species found in that zone. Maximum species richness is found at the wetter stations (Sunshine and Waterton) and in the transition zone at Plateau Mountain (a former nunatak). Parker Ridge shows the minimum species diversity, probably due to its proximity to the icefields. In contrast, the treeline ecotone shows a progressive increase in species southwards. Eighty-four percent of the species present in the alpine meadow at Sunshine Village also occur in the adjacent alpine tundra, indicating that the meadow and tundra are differentiated primarily on the basis of density of vegetation cover.

Species present at all sites

Only eight species (2.2%) occur at all six sites in the alpine zone. These include the North American endemics (*Carex albo-nigra*; *Claytonia lanceolata*; *Arenaria capillaris* ssp.

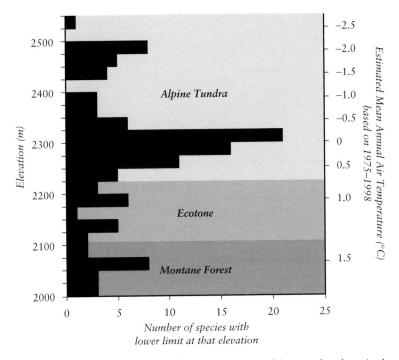

Figure 4.4 Histogram of the lower limits of elevation of the vascular plants in the alpine zone of Plateau Mountain. Also shown are the mean annual air temperatures at the various elevations based on 24 years of weather data collected at adjacent weather stations

capillaris), four Circumarctic species (*Poa alpina*; *Silene acaulis* ssp. *acaulis*; *Oxyra digyna* and *Polygonium viviparum*), and one Beringian species (*Myosotis alpestris* ssp. *asiatica*). Only 32 species (8.9%) are present at three or more sites, including both Waterton and Jasper.

Unique species at the study sites

The numbers of unique species, occurring at only one of the six sites, ranged from 11% to 30% (Table 4.4). This indicates a high degree of variation in species composition in the alpine zone in southwest Alberta, unlike other mountain areas. Many species are potentially very rare.

Both Argus and White (1978) and Gould (1999) have produced lists of the rare vascular plants of Alberta, and their distribution in the alpine zone is similar. The rare taxa occur mainly in the transition zone at Plateau Mountain and Waterton National Park. However,

the numbers are increasing as botanists discover the rarity of many of the species.

Comparing the species present with the list of Canadian rare vascular plants (Argus and Pryer, 1990), the numbers of rare plants drop dramatically, being largely confined to Waterton National Park. One species, *Aquilegia jonesii*, also occurs on the American list (Ayensu and DePhillipps, 1978). This is consistent with the flora being an extension of the American timberline flora into southern Alberta.

Species distribution on mountain tops

Figure 4.4 shows the number of species with their lower distributional limit in 25-m classes through the timberline ecotone and alpine tundra at Plateau Mountain. The obvious break at 2325 m is probably due to the mean annual air temperature having risen 2.5 °C during the climatic optimum. It also demonstrates that the species present on mountain peaks are not necessarily representative of the flora at equivalent altitudes on adjacent higher mountains.

DISCUSSION AND CONCLUSIONS

The data in Table 4.2 strongly support the concept of a thorough mixing of the alpine floras in North America excluding those of the Appalachians, prior to the Late Wisconsin glaciation. The fact that 73% of the species in the alpine zone only occur in North America suggests that considerable evolution of the species occurred on this continent when it was not connected to Eurasia. This is consistent with the evidence for only six periods with connections across the Bering Strait during the last 3.5 Ma, based on migrations of rodents (Repenning, 1980). This is far fewer than the number of glaciations (Harris, 1994, 2000).

The Late Wisconsin ice sheets largely destroyed the pre-existing vegetation in southwest Alberta except for species that survived on nunataks, e.g. Plateau Mountain. These may have permitted the survival of a few local endemics within the arctic/alpine vegetation at a few isolated locations in the middle section of the Rocky Mountains, which now appear as the Middle Cordilleran endemic flora. As deglaciation commenced about 14 000 BP, the area was recolonized with alpine species from the unglaciated areas to the south and from the nunataks prior to the climate warming up about 10 200 BP (MacDonald, 1982). The warming caused the alpine vegetation to become stranded on the higher mountain tops before proper mixing of species had taken place, hence the high variability from site to site. During the climatic optimum, the mean annual temperature in the alpine at Plateau Mountain was 2.5 °C warmer than today, resulting in the elimination of many species that would otherwise be found on the summit of the mountain today. Studies of the effects of possible climatic changes on the alpine vegetation should take this problem into account.

References

A.E. S. (no date). *Canadian climatic normals; temperature and precipitation*. Prairie Provinces. Environment Canada, Atmospheric Environment Service, Downsview, Ontario, pp 429

Alley NF (1973). Glacial stratigraphy and the limits of Rocky Mountain and Laurentide Ice Sheets in southwestern Alberta, Canada. *Canadian Society of Petroleum Geologists*, 21:153–77

Argus GW and Pryer KM (1990). *Rare Vascular Plants in Canada*. Canadian Museum of Nature, Ottawa, pp 191

Argus GW and White DJ (1978). The rare vascular plants of Alberta. National Museum of Canada, Ottawa. *Syllogeus*, 17:47

Ayensu ES and DeFilipps RA (1978). *Endangered and threatened plants of the United States*. Smithsonian Institution, Washington, pp 403

Baptie JB (1968). *Ecology of the alpine soils of Snow Creek valley, Banff National Park, Alberta*. Unpubl MSc thesis, Dept Biology, Univ of Alberta. pp 135

Barendregt RW and Irving E (1998). Changes in the extent of the North American ice sheets during the Late Cenozoic. *Canadian Journal of Earth Sciences*, 35:504–09

Brewster GR (1974). *Soil development within a sub-alpine ecotone: Bow Pass, Banff National Park, Alberta*. Unpubl MSc thesis, Univ of Western Ontario. pp 237

Broad J (1973). *Ecology of alpine vegetation at Bow Summit, Banff National Park*. Unpubl MSc thesis, Dept Biology, Univ of Calgary. pp 93

Brockmann-Jerosch H (1919). *Baumgrenze und Klimcharakter*. Pflanzengeographische Kommission der Schweiz. Naturforsch Gesell, Beitrage zur geobotanischen Landesaufnahme 6. Rascher, Zürich

Bryant JP (1968). *Vegetation and frost activity in an alpine fell-field on the summit of Plateau Mountain, Alberta*. Unpubl MSc thesis, Dept of Biology, Univ of Calgary. pp 93

Bryant JP and Scheinberg E (1970). Vegetation and frost activity in an alpine fellfield on the summit of Plateau Mountain, Alberta. *Canadian Journal of Botany*, 48:751–71

Cody WJ (1996). *Flora of the Yukon Territory*. NRC Research Press, Ottawa

Crack (1977). Flora and vegetation of Wilcox Pass, Jasper National Park. Unpublished MSc thesis, Department of Biology, University of Calgary, 284 pp.

Däniker A (1923). Biologische studien über Baum- und Waldgrenze, insobesondere über die klimatischen Ursachen und deren Zusammenhänge. *Vierteljahresscher Naturforsch Ges Zürich*, 68:1–102

Darwin C (1883). *Origin of the Species*. D. Appleton and Co., London

Gould J (1999). *Plant species of special concern*. Alberta Heritage Information Centre, Alberta Environmental Protection, Edmonton, Alberta. pp 27

Hare FK and Thomas MK (1979). *Climate of Canada*. J. Wiley and Sons, Toronto. 2nd Edition

Harris SA (1991). Vascular plant checklist – the Plateau Mountain area. *Pica*, 11(3):14–22

Harris SA (1994). Chronostratigraphy of glaciations and permafrost episodes in the Cordillera of western North America. *Progress in Physical Geography*, 18:366–95

Harris SA (1998). Nonsorted circles on Plateau Mountain, S. W. Alberta, Canada. Proceedings, 7th International Permafrost Conference, Yellowknife. *Collection Nordicana*, pp 441–8

Harris SA (2000). Pliocene and Pleistocene glaciations and permafrost events proven to date in the Cordillera of western North America. *Earth Cryology*, 4(1):24–43 [In Russian]

Harris SA, Pip E (1973). Molluscs as indicators of Late and Post-glacial history in Alberta. *Canadian Journal of Zoology*, 51:209–15

Hermes K (1955). Die Lage der oberen Waldgrenze in den Gebirgen der Erde und ihr Abstand zur Schneegrenze. *Kölner geographische Arbeiten*, 5

Hitchcock CL, Cronquist A, Ownbey M and Thomson JW (1969). *Vascular plants of the Pacific Northwest*. Univ Washington Press, Seattle and London. 5 volumes

Hooker JD and Gray A (1880). The vegetation of the Rocky Mountain region and a comparison with that of other parts of the world. *US Geological Survey, Terr.*, 6:1–62

Howell JD and Harris SA (1978). Soil-forming factors in the Rocky Mountains of southwestern Alberta, Canada. *Arctic and Alpine Research*, 10: 313–24

Hultén E (1937). *Outline of the history of arctic and boreal biota during the Quaternary period*. Okförlags Aktiebolaget Thule, Stockholm

Hultén E (1958). The Amphi-Atlantic plants and their phytogeographical connections. *Svensk Vetensk Akad Handlungen* 7, Stockholm

Hultén E (1962). The Circumpolar Plants. vol. 1. Vascular cryptogams, conifers, monocotyledons. *Kungl Svensk Vtenskapsakad Handlungen*, ser 4, 8:5. Stockholm

Hultén E (1971). The circumpolar plants. vol. 2. Dicotyledons. *Kungl Svensk Vetenskapsakad Handlungen*, volume 13, Stockholm

Hultén E (1968). *Flora of Alaska and Neighbouring Territories*. Stanford Univ Press, Stanford

King RH and Brewster GR (1976). Characteristics and genesis of some subalpine podsols (spadosols), Banff National Park, Alberta. *Arctic and Alpine Research*, 8:91–104

Knapik LJ (1972). *Soils of the Sunshine region, Banff National Park*. Canadian Wildlife Service Contract WR070/71, 37 Project 382–5206 (Scotter), 140 pp

Knapik L (1973). *Alpine soils of the Sunshine area, Canadian Rocky Mountains*. Unpubl MSc thesis, Dept of Soil Sci, Univ of Alberta. pp 213

Knapik L, Scotter GW and Pettapiece WW (1973). Alpine soils of the sunshine area, Banff National Park. *Arctic and Alpine Research*, 5:161–70

Körner C (1998). A re-assessment of high elevation treeline positions and their explanation. *Oecologia*, 115:445–59

Kuchar P (1973). *Habitat types of Waterton Lakes National Park*. Indian and Northern Affairs, Parks Canada. Unpubl report

Kuchar P (1975). *Alpine tundra communities and Dryas octapetala ssp. hookeriana in the Bald Hills, Jasper National Park*. Unpubl PhD thesis, Dept of Botany, Univ of Alberta

Kuijt J (1982). *A Flora of Waterton Lakes National Park*. University of Alberta Press, Edmonton

Longley RW (1972). *The Climate of the Prairie Provinces*. Environment Canada, Atmospheric Environment Service, Toronto

Löve D (1970). Subarctic and subalpine: where and what? *Arctic and Alpine Research*, 2:63–73

Luckman BH and Kearney MS (1986). Reconstruction of Holocene changes in alpine vegetation and climate in the Maligne Range, Jasper National Park, Alberta. *Quaternary Research*, 26:244–61

MacDonald GM (1982). Late Quaternary palaeo-environments of the Morley Flats and Kananaskis Valley of southwestern Alberta. *Canadian Journal of Earth Sciences*, 9:23–5

Major J and Bamberg SA (1968a). Comparison of some North American and Eurasian alpine systems. In Wright HE Jr and Osburn WE (eds), *Arctic and Alpine Environments*. Indiana Univ Press, Bloomington, pp 89–118

Major J and Bamberg SA (1968b). Some Cordilleran plants disjunct in the Sierra Nevada of California and their bearing on Pleistocene ecological conditions. In Wright HE Jr and Osburn WE (eds), *Arctic and Alpine Environments*. Indiana Univ Press, Bloomington, pp 171–88

Moss EH (1983). *Flora of Alberta*. 2nd Edition, revised by J.G. Packer, Univ of Toronto Press, Toronto

Mountjoy EW (1964). Rocky Mountain Front Ranges between Rocky River and Medicine Lake, Jasper National Park, Alberta. In Edmonton Geological Society, *6th Annual Field Trip Guide Book*, Medicine and Maligne Lakes, Jasper National Park

Ogilvie RT (1969). The mountain forest and alpine zones of Alberta. In Nelson JG and Chambers MJ (eds), *Vegetation, Soils, and Wildlife*. Methuen, Toronto. Chapter 2

Peschkova GA (various dates). *Flora Sibiriae*. Siberian Publishing firm Ran, Novosibirsk 14 volumes [in Russian]

Porsild AE and Cody WJ (1979). *Vascular plants of continental Northwest Territories, Canada*. National Museums of Canada, Ottawa

Price RA and Mountjoy EW (1970). Geological structure of the Canadian Rocky Mountains between Bow and Athabasca Rivers – a progress report. In Wheeler JO (ed), *Structure of the southern Canadian Cordillera*. Geol Assoc of Canada Special paper no. 6

Repenning CA (1973). Faunal exchanges between Siberia and North America. *Canadian Journal of Anthropology*, **1**:37–44

Roed MA (1975). Cordilleran and Laurentide multiple glaciation, west-central Alberta, Canada. *Canadian Journal of Earth Sciences*, **12**:1493–1515

Rydberg PA (1914). Phytogeography and its relation to taxonomy and other branches of Science. *Torreya*, **12**:73–85

Timoney K (1999). *Limber pine, whitebark pine, alpine heath, and terricolous alpine lichen vegetation alliances in Alberta*. Report for Alberta Environmental Protection Resource data Division, Edmonton

Tolmachev A (1959). *Sur l'origine de la flore arctique: Quant ou, et comment surgit la flore arctique?* 9th Int Bot Cong, Toronto. Toronto Univ Press, Toronto. vol. **2**, p 399

Wallace AR (1900). *Studies, Scientific and Social*. MacMillan, New York. vol. **1**, p 526

Weber WA (1965). Plant geography in the southern Rocky Mountains. In Wright, HE Jr and Frey DC (eds), *The Quaternary of the United States*. Princeton Univ Press, Princeton. pp 453–68

Plant Species Richness and Endemism of Upper Montane Forests and Timberline Habitats in the Bolivian Andes

5

Michael Kessler

INTRODUCTION

The eastern tropical Andes contain some of the world's richest plant communities (Gentry, 1982; Henderson *et al.*, 1991; Barthlott *et al.*, 1996; Myers *et al.*, 2000). However, knowledge of the magnitude and spatial distribution of this plant diversity is still fragmentary (Gentry, 1995; Webster, 1995). There are several reasons for this. First, it is very difficult and time-consuming to study complex vegetation types in the steep and inaccessible mountain terrain. Thus, there is a concentration of botanical collecting activities at a small number of sites and consequently the current knowledge of patterns of diversity and endemism reflect collecting intensity rather than natural patterns. Second, most vegetation studies in neotropical montane forests have focussed on woody plants, especially trees (Gentry, 1988, 1995; Lieberman *et al.*, 1996; Valencia *et al.*, 1998). However, the majority of vascular plant species in humid tropical forests belong to non-tree life forms (Gentry and Dodson, 1987 a, b; Ibisch, 1996; Balslev *et al.*, 1998; Galeano *et al.*, 1999). Third, few studies of tropical mountain vegetation have considered more than one measure of diversity, even though it is well-known that the detection of patterns and mechanisms of diversity depends considerably on spatial scale, sampling regime, and method of data analysis (Whittaker, 1972; Magurran, 1988; Pickett *et al.*, 1994; Lyon and Willig, 1999).

Even less is known about the elevational and regional distribution of plant endemism. Mountain regions are well known to harbour more endemic plant species than continental lowland areas (Balslev, 1988), but little is known about the exact elevational distribution of range size and about the patterns of different plant groups. Also, while isolated mountain ranges and habitats are generally characterised by high endemism, there also appears to be much variation of endemism in apparently continuous habitats on the main Andean slope. Among birds, variation of endemism along the Andean timberline from Ecuador to Bolivia has been linked to ecoclimatic stability, i.e. the local topographically determined moderation of climatic extremes such as droughts and cold air influxes (Fjeldså *et al.*, 1999b).

With the aim of describing in as much detail as possible the elevational and spatial distribution of plant diversity and endemism, botanical surveys have been conducted at 65 sites on the eastern Andean slope of the Bolivian Andes. As the study region is exceedingly species-rich and since many plant groups are poorly known, the study has been limited to vascular plant groups that are comparatively well-known, that are well-represented in the studied habitats, and that include a wide variety of life forms with different ecological requirements and adaptations, namely Acanthaceae, Araceae, Bromeliaceae, Melastomataceae, Palmae, and Pteridophyta (Kessler and Bach, 1999). Partial results of this project have already been published (Bach *et al.*, 1999; Fjeldså *et al.*, 1999a; Kessler, 1999a, 2000a-d, 2001a-c, in press a,b; Kessler and Beck, 2001; Kessler and Croat, 1999; Kessler and Helme, 1999; Kessler and Krömer, 2000; Kessler *et al.*, 1999, 2000, 2001 a,b; Krömer *et al.*, 1999; Smith *et al.*, 1999).

In the present Chapter, a synthesis is provided of the above studies, focussing on humid tropical montane forests at and above 3000 m and on the forest–grassland transition in the northern portion of the Bolivian Andes. The analysis is limited to the region north of the 'Andean knee' (about 18° S), where the ecological conditions are typical of the tropical eastern Andean slope north to Venezuela. Further south, the montane forest regions are climatically, topographically, biogeographically, and ecologically distinct and are generally referred to as 'Tucumano Bolivian forests'.

MATERIAL AND METHODS

A general description of the sampling regime and data analysis methods are provided. For details see Kessler (1999a, 2000d, 2001a-c, in press a,b) and Kessler and Bach (1999).

Field work was carried out in Bolivia for a total of 20 months from 1995 to 1997, covering 65 sites at 200–4050 m and at 13° 30'–20° S, including 47 sites north of 18° S, 13 of which are located above 3000 m (Figure 5.1). At each locality, the six botanical study groups (Acanthaceae, Araceae, Bromeliaceae, Melastomataceae, Palmae, Pteridophyta) were sampled in plots of 400 m². The total of 964 vegetation plots included 676 plots in natural zonal forest vegetation and 288 plots in azonal (ravines, ridges, rock faces) or secondary vegetation. An elevational transect in Carrasco National Park (Figure 5.1) was sampled particularly intensively, with 204 plots at 220–3950 m (Kessler 1999a, in press b; Kessler et al., 1999, 2001b) and is frequently referred to in this paper.

Plant diversity was quantified independently for the different study groups in several ways following Whittaker (1960, 1972): *Point diversity* was calculated as the mean number of species per zonal forest plot. *Alpha diversity* was expressed as the total species number recorded in zonal forest at a given site. *Pattern diversity*, the ratio of point diversity to alpha diversity, gives an indication of the local niche differentiation of the plant communities within a given habitat and elevation. *Beta diversity* was defined as the contribution of azonal or disturbed habitats to overall plant diversity at a given elevation.

Endemism of the plant communities at a given habitat or elevation was quantified through an index of *range size rarity*, calculated as the mean inverse range size of all species of each group recorded at each elevational step (Williams and Humphries, 1994; Fjeldså et al., 1999b). This index gives stronger weighting to narrowly distributed taxa while not being based on subjective cut-off limits (Usher, 1986; Fjeldså and Rahbek, 1997). The range sizes of all recorded species was quantified in a 1°-grid map based on extensive surveys of literature and herbaria and on information provided by the respective group specialists (see Acknowledgements).

To study the environmental correlates of species richness and endemism, each study site was characterised by the following variables: elevation, elevational range covered by the study area, latitude, overall area of sampled habitat, mean annual precipitation, mean annual temperature, number of frost days per year, distance to wetter and dryer habitats (as a proxy for habitat heterogeneity and for migration distances during climatic changes), canopy height (as a proxy for ecosystem productivity), bryophyte cover (as an index of air humidity), and a rough estimate of human impact. Additionally, to quantify erratic and anomalous climatic events that may be biologically critical (e.g. droughts, cold spells), coefficients of variation in remotely sensed NDVI and T_s values were calculated by E. Lambin and coworkers (University of Louvain, Belgium) on a month-by-month basis based on 10 years' data of the Global Area Coverage (GAC) data (see Fjeldså et al., 1999b for details). NDVI (Normalized Difference Vegetation Index) is highly correlated to vegetation parameters such as green leaf biomass, absorbed photosynthetically active radiation, and biomass green vegetation (Holben et al., 1980; Tucker et al., 1985). T_s (land surface 'skin' brightness temperature) is related, through the surface energy balance equation, to surface moisture availability and evapotranspiration, as a function of latent heat flux (Planet, 1988).

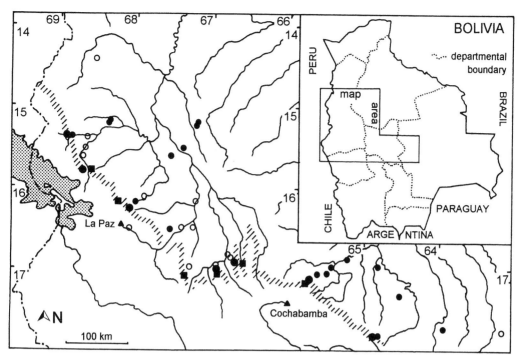

Figure 5.1 Location of the study areas in *Polylepis* forest (filled squares), in humid upper montane forest above 3000 m (large filled circles), in mixed humid forest below 3000 m (small filled circles), and in dry montane forest (open circles) on the eastern Andean slope in Bolivia. The hatched area shows the narrow and fragmented distribution of humid upper montane forest (including *Polylepis*) above 3000 m; triangles are cities. Nevado Huayna Potosí is located just northeast of La Paz, the elevational transect in Carrasco National Park corresponds to the series of six study sites to the northeast of Cochabamba

Geological substrate was not included as an ecological variable because of its homogeneity throughout the study region, corresponding mostly to red sandstones, lutites and quarcitic rocks of Devonian and Ordovician age (Pareja *et al.*, 1978) with fairly constant nutrient availability (P. Schad, pers. comm.).

Several analyses were conducted specifically for this publication: A graph of the elevational distribution of the number of vascular plant species per 0.1 ha by life form (Figure 5.2b) was compiled based on values in Gentry (1995), Seidel (1995), Ibisch (1996), and Smith and Killeen (1998). For non-forest habitats above 4000 m, unpublished data of J. Gonzales, Herbario Nacional de Bolivia were used, from a study on Nevado Huayna Potosí, departamento La Paz, Bolivia (Figure 5.1). T. Krömer (pers. comm.) provided epiphyte counts at 500 m and 700 m in departamento La Paz. As these sources did not cover all life forms at all elevations, ratios of species richness of the six study groups and of the four life form classes based on full floristic inventories in Andean and Amazonian forests were calculated (Young and León, 1990; Ibisch, 1996; Balslev *et al.*, 1999). These ratios were then used to extrapolate primarily species numbers of shrubs and herbs from the data in Figure 5.2a. To express species numbers at a constant spatial scale of 0.1 ha, species numbers obtained at differing plot sizes were corrected with the Arrhenius (1920, 1921) equation (see Barthlott *et al.*, 1996). To compare the elevational patterns of species richness at different spatial scales, point and alpha diversity of Bromeliaceae for elevational steps of 200 m in humid montane forest in Carrasco National Park were

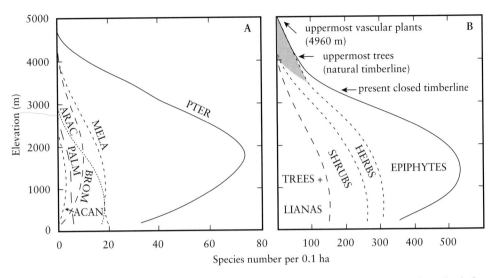

Figure 5.2 Elevational patterns of species richness (number of species per 0.1 ha) (a) independently by study group and (b) cumulative by life form in natural zonal vegetation on the humid eastern Andean slope in central Bolivia. In (a), values below 4000 m are from Carrasco National Park (Kessler, in press c) and values above 4000 m from the eastern slope of Nevado Huayna Potosí (J. Gonzales, pers. com.). In (b) values were compiled from a variety of sources from Bolivia and elsewhere in the Andes (see Material and Methods). In (b) the shaded area corresponds to non-forest zonal vegetation at and above timberline

calculated. All species recorded in the entire study region were added together, i.e. covering a distance of about 500 km, by elevational steps of 200 m in humid forests at 28 study sites and in all forest and non-forest habitats at 47 sites including dry forests. To study the regional variability of species richness and endemism in humid forests above 3000 m, values of point diversity and range size rarity at 13 study sites were calculated (Figure 5.1). These were then linearly correlated to the environmental parameters outlined above, subjecting overall probability values to sequential Bonferroni correction.

RESULTS AND DISCUSSION

Natural and present-day timberline positions

The closed upper timberline on the humid eastern Andean slope in Bolivia is currently located at about 3300–3600 m. Locally, isolated forest patches, mostly consisting of species of

Polylepis (Rosaceae), occur up to about 4000–4200 m. This has long been considered to represent a natural pattern (e.g. Troll, 1959; Koepcke, 1962; Walter and Medina, 1969; Seibert and Menhofer, 1991) caused by special, localised microclimatic conditions. However, recent studies have shown that these forests are the natural zonal vegetation at these elevations and that the natural zonal timberline would be located at around 4000–4200 m. These forests have been destroyed through millennia of human activities, mainly by the use of fire for hunting and for clearing pastures, and by livestock grazing (Ellenberg, 1958; Lægaard, 1992; Fjeldså, 1992; Kessler, 1995a,b; Chepstov-Lusty *et al.*, 1998). Fires are easily started in open grasslands and, while they frequently do not permanently damage adult *Polylepis* trees which are protected by their thick multilayered bark, they kill young trees, thus preventing forest regeneration (Lægaard, 1992; Kessler and Driesch, 1993; Hensen, 1995; Kessler, 1995a,b). Mixed humid montane forests composed of *Weinmannia, Clusia,*

Clethra, *Hedyosmum*, and *Oreopanax* species, typically occurring to about 3500 m elevation, are generally too wet to allow the spread of fires (though this may occur in exceptionally dry periods) but are affected at their margins and slowly pushed back by consecutive burning (Kessler, 2000b). As a result, the closed upper timberline is gradually pushed down, while remnant *Polylepis* forest patches survive in sheltered sites, such as ravines, rock faces, and boulder slopes (Young, 1993). These sites are not microclimatically favoured. For example, boulder slopes have colder, rather than warmer, soils than adjacent fine-grained soil (Kessler and Hohnwald, 1998). Overall, timberlines on the eastern Andean slope in Bolivia have been pushed down by around 500 elevational metres.

Timberline forest destruction probably started long before the Spanish conquest (Kessler and Driesch, 1993; Kessler, 1995a; Chepstov-Lusty *et al.*, 1998). Currently, less than 5% of the original forest cover above about 3500 m remains, most of it in small, scattered patches and in a degraded state (Kessler, 1995a,b; Fjeldså and Kessler, 1996). In Bolivia, only two or three natural timberline habitats larger than 1 km^2 (only one > 10 km^2) are known to survive (Kessler and Herzog, 1999). As a result, very little is known about the natural aspect of the humid Andean timberline. Based on observation at a few isolated and very wet mountain ridges, the natural forest–grassland ecotone probably consists of a mosaic of forest patches on deep, well-drained soils and of scrub and grassland on flat, water-saturated or on steep, shallow soils (Kessler, 1995a,b). Unless soil gradients are abrupt, the transition from forest to grassland is gradual. The current situation, with an abrupt transition, is generally due to fire.

Given the lack of knowledge on the natural timberline situation and the paucity of intact sites on which to study this ecotone, knowledge of the natural patterns of plant diversity and consequently on the impact of human activities is almost nil. In fact, most studies on tropic-alpine vegetation near the Andean timberline have assumed only recent, short-term human activities and have not considered the long-term impact of humans (see also Lutyen, 1999). One of the very few examples of a natural timberline situation can be found in Podocarpus National Park, southern Ecuador, but this is located on a rather low, very wet, and windswept ridge (Keating, 1999) and may not be typical of the situation on the main Andean chain. The destruction of *Polylepis* forests and timberline habitats is so extensive and little-known that it has frequently been overlooked in neotropical conservation assessments (e.g. Dinerstein *et al.*, 1995; Stotz *et al.*, 1996; Davis *et al.*, 1997).

Open questions with regard to the Andean timberline include: (a) the ecological and eco-physiological causes of timberlines (Körner, 1998), (b) the role of natural fires, e.g. those ignited by thunderstorms, and (c) the natural distribution of *Polylepis* in Colombia and Venezuela, where few ecologically rather specialised species occur, and their relationship to the páramo vegetation (Kessler, 1995a; Luteyn, 1999).

Elevational patterns of species richness

Along an elevational transect from the lowlands to the snowline, the number of species per 0.1 ha varies strongly between the study groups (Figure 5.2a). However, the patterns themselves are rather homogeneous. Species richness either remained roughly constant up to about 1500 m and then declined sharply (Araceae, Melastomataceae, Palmae) or revealed a more or less pronounced hump-shaped curve (Acanthaceae, Bromeliaceae, Pteridophyta). This coincides with patterns of woody plant diversity throughout the tropical Andes (Gentry, 1995), of vascular epiphytes in Peru and Bolivia (Ibisch, 1996; Ibisch *et al.*, 1996), and of bryophytes and lichens in Colombia and Peru (Gradstein and Frahm, 1987; Wolf, 1993). Overall, these observations support Rahbek's (1995, 1997) assumption that the hump-shape is the prevalent elevational pattern of plant species richness.

The above pattern is supported by the elevational distribution of species richness by life form in natural zonal vegetation (Figure 5.2b). Woody plants with > 2.5 cm dbh have roughly constant values up to 1500 m, with a linear decline at higher elevations (Gentry, 1995). Shrubs presumably show a roughly parallel pattern, though this is poorly documented. Even though data is completely absent, terrestrial herbs apparently show two maxima of species richness, one at about 2000 m corresponding to forest undergrowth herbs, and a second around the timberline corresponding to species of open non-forest habitats. Epiphytes have a well-marked hump-shaped distribution of species richness (Ibisch, 1996; Ibisch *et al.*, 1996).

Total vascular plant species richness shows a hump-shaped curve with a maximum at around 1500 m. This pattern is mostly due to the distribution of epiphytes. The inclusion of bryophytes and lichens would further strengthen the hump-shaped pattern of plant species richness (Gradstein and Frahm, 1987; Wolf, 1993). The values shown in Figure 5.2b are preliminary estimates applicable to the Bolivian Andes and probably differ somewhat from numbers further north in the Andes. The lowlands are particularly subject to less equitable climatic conditions in Bolivia than the Amazonian region further north. Thus, forests below about 700 m are floristically somewhat depauperate in Bolivia in comparison to e.g. Ecuador (Balslev *et al.*, 1999).

Above 2000 m, forest species richness declines by about 10% per 100 m elevational increase. Around the timberline elevation, non-forest habitats are more species-rich than forests. This has previously been noted in the *Polylepis* forest–páramo ecotone in the northern Andes (Luteyn, 1999). Presumably, the combination of low temperatures and low light levels inside timberline forests is too stressful for most plant species.

Elevational patterns of alpha, beta, and gamma diversity

A comparison of the elevational patterns of species richness among Bromeliaceae shows that from scales of 0.04 ha to about 2500 km^2,

i.e. across seven orders of magnitude, patterns remain remarkably constant (Figure 5.3). At four different scales, richness curves are highly correlated ($r = 0.89$ to 0.96). Similar observations apply to the other study groups, albeit with differing patterns and absolute values (data not shown). In Carrasco National Park, all six study groups show a remarkable elevational constancy of pattern diversity, i.e. of the ratio of point diversity to alpha diversity (Kessler, in press c). Likewise, there is no elevational trend in beta diversity, i.e. the contribution of different habitats (ravines, rock faces, etc.) to total species richness: independently of elevation, about 20–25% of the vascular plants are restricted to such habitats (Kessler, 1999a, b, in press b). Thus, elevation does not appear to influence the degree of niche differentiation in plant communities.

It is striking that the patterns at large spatial scales do so closely reflect those at a small scale, since endemism in Bromeliaceae increases with elevation (Figure 5.3a). Thus, it should be expected that, as area increases, total species numbers rise more strongly at high elevations. This is because as one moves along the Andean slope, at elevations with high endemism one does more frequently encounter additional species, whereas at elevations dominated by widespread taxa, one should always find the same selection of species. This expected pattern is indicated very slightly by the somewhat higher values of overall species richness in forests at 2000–3000 m and of overall species in all habitats above 3000 m (Figure 5.3b). The reason why this pattern is not more pronounced apparently involves the fact that even widespread species are frequently local in their distribution. Thus, many wide-ranging species were only found at one or a few sites in the present study. More intensive sampling and a further increase of the study area northwards along the Andes would presumably lead to a relative increase of bromeliad species richness at higher elevations.

The above considerations also apply to the other study groups. For example, among the 755 pteridophyte species recorded at the

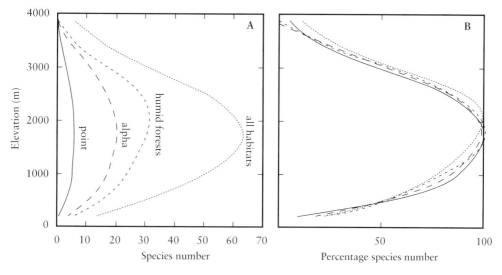

Figure 5.3 Elevation patterns of (a) absolute and (b) relative bromeliad species richness at four spatial scales: (a) point diversity, here calculated as the mean number of species per 400 m² plot along a transect in humid montane forest in Carrasco National Park; (b) alpha diversity, i.e. the total number of species recorded in steps of 200 m in humid montane forest in Carrasco National Park; (c) overall species richness in humid montane forests, i.e. the total number of species recorded in humid forests in steps of 200 m at 28 study sites north of 18° S on the eastern Andean slope in Bolivia; (d) overall species richness in all habitats, i.e. the total number of species recorded in all habitats in steps of 200 m at 47 study sites north of 18° S on the eastern Andean slope in Bolivia. To allow direct comparison of patterns, in (b) all curves are expressed as the percentage relative to the highest value of each curve

65 study sites, no less than 255 species (34%) were found at only one study site, and 119 species (16%) even only in one of the 676 study plots (Kessler, 2001b). Obviously, many plant species are so rare and/or local in their occurrence in montane forests that much more intensive sampling is needed to adequately document their distribution patterns and to separate the effects of this noncontinuous distribution of species within their ranges from the geographical turnover in species composition (gamma diversity).

A different aspect of the scale dependence of diversity patterns involves the comparison of life forms (Ibisch, 1996; Ibisch *et al.*, 1996). Because of different species–area curves of the different life forms, the relative importance of terrestrial species, particularly trees, increases at larger sample size, while that of epiphytes decreases. As a result, at scales of 1–10 km², the hump-shaped pattern shown in Figure 5.2 is likely to be less pronounced.

Regional patterns of species richness

Among the 13 study sites above 3000 m elevation, point diversity of Pteridophyta varies regionally by a factor of about 4 (6.5–24.8 species per 0.04 ha), while among Bromeliaceae and Melastomataceae species numbers range from 0 to 2.6 and 3.0 species, respectively. With respect to the 12 studied environmental parameters (see Material and Methods), variance of point diversity is significantly correlated to number of humid months among Bromeliaceae ($r = -0.75$, $p < 0.05$) and Pteridophyta ($r = 0.71$, $p < 0.05$). Species richness of Melastomataceae is significantly correlated to habitat area ($r = 0.84$, $p < 0.01$) and to bryophyte cover on canopy branches ($r = 0.72$, $p < 0.05$). Thus, the main factor determining the variation of species richness within the narrow elevational belt studied appears to be humidity, here captured either by number of humid months and by bryophyte

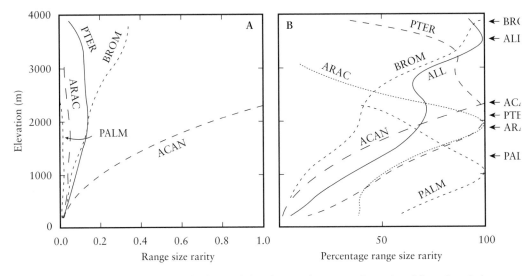

Figure 5.4 Elevational patterns of (a) absolute and (b) relative values range size rarity of five selected plant groups in humid montane forest in Carrasco National Park, Bolivia (Kessler, in press c). Higher values correspond to smaller ranges, i.e. to higher endemism. To allow direct comparison of patterns, in (b) the highest value of every group was equalled to 100%. Arrows show the elevation of the range size rarity maxima of each group. The full line in (b) shows the mean of all five study groups

cover. The relationship of Melastomataceae species richness to area is striking but cannot yet be meaningfully interpreted.

These results correspond to analyses of all 65 study sites (Kessler, in press a, 2001b), where species richness of the different study groups varies primarily in relation to elevation and to humidity, with highest overall diversity in the wettest regions.

Elevational patterns of endemism

Along the elevational transect in Carrasco National Park, patterns of range size rarity of five study groups reveal highly differing absolute values (Figure 5.4a). Species of Palmae and Araceae tend to be most widespread, while Acanthaceae have the highest absolute values with several species known only from the study region. When the patterns of range size rarity are expressed relative to the highest values of every single group, all groups show quite similar patterns (Figure 5.4b). Range size rarity is low in the lowlands, mostly with values less than 50% of the maxima, and increases either linearly with

elevation (Acanthaceae, Bromeliaceae) or shows a hump-shaped distribution relative to elevation (Araceae, Palmae, Pteridophyta). Averaging the indices of all study groups reveals a relative value of range size rarity of < 10% in the lowlands, the absolute maximum at 3500 m, and a decline at even higher elevations (Figure 5.4b). Similar patterns, though with regionally differing absolute values, have been obtained along partial elevational transects elsewhere in the Bolivian Andes (Kessler, 1999b, unpubl. data).

The peak of plant endemism at the timberline elevation coincides with previous studies on birds (Graves, 1988). In Ecuador, Balslev (1988) studied plant endemism in three broad elevational belts and found maximum endemism at 900–3000 m and somewhat lower values above 3000 m. Given the poor elevational resolution of this study, the actual pattern may well correspond to that detected here. The decline of plant endemism above timberline has also been documented in the Ecuadorian páramo vegetation (Sklenà and Jørgensen, 1999). A general explanation for the overall elevational pattern of endemism probably involves the spatial extent and discontinuity of the different elevational belts. The timberline on the eastern Andean slope is located in the region of steepest slopes. Correspondingly, vegetation belts are very narrow and fragmented by valleys and ridges. At higher and lower elevations, habitats are distributed more continuously. As a result, species of upper montane forests have small, fragmented populations which are subject to local differentiation and speciation (Graves, 1988). That the decline of endemism above the timberline is primarily a result of the topographical configuration of the high mountain region is exemplified by the Venezuelan Andes. Unlike the situation in Bolivia and Ecuador, in Venezuela there is no extensive tropic–alpine region. Rather, mountains are small, isolated and the surface area declines continuously with elevation. Correspondingly, plant endemism in the páramo zone increases constantly with elevation (A. Berg, pers. comm.).

Regional patterns of endemism

Of the study groups, only Bromeliaceae and Pteridophyta are represented well enough among the 13 forest sites above 3000 m to allow an analysis of the regional variance of range size rarity. Values of range size rarity vary by a factor of 17 among both Bromeliaceae and Pteridophyta. For Bromeliaceae, this variation is not significantly correlated to any of the 12 studied environmental parameters, with the strongest relationships shown to latitude ($r = 0.54$) and to distance of the study sites to habitats with at least 30% less precipitation ($r = 0.43$). Among Pteridophyta, variance of range size rarity is significantly correlated to mean annual precipitation ($r = 0.77$, $p < 0.05$) and bryophyte cover on canopy branches ($r = 0.70$, $p < 0.05$), i.e. two measures of environmental humidity. This latter pattern, which is also shown among all 65 study sites, may be due to increased speciation by allopolyploidy in regions with many co-occurring species (Kessler, 2001b).

These results coincide well with a more inclusive analysis of six study groups (Acanthaceae,

Araceae, Bromeliaceae, Cactaceae, Palmae, Pteridophyta) at all 65 study sites (M. Kessler and E. Lambin, unpubl. data). Here too, different groups show independent patterns of endemism which are rarely correlated to species richness. Neither is there a clear correlation of range size rarity to ecoclimatic stability, a pattern detected by Fjeldså et al. (1999b) among upper montane birds and hypothesised to represent a general mechanism that also applies to other groups of organisms.

In summary, with the present state of knowledge, the regional variability of plant endemism cannot be explained by one or a few environmental variables or underlying mechanisms. This is unfortunate for conservation purposes, since many Andean regions are biologically little known. An assessment of the regional distribution of plant endemism at the timberline has to be achieved through labour-, time- and funding-intensive field surveys.

Impact of natural and human disturbances on species richness and endemism

Along a successional vegetation gradient on landslides in humid montane forest at 1900–2800 m in Carrasco National Park, Bromeliaceae, Melastomataceae and Pteridophyta show an increase in species richness from early successional stages to late successional stages and mature forest, and a subsequent decline in senescent forest (Kessler, 1999a). Terrestrial ferns, terrestrial bromeliads, and melastomes have 67%, 50%, and 10%, respectively, of their species restricted to early successional stages, while almost no epiphytic species show this pattern. Overall, about 20% of the total vascular plant flora depends on early successional vegetation types induced by landslides. In all study groups, range size rarity is most pronounced in very early successional stages and in mature forest, perhaps as a result of less stable conditions at mid-succession and due to competitive displacement of endemic species in senescent forest.

Along gradients of human disturbance, species richness of the study groups declines

67

only slightly in somewhat degraded forests compared to mature natural forests (Kessler, 1999b; 2001c). Species richness declines strongly in more heavily disturbed habitats such as secondary forest, scrub, and pastures. Range size rarity shows a roughly similar pattern, but with one important exception: in 21 of 25 studied cases, range size rarity was somewhat higher in anthropogenically disturbed than in natural forests (Kessler, 2001c).

The relationship between plant ranges size and habitat disturbance has been studied only superficially and still lacks a coherent theoretical background (Huston, 1994). Endemic plants are frequently given high priority in conservation strategies (Bibby *et al.*, 1992; Williams *et al.*, 1996; Davis *et al.*, 1997) because their small ranges render them particularly vulnerable to habitat loss (Balmford and Long, 1994) and because they are assumed to be more susceptible to anthropogenic habitat disturbance than widespread taxa (Moolman and Cowling, 1994; Samways, 1994; Andersen *et al.*, 1997). However, the high values of range size rarity at low levels of human disturbance and at early successional stages observed here suggest that endemic humid montane forest plants may not be quite so sensitive to habitat disturbance as currently assumed. This is not surprising, considering that montane forests are subject to intensive natural disturbances by landslides and tree falls (Garwood *et al.*, 1979; Stern, 1995; Ibisch, 1996; Kessler, 1999a) and that plant species may therefore be adapted to disturbances.

I propose that endemic species are somewhat competitively inferior to other co-occurring taxa, hindering the establishment and maintenance of new populations following dispersal, and thus limiting their range expansion ability (Kessler, 2001c). Within their established ranges, such endemic species depend on natural habitat disturbances to prevent their competitive exclusion by other species and therefore also profit from a certain level of anthropogenic disturbance. While this pattern certainly does not apply to every single endemic tropical plant species, it indicates that the conservation of part of the endemic tropical

forest flora may be achieved in forest areas subject to sustainable forest use without the need to completely exclude human activities. The crucial question concerns which types and intensities of disturbances, i.e. uses, are compatible with species conservation.

CONCLUSIONS AND CONSERVATION IMPLICATIONS

Several general conclusions can be drawn from the present study for optimising biodiversity conservation and management in upper montane forests and timberline habitats of Bolivia and elsewhere in the tropical Andes:

(1) Upper montane forests, *Polylepis* forests, and timberline ecosystems are more severely threatened in Bolivia than humid montane forests at mid-elevation (1000–3000 m), which are still extensively preserved, and than tropic–alpine vegetation types above the timberline, which are largely man-made habitats that have been subject to human uses for millennia.

(2) The widespread and continuing destruction and lowering of timberline habitats requires immediate conservation attention. Intact timberline habitats should be identified and protected throughout the tropical Andes.

(3) For the *Polylepis* forest remnants, a conservation strategy has already been designed by Fjeldså and Kessler (1996). This strategy takes into account that the remnants are mostly located on communal lands, that they are subject to human uses, and that their conservation can only be implemented in collaboration with the local inhabitants in the scope of more inclusive land management plans. While several local projects are already implementing the strategy, many important areas receive no attention at present (Kessler, 1998).

(4) Increasing endemism with elevation calls for a large number of conservation areas at high elevations, particularly around the timberline elevation.

(5) However, the optimal location and design of such a network of timberline reserves is not yet known because of the paucity of biological data from many timberline regions, the limited overlap of patterns of richness and endemism between different organism groups, and the difficulty of modelling such patterns from environmental data. Clearly, extensive, well-planned additional field work is needed here.

(6) Given the rather low plant species richness above 3000 m, timberline conservation units need not be particularly large in order to representatively cover local vegetation types and to maintain viable plant populations. However, conservation management may not be practicable below a certain minimum unit size and some animals, e.g. the Andean deer *Hippocamelus antisensis*, may also require larger areas (Young, 1997).

(7) Contrary to the recommendations for areas above 3000 m, at elevations below 2000 m, high plant species richness, low population density of many species (Pitman *et al.*, 1999), and comparatively low endemism all call for a more limited number of large conservation areas.

Important questions for optimising habitat and species conservation and management include: What is the relationship of plant community composition, diversity and endemism in upper montane forests to human activities? Which uses are sustainable at what intensity? Is the conservation of upper montane forests dependent on reserves or is it more efficiently conducted in collaboration with the local people? Considering that non-forest vegetation types between the current and natural timberline elevations are mostly man-made, are these habitats worthy of conservation actions? Or should they be considered to represent artificial ecosystems that should be managed primarily with a view to providing services to humans?

ACKNOWLEDGEMENTS

I thank A Acebey, K Bach, JA Balderrama, J Bolding, J Fjeldså, J Gonzales, A Green, SK Herzog, B Hibbits, S Hohnwald, I Jimenez, E Kessler, T Krömer, J-C Ledesma, M Olivera, A Portugal, J Rapp, J Rodriguez, and M Sonnentag for help during fieldwork; TB Croat, PL Ibisch, B León, H Luther, JT Mickel, M Moraes, RC Moran, G Navarro, B Øllgaard, AR Smith, I Valdespino, D Wasshausen, and KR Young for specimen identification, advice, and hospitality during range map compilation; K Bach and E Kessler for database management; and the Herbario Nacional de Bolivia, La Paz, especially SG Beck, M Cusicanqui, A de Lima, R de Michel; and M Moraes, and the Dirección Nacional de Conservación de la Biodiversidad (DNCB), La Paz, for logistic support and permits. This study was supported by the Deutsche Forschungsgemeinschaft, Germany, the DIVA project under the Danish Environmental Programme, and the A.F.W.-Schimper Stiftung, Germany.

References

Ackery PR and Vane-Wright RI (1984). *Milkweed Butterflies*. British Museum (Natural History), London

Andersen M, Thornhill A and Koopowitz H (1997). Tropical forest disruption and stochastic biodiversity losses. In Laurance WF and Bierregaard RO (eds), *Tropical forest remnants: ecology management and conservation of fragmented communities*. University of Chicago Press, Chicago, pp 281–91

Arrhenius O (1920). Distribution of the species over the area. *Meddeland Vetenskapsakad Nobelinst*, 4:1–6

Arrhenius O (1921). Species and area. *Journal of Ecology*, 9:95–9

Bach K, Kessler M and Gonzales J (1999). Caracterización preliminar de los bosques deciduos andinos de Bolivia en base a grupos indicadores botánicos. *Ecología en Bolivia*, 32:7–22

Balmford A and Long A. (1994). Avian endemism and forest loss. *Nature*, 372:623–4

Balslev H (1988). Distribution patterns of Ecuadorean plant species. *Taxon*, 37:567–77

Balslev H, Valencia R, Paz y Miño G, Christensen H and Nielsen I (1999). Species count of vascular plants in one hectare of humid lowland forest in Amazonian Ecuador. In Dallmeier F and Comiskey JA (eds), *Forest Biodiversity in North Central and South America and the Caribbean. Research and Monitoring*. Man and the Biosphere series 21. UNESCO Paris and The Parthenon Publishing Group, New York, pp 585–94

Barthlott W, Lauer W and Placke A (1996). Global distribution of species diversity in vascular plants: towards a world map of phytodiversity. *Erdkunde*, 50:317–27

Bibby CJ, Collar NJ, Crosby MJ, Heath MF, Imboden C, Johnson TH, Long AJ, Stattersfield AJ and Thirgood SJ (1992). *Putting biodiversity on the map: priority areas for global conservation*. ICBP, Cambridge, UK

Chepstov-Lusty AJ, Bennet KD, Fjeldså J, Kendall A, Galiano W and Tupayachi Herrera A (1998). Tracing 4.000 years of environmental history in the Cuzco area, Peru, from the pollen record. *Mountain Research and Development*, 18:159–72

Davis SD, Heywood VH, Herrera-MacBryde O, Villa-Lobos J and Hamilton AC (eds) (1997). *Centres of plant diversity: A guide and strategy for their conservation*. WWF-IUCN, Gland

Dinerstein E, Olson DM, Graham DJ, Webster AL, Primm SA, Bookbinder MP and Ledec G (1995). *A conservation assessment of the terrestrial ecoregions of Latin America and the Caribbean*. The World Bank, Washington, DC

Ellenberg H (1958). Wald oder Steppe? Die natürliche Pflanzendecke der Anden Perus I, II. *Umschau Wiss Technol.*, 21:645–8; 22:679–81

Fjeldså J (1992). Biogeographic patterns and evolution of the avifauna of relict high-altitude woodlands of the Andes. *Steenstrupia*, 18:9–62

Fjeldså J and Kessler M (1996). *Conserving the biological diversity of Polylepis woodlands of the highlands of Peru and Bolivia: A contribution to sustainable natural resource management in the Andes*. NORDECO, Copenhagen, Denmark

Fjeldså J and Rahbek C (1997). Species richness and endemism in South American birds: implications for the design of networks of nature reserves. In Laurance WF and Bierregaard R (eds), *Tropical Forest Remnants: Ecology Management and Conservation of Fragmented Communities*. Chicago University Press, Chicago, pp 466–82

Fjeldså J, Kessler M and Swanson G (eds) (1999a). *Cocapata and Saila Pata: People and biodiversity in a Bolivian montane valley*. DIVA Technical Report 7, Rønde, Denmark

Fjeldså J, Lambin E and Mertens B (1999b). Correlation between endemism and local ecoclimatic stability documented by comparing Andean bird distributions and remotely sensed land surface data. *Ecography*, 22:63–78

Galeano G, Suárez S and Balslev H (1999). Vascular plant species count in a wet forest in the Chocó area on the Pacific coast of Colombia. *Biodiversity Conservation*, 7:1563–75

Garwood NC, Janos DP and Brokaw N (1979). Earthquake-caused landslides: A major disturbance to tropical forests. *Science*, 205:997–9

Gentry AH (1982). Patterns of neotropical plants species diversity. *Evolutionary Biology*, 15:1–84

Gentry AH (1988). Changes in plant community diversity and floristic composition on environmental and geographical gradients. *Annals of the Missouri Botanical Gardens*, 75:1–34

Gentry AH (1995). Patterns of diversity and floristic composition in neotropical montane forests. In Churchill SP, Balslev H, Forero E and Luteyn JL (eds), *Biodiversity and Conservation of Neotropical Montane Forests*. The New York Botanical Garden, Bronx, pp 103–26

Gentry AH and Dodson CH (1987a). Contribution of nontrees to species richness of a tropical rain forest. *Biotropica*, 19:149–56

Gentry AH and Dodson CH (1987b). Diversity and biogeography of neotropical vascular epiphytes. *Annals of the Missouri Botanical Gardens*, 74:205–33

Gradstein SR and Frahm JP (1987). Die floristische Höhengliederung der Moose entlang des BRYTROP-Transektes in NO-Peru. *Beih Nova Hedwigia*, 88:105–13

Graves GR (1988). Linearity of geographical range and its possible effect on the population structure of Andean birds. *Auk*, 105:47–52

Henderson A, Churchill SP and Luteyn JL (1991). Neotropical plant diversity. *Nature*, 351:21–2

Hensen I (1995). Die Vegetation von *Polylepis*-Wäldern der Ostkordillere Boliviens. *Phytocoenologia*, 25:235–77

Holben BN, Tucker CJ and Fan, CF (1980). Spectral assessment for soybean leaf area and leaf biomass. *Photographic Engineering and Remote Sensing*, 46:651–6

Hueck K and Seibert P (1981). *Vegetationskarte von Südamerika*. 2nd ed. Gustav Fischer Verlag, Stuttgart

Huston MA (1994). *Biological diversity. The coexistence of species on changing landscapes*. Cambridge University Press, Cambridge, UK

Ibisch PL (1996). *Neotropische Epiphytendiversität – das Beispiel Bolivien*. Archiv naturwissenschaftlicher Dissertationen 1. Martina Galunder-Verlag, Wiehl, Germany

Ibisch PL, Boegner A, Nieder J and Barthlott W (1996). How diverse are neotropical epiphytes?

An analysis based on the 'Catalogue of the flowering plants and gymnosperms of Peru'. *Ecotropica*, **1**:13–28

Keating PL (1999). Changes in páramo vegetation along an elevation gradient in southern Ecuador. *Journal of the Torrey Botanical Society*, **126**:159–75

Kessler M (1995a). Polylepis – *Wälder Boliviens: Taxa, Ökologie, Verbreitung und Geschichte*. Dissertationes Botanicae 246, J. Cramer, Berlin

Kessler M (1995b). Present and potential distribution of *Polylepis* (Rosaceae) forests in Bolivia. In Churchill SP, Balslev H, Forero E and Luteyn JL (eds), *Biodiversity and conservation of neotropical montane forests*. New York Botanical Gardens, Bronx, pp 281–94

Kessler M (1998). Land use, economy and the conservation of biodiversity of high-Andean forests in Bolivia. In Barthlott W and Winiger M (eds), *Biodiversity. A Challenge for Development Research and Policy*. Springer, Berlin, pp 339–51

Kessler M (1999a). Plant species richness and endemism during natural landslide succession in a perhumid montane forest in the Bolivian Andes. *Ecotropica*, **5**:123–36

Kessler M (1999b). Plant diversity. In Fjeldså J, Kessler M and Swanson G (eds), *Cocapata and Saila Pata: People and biodiversity in a Bolivian montane valley*. DIVA Technical Report 7, Rønde, Denmark, pp 105–15

Kessler M (2000a). Altitudinal zonation of Andean cryptogam communities. *Journal of Biogeography*, **27**:275–82

Kessler M (2000b). Observations on a human-induced fire a humid timberline in the Bolivian Andes. *Ecotropica*, **6**:89–93

Kessler M (2000c). Upslope-directed mass effect in palms along an Andean elevational gradient: a cause for high diversity at mid-elevations? *Biotropica*, **32**:756–9

Kessler M (2000d). Species richness, endemism, and altitudinal zonation of selected plant groups along an elevational transect in the central Bolivian Andes. *Plant Ecology*, **149**:181–93

Kessler M (2001a). Diversidad y endemismo de grupos selectos de plantas en la Serranía de Pilón Lajas, depto. Beni, Bolivia. *Rev Sociedad Boliviana de Botánica*, **3**:124–45

Kessler M (in press a). Environmental patterns of species richness, life-form and ecophysiological type among bromeliad communities of Andean forests in Bolivia. *Biodiversity and Conservation*

Kessler M (in press b). Patterns of diversity and range size of selected plant groups along an elevational transect in the Bolivian Andes. *Biodiversity and Conservation*

Kessler M (2001 b). Pteridophyte species richness in Andean forests in Bolivia. *Biodiversity and Conservation* **10**, 1473–95

Kessler M (2001c). Maximum plant community endemism at intermediate intensities of anthropogenic disturbance in Bolivian montane forests. *Biodiversity and Conservation*, **15**:634–41

Kessler M and Bach K (1999). Using indicator groups for vegetation classification in species-rich Neotropical forests. *Phytocoenologia*, **29**:485–502

Kessler M and Beck SG (2001). Bolivia. In Kapelle M and Brown AD (eds), *Bosques Húmedos Montanos Neotropicales*. IUCN INBio Fundación ANA, Instituto Nacional de Biodiversidad, Santo Domingo de Heredia, Costa Rica

Kessler M and Croat T (1999). State of knowledge of Bolivian Araceae. *Selbyana*, **20**:224–34

Kessler M and Driesch P (1993). Causas e historia de la destrucción de bosques altoandinos en Bolivia. *Ecología en Bolivia*, **21**:1–18

Kessler M and Helme, N (1999). Floristic diversity and phytogeography of the central Tuichi Valley, an isolated dry forest locality in the Bolivian Andes. *Candollea*, **54**:341–66

Kessler M and Herzog SK (1998). Conservation status in Bolivia of timberline habitats, elfin forest and their birds. *Cotinga*, **10**:50–4

Kessler M and Hohnwald S (1998). Bodentemperaturen innerhalb und außerhalb bewaldeter und unbewaldeter Blockhalden in den bolivianischen Hochanden. Ein Test der Hypothese von Walter und Medina (1967). *Erdkunde*, **52**:54–62

Kessler M and Krömer T (2000). Patterns and ecological correlates of pollination modes among bromeliad communities of Andean forests in Bolivia. *Plant Biology*, **2**:659–69

Kessler M, Bach K, Helme N, Beck SG and Gonzales, J (2000). Floristic diversity of Andean dry forests in Bolivia – an overview. In Breckle SW, Schweizer B and Arndt U (eds), *Ergebnisse weltweiter ökologischer Forschung*. Beiträge des 1. Symposiums der A.F.W. Schimper Stiftung von H. und E. Walter, Hohenheim, October 1998. Günter Heimbach Verlag, Stuttgart, pp 219–34

Kessler M, Herzog SK, Fjeldså, J and Bach K (2001a). Species richness and endemism of plant and bird communities along two gradients of elevation, humidity, and land use in the Bolivian Andes. *Diversity Distributions*, **7**:61–77

Kessler M, Krömer T and Jimenez I (2000). Inventario de grupos selectos de plantas en el Valle de Masicurí (Santa Cruz – Bolivia). *Rev Boliviana de Ecología y Conservación Ambiental*, **8**:3–15

Kessler M, Smith AR, Acebey A and Gonzales J (2001b). Registros adicionales de pteridófitos del Parque Nacional Carrasco, dpto. Cochabamba, Bolivia. *Revue de la Sociedad Boliviana de Botánica*, **3**:146–50

Kessler M, Smith AR and Gonzales J (1999). Inventario de pteridófitos en una transecta altitudinal del Parque Nacional Carrasco, dpto.

Cochabamba, Bolivia. *Revue de la Sociedad Boliviana de Botánica*, 2:227–50

Koepcke HW (1962). *Synökologische Studien an der Westseite der peruanischen Anden*. Bonner Geogr Abh 29, S Ferd Dümmler Verlag, Bonn

Körner C (1998). A re-assessment of high elevation treeline positions and their explanation. *Oecologia*, 115:445–9

Krömer T, Kessler M, Holst BK, Luther HE, Gouda E, Till W, Ibisch PL and Vásquez R (1999). Checklist of Bolivian Bromeliaceae with notes on species distribution and levels of endemism. *Selbyana*, 20:201–23

Lægaard S (1992). Influence of fire in the grass páramo vegetation of Ecuador. In Balslev H and Luteyn JL (eds), Páramo. *An Andean ecosystem under human influence*. Academic Press, London, pp 151–70

Lieberman D, Lieberman M, Peralta R and Hartshorn GS (1996). Tropical forest structure and composition on a large-scale altitudinal gradient in Costa Rica. *Journal of Ecology*, 84:137–52

Luteyn JL (ed.) (1999). *Páramos. A checklist of plant diversity, geographical distribution, and botanical literature*. The New York Botanical Garden, Bronx

Lyon SK and Willig MR (1999). A hemispheric assessment of scale dependence in latitudinal gradients of species richness. *Ecology*, 80:2483–91

Magurran AE (1988). *Ecological diversity and its measurement*. Chapman and Hall, London

Moolman HJ and Cowling RM (1994). The impact of elephant and goat grazing on the endemic flora of South African succulent thicket. *Biological Conservation*, 68:53–61

Myers N, Mittermeier RA, Mittermeier CG, da Fonseca GAB and Kent J (2000). Biodiversity hotspots for conservation priorities. *Nature*, 403:853–8

Pareja J, Vargas C, Suárez R, Ballón R, Carrasco R and Villarroel C (1978). *Mapa geológico de Bolivia*. YPFB – Servicio Geológico de Bolivia, La Paz

Pickett STA, Kolasa J and Jones CG (1994). *Ecological understanding. The nature of theory and the theory of nature*. Academic Press, San Diego

Pitman NCA, Terborgh J, Silman MR and Nuñez P (1999). Tree species richness in an upper Amazonian forest. *Ecology*, 80:2651–61

Planet WG (1988). *Data extraction and calibration of TIROS-N-NOAA radiometers*. NOAA Tech. Mem. NESS 101-Rev. 1. US Dept of Commerce – Nat. Oceanic and Atmos. Adm., Washington, DC

Rahbek C (1995). The elevational gradient of species richness: a uniform pattern? *Ecography*, 18:200–05

Rahbek C (1997). The relationship among area elevation and regional species richness in neotropical birds. *American Naturalist*, 149:875–902

Samways MJ (1994). *Insect conservation biology*. Chapman and Hall, London

Scheiner SM (1993). Introduction: Theories hypotheses and statistics. In Scheiner SM and Gurevitch J (eds), *Design and analysis of ecological experiments*. Chapman and Hall, New York, pp 1–13

Seibert P and Menhofer X (1991). Die Vegetation des Wohngebietes der Kallawaya und des Hochlandes von Ulla-ulla in den bolivianischen Anden. I. *Phytocoenologia*, 20:145–276

Seidel R (1995). Inventario de los árboles en tres parcelas de bosque primario en la Serranía de Marimonos Alto Beni. *Ecología en Bolivia*, 25:1–35

Sklenà P and Jørgensen PM (1999). Distribution patterns of páramo plants in Ecuador. *Journal of Biogeography*, 26:681–91

Smith AR, Kessler M and Gonzales J (1999). New fern records from Bolivia. *American Fern Journal*, 89:244–66

Smith DN and Killeen TJ (1998). A comparison of the structure and composition of montane and lowland tropical forest in the Serrania Pilón Lajas, Beni, Bolivia. In Dallmeier F and Comiskey JA (eds), *Forest Biodiversity in North, Central and South America, and the Caribbean. Research and Monitoring*. UNESCO, Paris and The Parthenon Publishing Group, New York, pp 681–700

Stern MJ (1995). Vegetation recovery on earthquake-triggered landslide sites in the Ecuadorian Andes. In Churchill SP, Balslev H, Forero E and Luteyn JL (eds), *Biodiversity and Conservation of Neotropical montane forests*. New York Botanical Garden, Bronx, pp 207–20

Stotz DF, Fitzpatrick JW, Parker III TA and Moskovitz DK (1996). *Neotropical birds: Ecology and Conservation*. Chicago University Press, Chicago

SYSTAT (1997). *SYSTAT for Windows statistics version 7.0*. SPSS Inc, Chicago

Terborgh J and Winter B (1983). A method for siting parks and reserves with special reference to Colombia and Ecuador. *Biological Conservation*, 27:45–58

Troll C (1959). Die tropischen Gebirge. *Bonner Geogr Abh.*, 25:1–93

Tucker CGF, Townshend JRG and Goff TR (1985). African land-cover classification using satellite data. *Science*, 227:369–75

Usher MB (1986). *Wildlife Conservation Evaluation*. Chapman and Hall, London

Valencia R, Balslev H, Palacios W, Neill D, Josse C, Tirado M and Skov F (1998). Diversity and family composition of trees in different regions of Ecuador: a sample of 18 one-hectare plots. In Dallmeier F and Comiskey JA (eds), *Forest Biodiversity in North Central and South America and the Caribbean. Research and Monitoring*. Man and the Biosphere series 21. UNESCO Paris and The Parthenon Publishing Group, New York, pp 569–84

Walter H and Medina E (1969). Die Boden-temperatur als ausschlaggebender faktor für die Gleiderung der subalpinen und alpinen Stufe in den Anden Venezuelas. *Ber Deutsche Bot Ges.*, **82**:275–81

Webster GL (1995). the panorama of Neotropical cloud forests. In Churchill SP, Balslev H, Forero E and Luteyn JL (eds), *Biodiversity and Conservation of Neotropical montane forests*. New York Botanical Garden, Bronx, pp 53–77

Whittaker RH (1960). Vegetation of the Siskiyou Mountains Oregon and California. *Ecological Monographs*, **30**:279–338

Whittaker RH (1972). Evolution and measurement of species diversity. *Taxon*, **21**:213–51

Williams PH and Humphries CJ (1994). Biodiversity taxonomic relatedness and endemism in conservation. In Forey PL, Humphries CJ and Vane-Wright, RI (eds), *Systematics and Conservation Evaluation*. Clarendon Press, Oxford, pp 269–87

Williams PH, Prance GH, Humphries CJ and Edwards KS (1996). Promise and problems in applying quantitative complementary areas for presenting the diversity of some Neotropical plants (families Dichapetalaceae, Lecythidaceae, Caryocaraceae, Chrysobalanaceae and Proteaceae). *Biological Journal of the Linnean Society*, **58**:125–57

Wolf JHD (1993). Diversity patterns and biomass of epiphytic bryophytes and lichens along an altitudinal gradient in the northern Andes. *Annals of the Missouri Botanical Gardens*, **80**:928–60

Young KR (1993). Tropical timberlines: changes in forest structure and regeneration between two Peruvian timberline margins. *Arctic and Alpine Research*, **25**:167–74

Young KR (1997). Wildlife conservation in the central landscapes of the central Andes. *Landscape Urban Planning*, **38**:137–47

Young KR and León B (1990). Catálogo de las plantas de la zona alta del Parque Nacional Río Abiseo Peru. *Publicaciones del Museo de Historia Natural UNMSM (B)*, **34**:1–37

Biotope Patterns, Phytodiversity and Forestline in the Andes, based on GIS and Remote Sensing Data

6

Gerald Braun, Jens Mutke, Andreas Reder and Wilhelm Barthlott

INTRODUCTION

On a continental or global scale, it is near to impossible to adequately assess biodiversity and monitor its losses solely by collecting ground data. Even at the regional scale, the difficulties are so immense that we hardly have any data for tropical and subtropical high mountains. In this paper, we illustrate how a combination of biological field data with remote sensing data and climatological, pedological and topological GIS data can be used to obtain large-scale phytodiversity patterns.

The tropical Andes are one of the hot spots of biological diversity (Barthlott *et al.*, 1996, 1999; Davis *et al.*, 1997; Myers *et al.*, 2000; Groombridge and Jenkins, 2000). Together with their foothills, the Andes cover an area of nearly 8.1 million km^2 and stretch from about 11° 30′ N to 55° 05′ S. Elevations above 4000 m reach a maximum in Peru, Bolivia and Argentina. The Andean complex north of the equator forms two or three ranges nearly reaching 10° N in Venezuela. A remarkable decline in altitude can be observed between Ecuador and Peru, where elevation does not exceed 4000 m. Mittermeyer *et al.* (1999) estimate that 45 000 species of vascular plants (a sixth of the global flora) live in the Andean region. For example, in the Peruvian Andes above 500 m, more than 200 families of flowering plants can be found according to the checklist of Brako and Zarucchi (1993). This is 20% more than in the Peruvian Amazon and 20% more than in the whole flora of Europe (Tutin *et al.*, 1968–1993) which covers an area more than 20 times as large. A major cause of this enormous regional biological richness is the compression of climatic life zones along elevational gradients.

On a global scale, a distinct increase of vascular plant diversity can be observed as one approaches the equator (Figure 6.1). Phytodiversity increases with increasing warmth and humidity as one moves from the poles to the equator, except for continental and coastal deserts. Secondary maxima of species richness can be found in both hemispheres between 30° and 35°, due to Mediterranean type ecosystems and subtropical evergreen forests. On a regional scale, topographic heterogeneity becomes a second important driver. Mountains act as barriers for advective atmospheric moisture and front ranges trap much of the water with often little left for inner chains or leeward slopes. As temperatures decrease with elevation, the well known altitudinal and vertical arrangement of ecoclimatological zones emerges (Lauer, 1986). Many authors have investigated the altitudinal zonation of the Andean vegetation and flora (e.g. Humboldt, 1805–1834; Weberbauer, 1911; Acosta-Solís, 1968; Ellenberg, 1975; Cleef *et al.*, 1984). However, it was only in the last decade that large floristic databases (e.g. Brako and Zarucchi, 1993; Jørgensen and León-Yánez, 1999) allowed detailed quantitative comparisons of floristical zones and altitudinal gradients. In addition, data scattered in the literature regarding small but very detailed

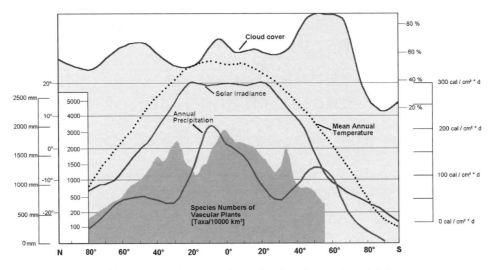

Figure 6.1 Pole to pole profile of climatic elements and vascular plant diversity in a global mean

floristic inventories, especially for woody plants, were collected in databases and synoptically analysed by different authors (e.g. Clinebell *et al.*, 1995; Terborgh and Andresen, 1998; Mutke, unpubl.).

Nested in these larger climatic gradients, steepness and aspect of slopes and micro-relief create a topographical diversity which interacts with solar irradiance, wind and precipitation, creating a multitude of habitats, here addressed as 'geodiversity'. The geodiversity of mountains by far exceeds that of lowlands and in part explains the high mountain biodiversity. Geodiversity is defined herein as the spatial heterogeneity of atmospheric and geospheric conditions (e.g. petrography, soils, topography and associated microclimate). In a given area, geodiversity reflects the multitude of life conditions and thus affects total regional biological diversity. The total diversity of an area, the ecodiversity, consists of biodiversity and geodiversity (Barthlott *et al.*, 1996, 2000). We prefer the term ecodiversity rather than landscape diversity, as the latter is not free of scale. Geodiversity may become ecologically effective at the centimetre scale. Hence, it is possible to describe the ecodiversity of, e.g. a cubic centimetre of soil, including highly diverse

microbial communities – a scale not covered by the term landscape diversity. Additionally, in biodiversity research the term landscape diversity has long been used as a synonym for the gamma diversity (Whittaker, 1972, 1977), the mere biodiversity at landscape level, without taking into account geodiversity.

MATERIAL AND METHODS
GIS and remote sensing data

The BIO-GIS-Project was founded in 1996 by the German Remote Sensing Data Centre, Environment and GIS Unit. It had been designed as an open frame for multiscale biogeography-related data and models. The project has been laid out complementary to the BIOMAPS project of the Dept. of Botany, University of Bonn. The geospatial and biological data were processed on a continental scale with a spatial resolution between 1 and 5 km, currently covering South America and Africa. Table 6.1 gives an overview on the basic datasets incorporated in BIO-GIS.

Topographical and topological data (DEM, DCW) have been included without any further adjustments. For the FAO soil map, we used a

Table 6.1 Primary data sets, available in the BIO-GIS

Dataset	Resolution	Source
DEM	1 km	USGS, GTOPO 30
NOAA–AVHRR–NDVI	1 km/1 month	1–km AVHRR global land data set
Soil map	–	FAO, processed to 5 km raster data
Climatic data	2–5 km	CRU, global datasets of station time series
Roads, rivers, cities	–	Digital chart of the world
Vascular plant species richness after Barthlott *et al.* (1999)	–	BIOMAPS Group, Botanical Institute, Univ. Bonn
Geographical units with known species numbers	–	BIOMAPS Group, Botanical Institute, Univ. Bonn
Small-scale inventories of woody plants in the Neotropics	point data	Jens Mutke, Botanical Institute, Univ. Bonn
Vegetation	1 km	Supervised classification of multi-temporal AVHRR–NDVI data

5 km resolution and differentiated soil genesis and texture classes.

The BIO-GIS climate database was derived from a standardised global climate dataset in a 0.5° spatial resolution compiled by New *et al.* (1999). The set covers the period 1960 to 1990, with monthly data derived from global terrestrial observations. For monthly precipitation, 19 295 stations had been included, and monthly mean temperatures were based on 12 092 locations. To meet the needs of a plant ecological analysis in the patchy high mountain terrain, the data had to be downscaled using high resolution topographic data. In a multiple regression analysis, the monthly climate data represented the dependent variable, whereas elevation, latitude and longitude in a 2.5° by 2.5° grid were the independent variables. For each month, about 10,000 equations were solved and the resultant coefficients were assigned to the GIS matrix. Additional other climate data such as climatic parameters, e.g. vapour pressure and atmospheric pressure at msl (National Centre for Atmospheric Prediction, National Centre for Atmospheric Research), were integrated to facilitate the calculation of the climatological water balance. In order to discuss potential vegetation change, we used modelled climate data for the year 2049 (derived from an ECHAM4 global climatic model) provided by the German Climate Computing Centre.

The vegetation map of South America by Hueck and Seibert (1972) had been digitised and referenced to the common Lambert Azimuthal Equal Area projection. A set of continental multi-temporal NOAA–AVHRR–NDVI data, covering the period April 1992 to March 1993 in 10 day composites, had been acquired. Due to the poor data quality, post processing had to be applied to correct for geometric deflections and cloud 'contamination'. Geometry was corrected manually in a stepwise approach (Braun *et al.* in preparation). By reaggregation of three 10-day composites per month, maximum composites were derived. A single path gradient analysis had been carried out to perform statistical cloud detection and the consecutive interpolation. The final data set, consisting of 12 monthly images, was added to the BIO-GIS data stack. Once the data were corrected for geometry and cloud gaps, clear phenological signals could be deduced. For the current purpose, the vegetation map of South America was used to extract and analyse the NDVI-derived phenological signatures for single vegetation units, as represented in conventional vegetation maps. If the phenological signals from the NDVI data showed distinct heterogeneities which could not be explained from the differentiation in the basic vegetation map, additional cartographic information was involved to differentiate and subdivide the vegetation categories. Step by step, the NDVI signal for 78 vegetation units of the Hueck-Seibert map have been examined and carefully reclassified, to derive a spatially more accurate and thematically deeper vegetation map, now differentiating nearly 180 vegetation units (Braun *et al*. in preparation).

The data sets, shown in Table 6.1 form the basis for calculation of numerous derivatives or more integrated GIS layering. The DEM was used to calculate slope inclination, aspect and soil moisture indicator values, described in Braun and Hörsch (1999). Climate data were used to calculate potential evaporation and the thermal vegetation period for each pixel. Additionally, the data were used to estimate large-scale topodiversity, climate diversity and geodiversity.

Phytodiversity data

The phytodiversity data of the BIOMAPS Group at the Botanical Institute, University of Bonn, consist of four different datasets for South America: The first data layer comprises the world map of vascular plant species richness published by Barthlott *et al*. (1996, 1999). The second dataset, the corresponding raw data with figures for total species richness, endemism and, if available, family numbers, comprises some 340 records, covering 271 different geographical units in South America. Some 150 of these are located in the Andes or the adjacent forelands (Mutke *et al*. in preparation). For a general description of the data structure, consult Mutke *et al*. (in press).

To study the altitudinal zonation, we mainly used the third dataset, the digital version of the *Catalogue of the Flowering Plants and Gymnosperms of Peru* (Brako and Zarucchi, 1993, compare www.mobot.org). We extracted species and family numbers of different life forms in 10 elevational zones. The fourth dataset was used to analyse spatial patterns of tree species diversity. A database with complete species lists of almost 250 small-scale

inventories of woody plants in South and Central America and the Caribbean (but with a focus on the tropical Andes) was established, based on data taken from published literature and a few personal field studies. In addition, 50 records with species numbers, but without complete species lists, were included. The 188 inventories with known geographical coordinates in the dataset comprise a total number of 155 vascular plant families. These records are analysed using GIS and statistical software regarding their spatial patterns of plant richness as well as the underlying biogeographical patterns. The Andean treeline/ forestline analysis is based on information obtained from the Tropicos database of the Missouri Botanical Garden, herbarium records by Kessler (1995) and Fjeldså and Kessler (1996).

Quantifying geodiversity

On the basis of climatic, soil geographical and topographical data, diversity was determined in a 50 by 50 km window. From the single diversity layers, e.g. precipitation, temperature, soil texture, topographic diversity, an unweighted mean has been calculated and aggregated to geodiversity.

We propose a new normalised diversity algorithm which also accounts for the frequency of class occurrence. Based on the assumption that, for a given area, class diversity is highest (here defined as 1) if the class frequency is even, a maximum diversity is set for equal distribution of class frequencies. In contrast, lowest diversity will be reached if the class frequency has a maximum unevenness, e.g. if in a given area one class dominates the pattern (class diversity = 0). Between these two extremes we have to scale the diversity depending on the frequency of the different classes present. For a detailed description of the index see Braun (in preparation, compare Barthlott *et al*., 2000). Implementing a normalised diversity index into spatial information systems requires a sliding or stepwise window moving across the database. The size of the moving window should be equal or larger than the maximum number of classes. The microclimatic conditions were derived from topodiversity which is the key factor increasing the climatic diversity of the land-scape. In this first approach, we calculated topodiversity from the basic autochthone relief factors altitude, aspect and inclination of a single tope. A prerequisite for the GIS-based calculation of topodiversity is a set of conventions. For instance, it is necessary to define a lower inclination threshold for the calculation of slope aspect (here set at 7 ° inclination). Not yet accounted for are allochtone factors such as cold air drainage, wind exposure, and solar horizon of the tope.

Analysing the Andean forest line

The BIO-GIS database was also used to analyse the forestline/treeline in the South American Andes. The objectives of the project are:

(a) To find ecoclimatological key parameters or indices which support a physiological explanation of the treeline/forestline altitude,

(b) To identify anthropogenous depressions of treeline/forestline across the Andes, and

(c) To model the impact of climate change on the Andean forestline/treeline on the basis of GCM data.

About 150 000 treeline/forestline elevation findings had been extracted from the 1 km South American vegetation map and a DEM. An additional 2000 findings for specific taxa were extracted from floristic databases (compare Phytodiversity Data). Treeline positions were correlated with calculated water balance and thermal indicators (season length defined by monthly mean temperature > 4.5 °C). The water supply during the thermal growing season was estimated from the difference between precipitation and potential evaporation for each month (Lauer and Frankenberg, 1981). By superimposing the forestline maps and species findings onto the climatic data, it was possible to extract the temperature and water balance corresponding to the position of the whole Andean forestline, as well as the altitude of the different treeline taxa. Using

minimum distance classification procedures, the potential Andean forest cover could be estimated. A re-mapping of forest positions for the year 2049 climate projection was also done.

Comparing altitudinal trends of Andean phytodiversity

To analyse altitudinal gradients of Andean vascular plant diversity, we analysed two independent datasets. Using the digital version of *the Catalogue of the Flowering Plants and Gymnosperms of Peru* (Brako and Zarucchi, 1993), we extracted species and family numbers of different life forms in 10 elevational zones. From these data, species per family ratios for the different life forms and elevational zones were computed. In addition, we extracted total species and family richness in coastal Peru, the northern and the southern Peruvian Andes above 500 m, and the Peruvian Amazon. Furthermore, we studied the altitudinal trends found in small-scale tree species inventories of 0.1 and 1.0 ha in size.

RESULTS

Andean geodiversity

The geodiversity of the whole Andean range was differentiated into eight grey scale levels (Figure 6.2). Composed of four abiotic parameters – topography, soil texture, precipitation and temperature – the map shows geodiversity maxima for the Colombian Andes, around 5° N latitude and along the Peruvian and Bolivian Eastern Cordilleras. These maxima coincide almost completely with the Andean phytodiversity maxima shown in the global map of species numbers of vascular plants from Barthlott *et al.* (2000).

Present and future patterns of the Andean forest and treeline

Even though the data used for the analyses have not been filtered or corrected, Figure 6.3 and 6.4 illustrate an obvious pattern. Both, the topographical elevation depression around 5° S and the corresponding lack of a forestline can be clearly seen for the eastern Andean

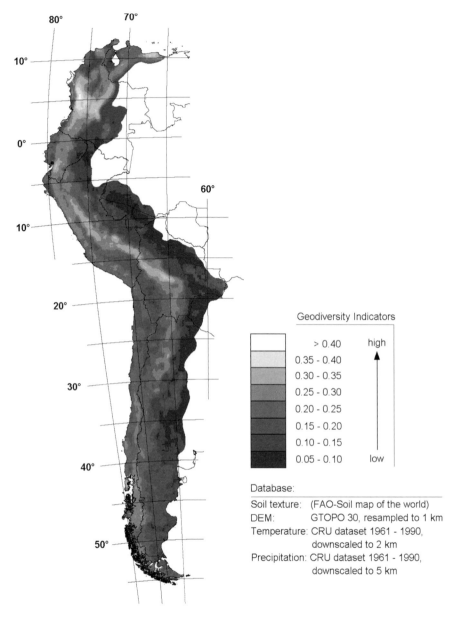

Figure 6.2 Andean geodiversity as calculated on the basis of the above-described parameters

ranges. Also, the depression and subsequent disappearance of forest between 25 and 30° S due to aridity becomes visible (Figures 6.3 and 6.4, eastern ranges). A remarkable gap in the Andean forestline can be observed on the western Andean ranges between 22° and 33° S. As a result of the increasing aridity at the western edge of the continent, the forestline/ treeline moves to more humid habitats in higher elevations and finally disappears around 24° S. Outposts of forest in the dry Altiplano are composed of a single tree species, *Polylepis tarapacana*, which can be found up to 5100 m, occupying azonal warm habitats on the

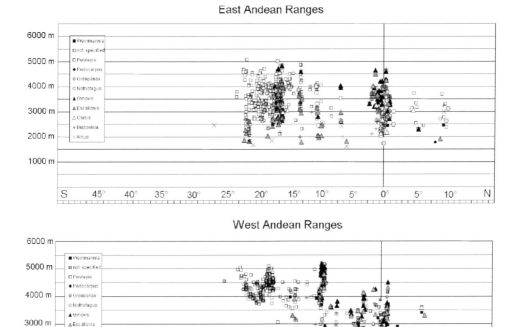

Figure 6.3 Altitudinal distribution of treeline and forestline species derived from floristic databases

north-facing slopes of large volcanoes (Jordan, 1979; Braun, 1997).

The forestline in Figure 6.5 was mapped from NOAA–AVHRR–NDVI data and the local climate data were derived from downscaling, as described earlier. For the whole Andean region, the mean temperature during the growing season at the treeline was calculated following Körner (1998). Temperatures were stratified into three ranges, namely 3.0–5.5, 5.5–7.5 and 7.5–10.5 °C. The last class has only been plotted into the upper right subset of the map, showing the central Andes of Bolivia and Peru. In the lower right part of Figure 6.5, the fractions of 'forestline pixels' falling into one of these temperature ranges was calculated.

The predominant part of the Andean forestline and the mesophytic forest and shrubs of the western Peruvian and Bolivian Andes, as mapped from the 1992 remote sensing data, occur at temperatures between 5.5–7.5 °C. A total of 30.7% of the mesophytic forestline occurs in habitats with 3.5–5.5 °C, which can be corroborated by field findings from the western Bolivian highlands. Only a minor part, 28.9%, occurs at temperatures between 7.5 and 10.5 °C (Figure 6.5). On the eastern Andean ranges the findings are reversed. In this predominantly humid environment, with a remarkably higher degree of human impact on the forestline, such as burning and clear cutting, 43.8% of forestline observations were found in a temperature range between 7.5–10.5 °C. If this effect was a result of geometric disturbances or inaccuracies in the climatic data sets instead of human impact, a larger fraction of forestline pixels than the observed 9.2% would fall into the 3.5–5.5 °C range.

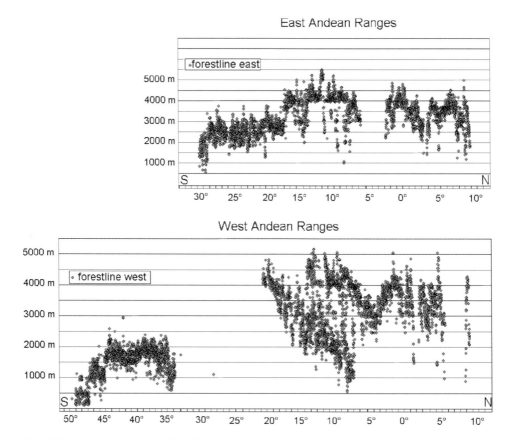

Figure 6.4 Altitude of Andean treeline/forestline mapped from remote sensing data

Provided that the hypothesis is correct, that forestlines on a large scale reflect a thermal limit, the analysis shown in Figure 6.5 suggests that human activities caused a much higher impact on the forestline in the humid eastern Andes than in the subhumid and semiarid ranges. Furthermore, Figure 6.6 indicates possible changes of forestline altitude due to climate change. On the basis of modelled future climate data, the forestline for the whole Andean region indicates either no changes for 2049 (green line), a retraction (red line), or an advance (blue).

Altitudinal gradients of Andean phytodiversity

Comparing different zones within Peru and Ecuador, the Andes region has the highest species and family numbers of vascular plants (Figure 6.7). However, when analysing full altitudinal gradients of Andean plant richness, the highest species numbers per 500-m altitudinal interval occur in the lowlands. The same is true for family numbers (Figure 6.8). Analysing such altitudinal gradients by life form, the sharpest reduction in species richness from the lowland to the higher elevations occurs in trees. In contrast, much of the diversity of herbs or shrubs can be found at mid elevations. For these groups, there are also higher numbers of species per family at these elevations.

Regarding the diversity of trees (dbh $\geq$ 10 cm) in 1.0-ha plots, species richness also sharply declines from the lowland to the montane forests (Figure 6.9). The Gentry transect data (Gentry, 1982b; Clinebell *et al.*, 1995) on

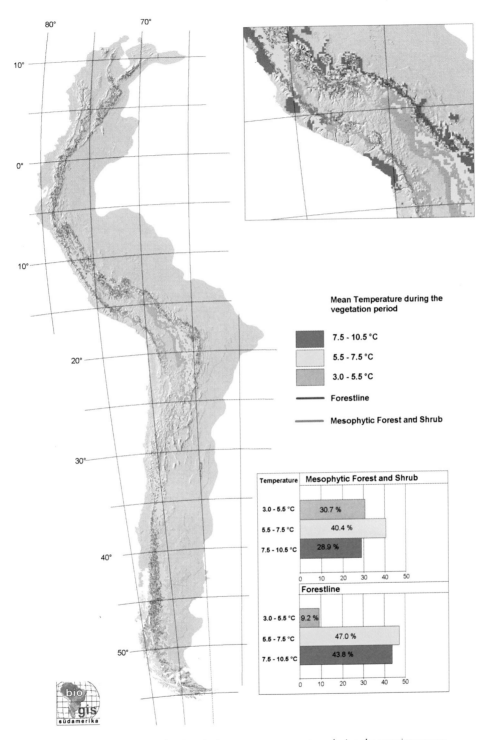

Figure 6.5 Position of the Andean forestline in relation to mean temperature during the growing season

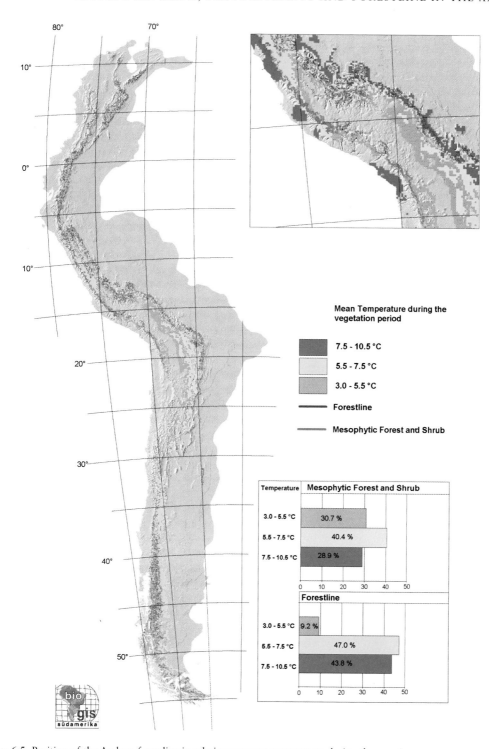

Figure 6.5 Position of the Andean forestline in relation to mean temperature during the growing season

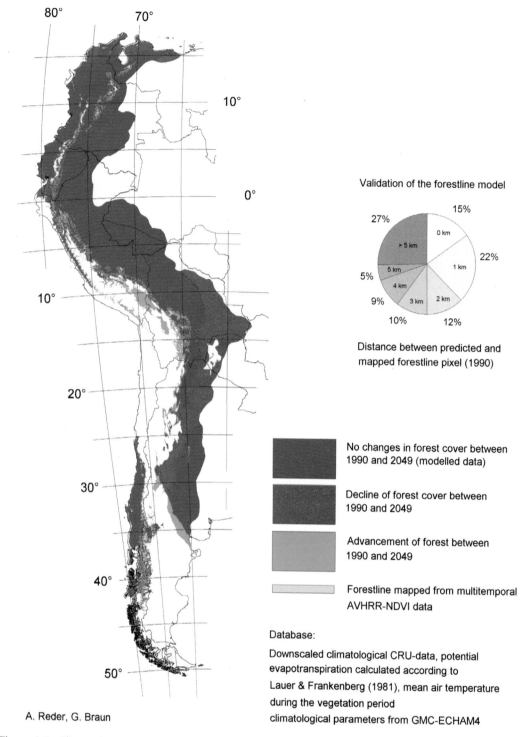

Validation of the forestline model

Distance between predicted and
mapped forestline pixel (1990)

No changes in forest cover between
1990 and 2049 (modelled data)

Decline of forest cover between
1990 and 2049

Advancement of forest between
1990 and 2049

Forestline mapped from multitemporal
AVHRR-NDVI data

Database:

Downscaled climatological CRU-data, potential
evapotranspiration calculated according to
Lauer & Frankenberg (1981), mean air temperature
during the vegetation period
climatological parameters from GMC-ECHAM4

A. Reder, G. Braun

Figure 6.6 Climate-driven changes in the Andean forest cover. The modelling is based on actual and predicted climatological data as well as a satellite-based mapping of the Andean forestline

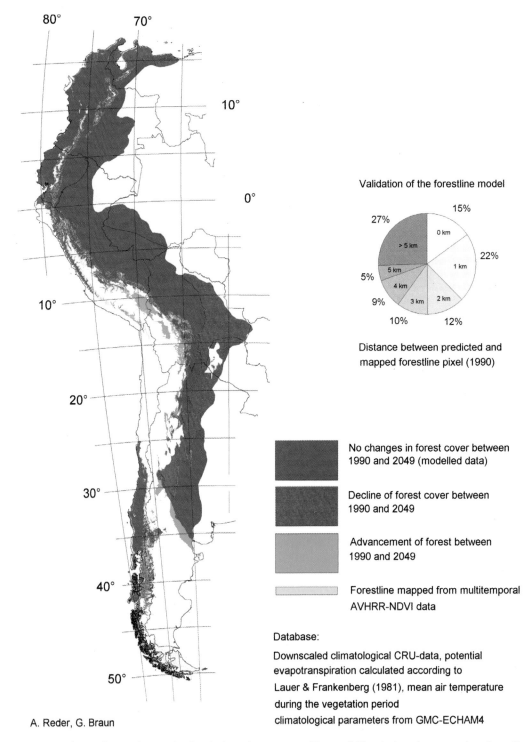

Validation of the forestline model

Distance between predicted and
mapped forestline pixel (1990)

No changes in forest cover between
1990 and 2049 (modelled data)

Decline of forest cover between
1990 and 2049

Advancement of forest between
1990 and 2049

Forestline mapped from multitemporal
AVHRR-NDVI data

Database:

Downscaled climatological CRU-data, potential
evapotranspiration calculated according to
Lauer & Frankenberg (1981), mean air temperature
during the vegetation period
climatological parameters from GMC-ECHAM4

A. Reder, G. Braun

Figure 6.6 Climate-driven changes in the Andean forest cover. The modelling is based on actual and predicted climatological data as well as a satellite-based mapping of the Andean forestline

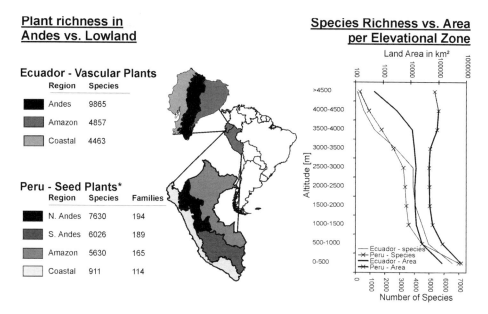

Figure 6.7 Documented species and family numbers of higher plants in different regions of Peru and Ecuador vs. altitudinal gradients of species richness and land area in 500-m height intervals in both countries. (Figures for Peru extracted from the online version of Brako and Zarucchi, 1993; family delimitation after ITIS (USDA), figures for Ecuador after Jørgensen and León-Yánez, 1999)
*Boundaries of the regions in Peru follow the boundaries of the provinces due to the structure of the data

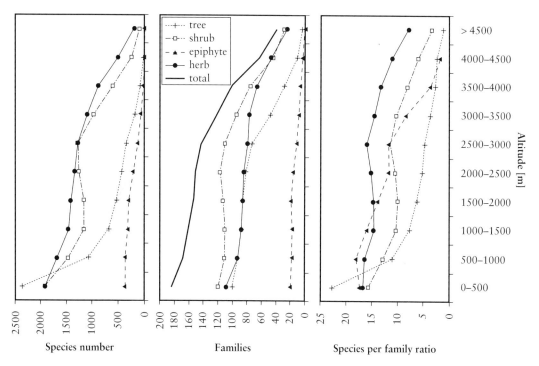

Figure 6.8 Altitudinal gradients in Peruvian plant diversity analysed on the basis of the *Catalogue of Flowering plants and Gymnosperms of Peru* (extracted from the online version of Brako and Zarucchi, 1993)

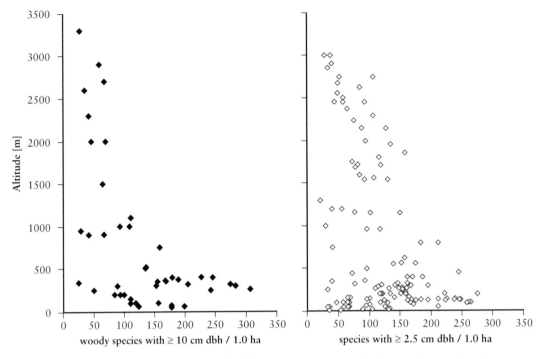

Figure 6.9 Species numbers of woody plants with dbh ≥ 10 cm in neotropical 1.0-ha study plots (data from various sources) and of plants with dbh ≥ 2.5 cm in neotropical 0.1-ha study plots (data from different sources, mainly the Gentry transect data, compare Gentry, 1982b; Clinebell *et al.*, 1995)

richness of plants with dbh ≥ 2.5 cm (including shrubs and occasionally large herbs) in 0.1 ha inventories in the Neotropics shows a less pronounced decline. The present analysis reveals a second maximum of species numbers at mid elevations around 2000 m asl (Figure 6.9).

DISCUSSION

Ongoing analysis on Andean forestline and treeline ecotone data show that correlations between tree growth at high altitude and ecologically relevant climatic parameters (seasonal temperature, water balance) are valid even on very large scales. Looking at the forestline/treeline from a life form point of view, neglecting the taxonomic differentiation, the correlation analysis between climate variables and forestline/treeline findings had to be performed separately for humid and semiarid habitats. In the semiarid Andes the forest/tree

boundaries are not as well pronounced as in the more humid parts because forest stands open up with increasing aridity as water availability becomes the limiting factor for tree growth. Analysed separately for the different treeline-forming species, this humid/arid differentiation was not necessary in order to arrive at clear correlations.

A comparison between the mapped forestline and the forestline modelled on the basis of temperature and water balance shows a high coincidence across the whole Andes. Nevertheless, higher deviations occur in the Eastern Cordilleras of Ecuador, Peru and in the Eastern Cordillera of Bolivia, near Cochabamba. While historical deforestation in the Cordilleras west of Cochabamba is evident, the deviations between the modelled and the mapped forestline in the eastern Andes of Peru and Ecuador could probably be retraced to errors in the digital elevation model. Therefore, the assessment of the anthropogenic impact on the Andean forestline on these large scales is delicate. A carefully estimated proportion of anthropogenic forestline depression would be in the range of 32% of the whole Andean forestline.

Using the relationship between forestline and climate and the predicted climatological situation for 2049, trends in altitudinal changes of the Andean forestline can be regionalised and attributed to positive or negative changes in temperature or water balance. Despite some errors in the digital elevation model, the large-scale trends are obvious. The main responses of forestline on climate change can be expected around the dry axis on the Altiplano. A slightly improved water balance could lead to an advance of tree species into the basins of Lake Titicaca and Lake Poopo. The mesophytic montane forests on the western Cordillera of Peru show a slight constriction in their altitudinal range, but a tendency to advance further south.

In contrast to these results, approaches to predict patterns of Andean plant species richness on the basis of environmental parameters show only partly satisfying results. To some degree this can be traced back to the highly heterogeneous quality of the phytodiversity data available for South America. However, in line with the high geodiversity – and presumably due

to the high geodiversity (Figure 6.2) – the Andean flora shows a very high beta diversity, resulting in exceptional high species numbers on a landscape level. These high species numbers in the Andes are not only found in vascular plants, but in many other groups of organisms (e.g. Fjeldså and Rahbeck, 1998; Brooks et al., 2000; Groombridge and Jenkins, 2000).

Another interesting feature of Andean phytodiversity are the differences between the altitudinal gradients in species and family richness that can be observed for different groups of plants. Whereas by far the highest species numbers of trees can be found in the lowlands (Figures 6.8, 6.9), shrubs show a second peak at 2000–3000 m asl Looking at the species per family ratios for different life forms of the Peruvian flora, it seems that shrubs, herbs, and epiphytes, in contrast to the trees, underwent important speciation processes after the uplift of the Andes (compare Gentry, 1982a; Ibisch et al., 1996). The fact that highest species numbers of all 500-m height intervals can be found in the lowland might be a result of the larger area available (Figure 6.7), as discussed by Körner (2000) and others. However, this seems to be only one explanation for the sharp decline in species numbers above 3000 m asl This effect could, instead, be explained with the parameters discussed in the context of the Andean forestline.

References

Acosta-Solís M (1968). *Divisiones fitogeograficas y formaciones geobotanicas del Ecuador*. Edit. Casa de la cultura Ecuatoriana, Quito

Barthlott W, Biedinger N, Braun G, Feig F, Kier G and Mutke J (1999). Terminological and methodological aspects of the mapping and analysis of global biodiversity. *Acta Botanica Fennica*, 162: 103–10

Barthlott W, Lauer W and Placke A (1996). Global distribution of species diversity in vascular plants: towards a world map of phytodiversity. *Erdkunde*, 50(4):317–27

Barthlott W, Mutke J, Braun G and Kier K (2000). Die ungleiche globale Verteilung pflanzlicher

Artenvielfalt – Ursachen und Konsequenzen. *Ber d Reinh Tüxen-Ges*, 12:67–84

Brako L and Zarucchi JL (1993). *Catalogue of the Flowering Plants and Gymnosperms of Peru*. Monographs in Systematic Botany from the Missouri Botanical Garden, 45, 1286 pp

Braun G and Hörsch B (1999). Hochwasserschutz und Hochwasserprävention – Integration von Fernerkundungs-und GIS-Daten durch räumliche Modelle. *Wasserwirtschaft*, 89(6):298–304

Braun G, Barthlott W, Teschner O (in prep.). Vegetation of South America: A satellite based map of the continental vegetation and human impact on landscape.

Braun G (1997). Digital methods in assessing forest patterns in an Andean environment – the *Polylepis* example. *Mountain Research and Development*, 17(3):253–62

Brooks T, Jahn A, Limp F, Smith K, Mehlman D, Roca R and Williams P (2000). *Conservation Priority Setting for Birds in Latin America/ WORLDMAP*. CD-ROM. The Nature Conservancy, Arlington, VA

Cleef AM, Rangel Ch O, Van der Hammen T and Jaramillo R (1984). La vegetación de las selvas del transecto Buritaca. In Van der Hammen T and Ruiz PM (eds), *La Sierra Nevada de Santa Marta (Colombia) transecto Buritaca – La Cumbre. Estudios de ecosistemas tropandinos II*. J. Cramer, Berlin, pp 267–406

Clinebell RR II, Phillips OL, Gentry AH, Starks N and Zuuring H (1995). Prediction of neotropical tree and liana species richness from soil and climatic data. *Biodiversity and Conservation*, 4:56–90

Davis SD, Heywood VH, Herrera-Macbryde O, Villa-Lobos J and Hamilton AC (eds) (1997). *Centres of Plant Diversity. A guide and strategy for their conservation*. vol. 3: The Americas. WWF & IUCN, Cambridge

Ellenberg H (1975). Vegetationsstufen in perhumiden bis perariden Bereichen der tropischen Anden. *Phytocoenologia*, 2(3/4):368–87

Fjeldså J and Kessler M (1996). *Conserving the Biological Diversity of Polylepis Woodlands of the Highland of Peru and Bolivia. A Contribution to Sustainable Natural Resource Management in the Andes*. NORDECO, Copenhagen, 250 pp

Fjeldså J and Rahbeck C (1998). Priorities for conservation in Bolivia, illustrated by a continent-wide analysis of bird distributions. In Barthlott W and Winiger M (eds), *Biodiversity – A Challenge for Development Research and Policy*. Springer, Berlin, Heidelberg, New York, pp 313–27

Gentry AH (1982a). Neotropical floristic diversity: Phytogeographical connections between Central and South America, Pleistocene climatic fluctuations, or an accident of the Andean orogeny? *Ann Miss Bot Gard*, 69:557–93

Gentry AH (1982b). Patterns of neotropical plant species diversity. *Evolutionary Biology*, 15:1–84

Groombridge B and Jenkins MD (eds) (2000). *Global biodiversity: Earth's living resources in the 21st century*. World Conservation Press, Cambridge

Hueck K and Seibert P (1972). *Vegetationskarte von Südamerika*. G. Fischer, Stuttgart

Humboldt A von (1805–1834). *Voyage aux régions équinoxiales du Nouveau Continent fait en 1799, 1800, 1801, 1802, 1803 et 1804 par Alexandre de Humboldt et Aimé Bonpland*. Redigé par Alexandre de Humboldt. 30 vols. Paris

Ibisch PL, Boegner A, Nieder J and Barthlott W (1996). How diverse are neotropical epiphytes? An analysis based on the "Catalogue of the flowering plants and gymnosperms of Peru". *Ecotropica*, 2:13–28

Jordan E (1979). *Das durch Wärmemangel und Trockenheit begrenzte Auftreten von Polylepis am Sajama Boliviens mit den höchsten Polylepis-Gebüschvorkommen der Erde*. 42. Deutscher Geographentag, Göttingen. Wiesbaden, pp 303–05

Jørgensen PM and León-Yánez S (eds) (1999). Catalogue of the Vascular Plants of Ecuador. *Monogr Syst Bot Miss Bo. Gard*, 75:VIII + 1181 pp

Kessler M (1995). *Polylepis-Wälder Boliviens: Taxa, Ökologie, Verbreitung und Geschichte*. Dissertationes Botanicae 246. Cramer, Berlin Stuttgart, 303 pp

Körner Ch (1998). A re-assessment of high elevation treeline positions and their explanation. *Oecologia*, 115:445–59

Körner Ch (2000). Why are there global gradients in species richness? Mountains might hold the answer. *TREE*, 15(12):513–14

Lauer W (1986). Die Vegetationszonierung der Neotropis und ihr Wandel seit der Eiszeit. *Deutsche Botanische Gesellschaft*, 99:211–35

Lauer W and Frankenberg P (1981). *Untersuchungen zur Humidität und Aridität von Afrika – Das Konzept einer potentiellen Landschaftsverdunstung*. Bonner Geographische Abh. 66. Dümmler, Bonn, 127 pp

Mittermeier RA, Myers N, Gil PR and Mittermeier CG (1999). *Hotspots – Earth's Biologically richest and most endangered Terrestrial Ecoregions*. CEMEX, Mexico City

Mutke J, Kier G, Braun G, Schultz C and Barthlott W (in press). Patterns of African vascular plant diversity – a GIS based analysis. *Syst Geogr Pl*, 71

Myers N, Mittermeier RA, Mittermeier CG, da Fonseca GAB and Kent J (2000). Biodiversity hotspots for conservation priorities. *Nature*, 403:853–8

New M, Hulme M and Jones P (1999). Representing Twentieth-Century Space–Time Climate Variability. Part I: Development of a 1961–90 Mean Monthly Terrestrial Climatology. *Journal of Climate*, 12:829–56

Terborgh J and Andresen E (1998). The composition of Amazonian forests: patterns at local and regional scales. *J Trop Ecol*, 14:645–64

Tutin TG, Burges NA, Chater AO, Edmondson JR, Heywood VH, Moore DM, Valentine DH, Walters SM and Webb DA (eds) (1993). *Flora Europaea*. vol. 1, 2. (ed.) Cambridge University Press, Cambridge, New York, Melbourne

Tutin TG, Heywood VH, Burges NA, Moore DM, Valentine DH, Walters SM and Webb DA (eds) (1968–80). *Flora Europaea*. vol. 2–5. Cambridge University Press, Cambridge, London, New York

Weberbauer A (1911). *Die Pflanzenwelt der peruanischen Anden: in ihren Grundzügen dargestellt.* Engelmann, Leipzig, 353 pp

Whittaker RH (1972). Evolution and measurement of species diversity. *Taxon*, **21**:213–51

Whittaker RH (1977). Evolution of Species Diversity in Land Communities. *Evolutionary Biology*, **10**:1–67

Multi-scale Patterns in Plant Species Richness of European High Mountain Vegetation

7

Risto Virtanen, Thomas Dirnböck, Stefan Dullinger, Harald Pauli, Markus Staudinger and Georg Grabherr

INTRODUCTION

One of the early-perceived patterns in alpine plant diversity of Europe is that some mountains are richer in species than others (Wahlenberg, 1813, 1814; Kerner, 1863; Blytt, 1876; Jaccard, 1912; DuRietz, 1924). Many of the patterns in mountain plant richness had been obvious to botanists and naturalists, but had been quantified in only a few cases. On a global scale, plant species richness on mountains has been suggested to decline with increasing latitude (Eurola, 1974; Körner, 1995). Within a certain region and on very small scales (< 20 km^2), alpine species richness has been found to increase with the area surveyed (Wohlgemuth, 1993) but not at larger scales (> 20 km^2). Locally, species richness also depends on disturbance, grazing intensity, and spatial landscape heterogeneity at different scales (Scharfetter, 1909; Wohlgemuth, 1993; Körner, 1995; Olofsson *et al.*, 2001). The elevational decline in species richness above the treeline is a widely recognised phenomenon (Grabherr *et al.*, 1995; Chapter 1).

The importance of the evolutionary history of glaciations and post-glacial climatic fluctuations for mountain plant species richness has been emphasised by many botanists (Jerosch, 1903; Scharfetter, 1909; Kalliola, 1939; Hegi *et al.*, 1977; see also Chapin and Körner, 1995). The significance of the species–area relationship and the island nature of mountains has also been considered by many authors (MacArthur and Wilson, 1967; Riebesell, 1982; Williamson, 1988; Holt, 1993; Körner, 2000). Isolation has

modified the alpine flora particularly during post-glacial warmer climates, during which habitats suitable for alpine plants were reduced in size and extinction rates were high. Ongoing atmospheric warming is likely to enhance this process (Grabherr *et al.*, 1994).

In addition to climate, soil conditions have long been known to play an important role for alpine plant species diversity (Wahlenberg, 1814; Kerner, 1863; Nordhagen, 1928; Dullinger *et al.*, 2000), with calcareous substrates commonly supporting more species. Whether this holds for mountains across Europe is unknown. Similarly, patterns of snow distribution affect alpine vegetation (Gjærevoll, 1956; Körner, 1999). Distinct snow cover gradients range from exposed sites with little or no snow cover in winter to snowbeds with heavy snow pack until late summer. Following the intermediate productivity (or stress) theory (Grime, 1979), it could be expected that low productivity wind-exposed communities and snowbeds support less species than in communities with moderate snow cover.

Vegetation samples (quadrats or relevés) provide data on local-scale species richness (or species density) and, together with other data on alpine vegetation, these data can be used to compile regional species pools (see Zobel, 1997). Alternatively, relationships between local and regional species richness can be used to model diversity at larger scales (Ricklefs, 1987; Cornell and Lawton, 1992). Using such scaling techniques, alpine plant diversity has been assessed between 1997 and 2000 as part of ALPNET, one of the European Science

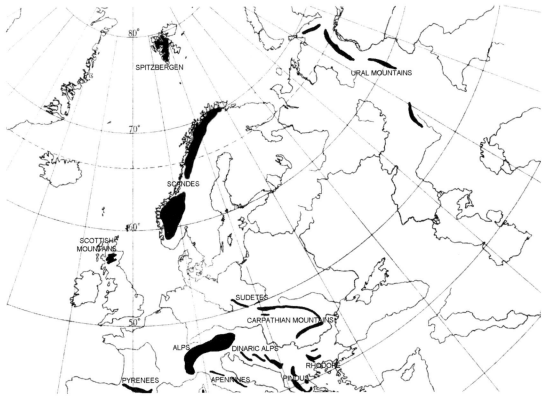

Figure 7.1 Map showing the main mountain areas in Europe. Black dots indicate sites where data were obtained for the present analyses

Foundation's research networks. It encompassed specialists on mountain environments from 14 European countries and institutes. The network aimed at fostering mountain biodiversity research across all the European mountain ranges (http://www.esf.org/life/ln/old/ALPNET/alpnet.htm; Grabherr *et al.*, 2002). One of the goals of ALPNET was to produce a synthetic picture of the plant diversity of the European alpine zone. Here, we report the resultant patterns in relation to theory and observations elsewhere. We seek to arrive at a general understanding of the drivers of mountain plant richness in the European mountain system (Figure 7.1).

MATERIAL AND METHODS

The analysis was based on data from ca. 100 published and unpublished vegetation studies in which the number of species was extractable and which were clearly defined as above the natural climatic treeline (varying from ca. 2300 m in the south to 700 m in the north; for full description see Virtanen *et al.*, 2002). The community types included wind-exposed (ridge) communities, snow-protected communities and snowbeds on both calcareous and siliceous soils. From each mountain area, data on similar vegetation types were selected (e.g. grasslands). Plot sizes varied between 0.25 and 10 m^2. Cumulative species–area curves were calculated for each data set and species richness for 10 and/or 100 m^2 was estimated by interpolation. We assumed that sampling errors were random across studies. The concept 'regional species pool' refers to the number of species available in a given area which could occupy a certain habitat type (see above), i.e. not the total flora of the area (see also Zobel, 1997). Species from steep rock walls, mires and water courses were excluded. The

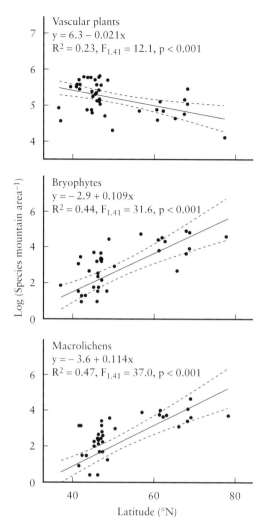

Vascular plants
$y = 6.3 - 0.021x$
$R^2 = 0.23$, $F_{1,41} = 12.1$, $p < 0.001$

Bryophytes
$y = -2.9 + 0.109x$
$R^2 = 0.44$, $F_{1,41} = 31.6$, $p < 0.001$

Macrolichens
$y = -3.6 + 0.114x$
$R^2 = 0.47$, $F_{1,41} = 37.0$, $p < 0.001$

Log (Species mountain area^{-1})

Latitude (°N)

Figure 7.2 Species richness of vascular plants, bryophytes and macrolichens in relation to latitude. The species richness values represent the estimates of regional species pools of mountain vegetation. Regression lines with 95% confidence limits are based on ordinary linear regression for log transformed data

estimates of the regional species pools for vascular plants are more reliable than those for bryophytes and macrolichens, which have not always been recorded exhaustively (Virtanen et al., 2002).

RESULTS AND DISCUSSION

Latitudinal patterns

The species richness of vascular plants declined from south to north, whereas that of bryophytes and macrolichens increased from south to north (Figure 7.2). The higher vascular plant species richness in southern mountains (e.g. the Alps) than in the north (e.g. the Scandes) is consistent with most latitudinal patterns in biodiversity reported elsewhere (Fischer, 1960; Pianka, 1966; Billings, 1973, 1992; Major, 1988; Malyshev et al., 1994; Rosenzweig, 1995) and can be largely explained by glaciation history. It seems obvious that extinction rates were smaller in mountains of southern Europe, where numerous unglaciated and periglacial refugia allowed species survival (e.g. Hegi et al., 1977; Stehlik et al., 2001), while the opposite was true in the northern mountains (e.g. Birks, 1993; Comes and Kadereit, 1998).

The opposite trend for cryptogams is rather surprising and new evidence. Our analysis clearly suggests that high latitude mountains are richer in bryophytes and lichens. The opposite trends in vascular plants and cryptogams suggest that these plant groups respond differently to the factors changing along this latitudinal gradient. Although the lower number of cryptogam species in the south may reflect insufficient census, this phenomenon has been noted before (Ochsner, 1954). This field require more research, possibly of the type pioneered by Geissler and Velluti (1996), with molecular markers opening new possibilities of detecting phytogeographic links (e.g. Printzen and Lumbsch, 2000).

Species richness in relation to the size of mountain area

There was a poor fit in the regression between the size of mountain areas and the number of vascular plant species (Figure 7.3). This implies, that independent of the mountain area, the mean vascular plant species richness is ca. 200–300 species which corresponds to values found elsewhere (Körner, 1995). The substantial scatter reflects the latitudinal variation in species richness. Unconfounded species richness–area relationships will require that the ecological conditions and the history among the studied systems are similar and that a sufficiently long time has elapsed since major

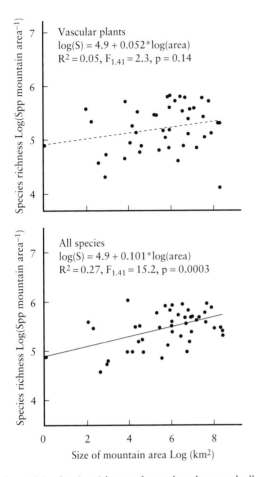

Figure 7.3 Species richness of vascular plants and all plant groups (including cryptogams) in relation to the size of the respective mountain region. Species richness values as in Figure 7.2. The regressions are based on linear regression for ln transformed data

extinctions (MacArthur and Wilson, 1967; Williamson, 1988). These conditions are clearly not met within the European mountain system, where there is little evidence of an equilibrium between immigration and extinction having been reached after glaciation (see also White and Miller, 1988). In addition, the different degrees of isolation, geological structure and elevational ranges of mountains introduce additional variability (Körner, 2000). For instance, the High Tatra mountains are relatively small in area but have a high vertical zonation from alpine to subnival zones. We also suspect that, for some regions, the species

pool has been overestimated by the inclusion of elements of lower altitudes and uncertainties about the natural treeline position.

Interestingly, a regression model fitted the data well when the richness of all species groups was used (Figure 7.3). This suggests that, considering the overall plant diversity, the species–area relationship has fairly high predictive power. We cannot offer any good explanation why the fit in this analysis is so much better than for vascular plants alone.

Species richness in relation to bedrock type

Plant species richness of alpine communities on calcareous soils was clearly higher than on siliceous soils, and this trend was consistent across all three types of communities with respect to snow cover (Figure 7.4; Table 7.1) – a trend already observed by Wahlenberg (1814) and further confirmed later (Domin, 1928; Nordhagen, 1943; Gjærevoll, 1956; Wohlgemuth, 1993). Perhaps this fits Grimes' (1979) 'reservoir hypothesis' by which he explained the richness of calcareous grasslands in Britain as having a generally larger pool of species adapted to this soil type (see also Grubb, 1987; Zobel, 1997). The majority of original alpine taxa may be adapted to calcareous soils because these were much more abundant during the early geosynclinal uplifting (Hegi *et al.*, 1977). Species adapted to acidic and/or siliceous soils may be of more recent evolutionary origin (Conti *et al.*, 1999). Alternatively, calcium-poor, acidic substrates require a higher degree of specialisation, thus narrowing the number of species which can cope with such situations. Calcium-rich substrates generally increase nutrient availability and may thus reduce competitive exclusion through resource depletion (Tilman, 1982).

The relationship between species richness and soil conditions was not found to be consistent across the mountains of Europe (Figure 7.5). Species richness was clearly higher on calcareous soils than on siliceous soils in northern Europe, as repeatedly noted in the literature for this region (e.g. Blytt, 1876;

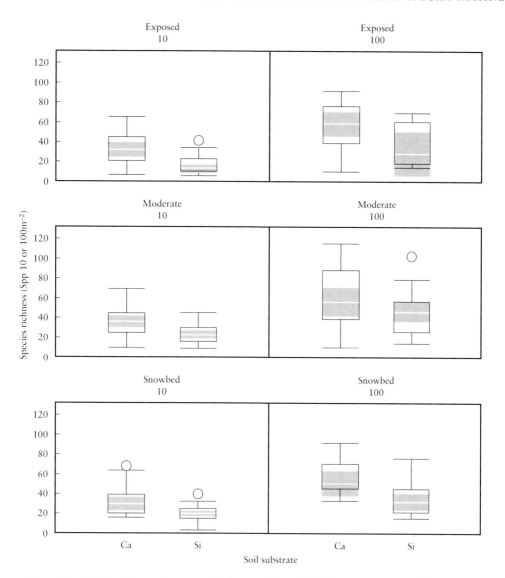

Figure 7.4 Species richness of communities in two soil type 8 differing in exposure. Calcareous soil (Ca), Siliceous soil (Si). The box plots show the median, the upper and lower quartiles, the fences indicate the extent of data points beyond the quartiles, the circles show outliers. Grey shadings indicate the 95% confidence interval of the median values

Table 7.1 Summary of a general linear model for vascular plant richness in alpine plant communities (Poisson regression, quasi link, dispersion parameter taken as 7.7). Significance according to deletion tests: Area ($F_{1,246}$ = 93.1, $p < 0.001$), Soil ($F_{1,246}$ = 52.9, $p < 0.001$), Snow ($F_{2,246}$ = 4.0, $p = 0.02$)

Explanatory variable	Estimate	s.e.	t
(Intercept)	3.5640	0.0311	114.6
Area (10 vs. 100 m²)	0.2796	0.0293	9.5
Soil type (calcareous vs. siliceous)	−0.2192	0.0306	7.2
Snow (level 1)	0.0760	0.0344	2.2
Snow (level 2)	−0.0324	0.0229	1.4

Nordhagen, 1928; Kalliola, 1939; Halvorsen and Salvesen, 1983; Moe, 1995; Birks, 1996). This is rather surprising, since siliceous bedrock type dominates the Scandes and the Scottish mountains, and the calcareous bedrock is fragmented in these areas. We suspect that the species-rich alpine vegetation that we find there today on calcareous substrates may represent remnants of a ubiquitous vegetation that prevailed after glacial retreat.

Species richness in relation to snow cover

Vascular plant species richness varied with snow cover duration (Figure 7.4; Table 7.1).

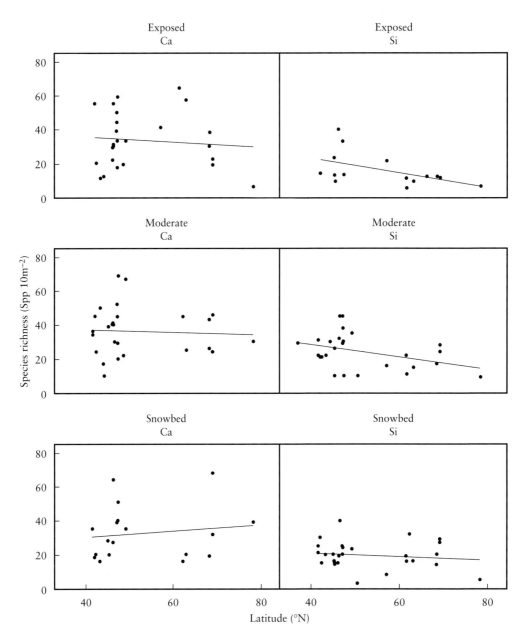

Continued

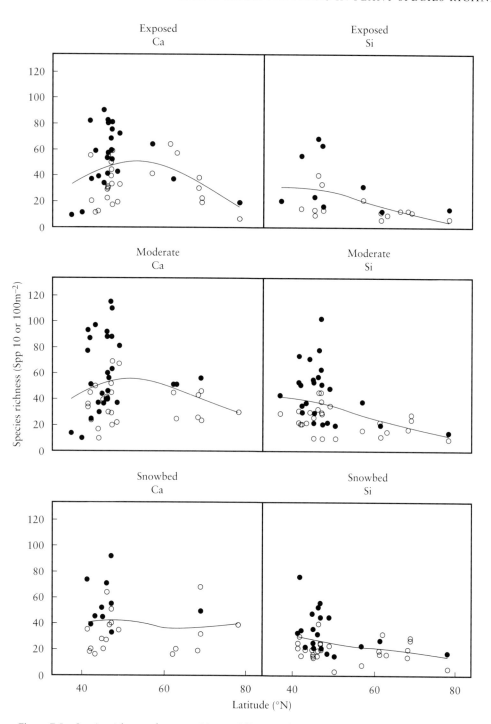

Figure 7.5 Species richness of communities on different soil types along a latitudinal gradient. First order regressions to show the dominant linear trends (a). In (b) cubic splines suggest some curvature in the relationship. In (b) open circles denote species richness estimates for 10 m² plots, solid circles for 100 m² plots

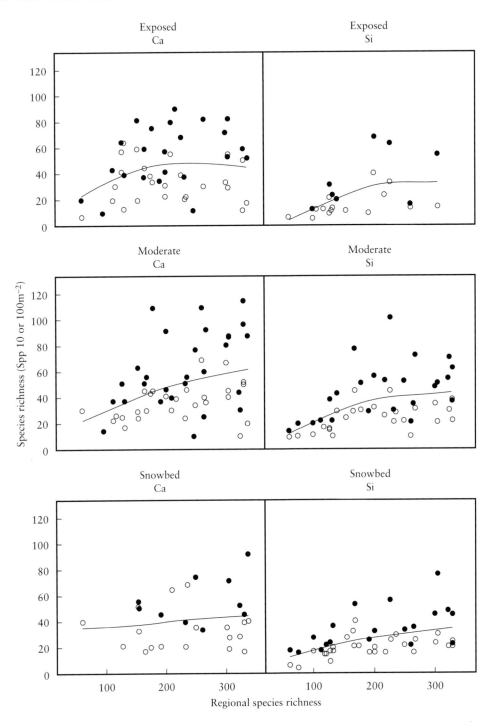

Figure 7.6 The relationship between local and regional vascular plant species richness. Open circles are for species richness estimates for 10 m² plots, solid circles for 100 m² plots. The regression lines result from a cubic spline smoothing

Richness was highest in communities with moderate snow cover and significantly lower on sites with very little or very high snow cover. This is compatible with earlier studies showing that the uneven distribution of snow cover strongly affects plant distribution and abundance relationships between plants. Vestergren (1902) noted that exposed habitats with poor snow cover and snowbeds support very few plant species. Obviously, sites requiring very specialised species also support less species (Grime, 1979). Our sample was too small to test whether the size of the species pool in extreme habitats depends on the size of the overall species pool. The great variation in snow duration and floristic richness across European mountain chains provides good opportunities for such research.

Regional versus local species richness

There were several types of relationships between local and regional species richness (Figure 7.6). Saturating curves were typical for vegetation on siliceous soil and for wind-exposed communities on calcareous soil. These relationships indicate that the local richness plateaus were far below the regional species richness. Physical limitations in space may lead to reduced numbers of species at the plot scale (Loreau, 2000). However, the fact that these trends are fairly similar for 10 and 100 m² plots (Virtanen et al., 2002) suggests that sampling scale is not the primary cause for the saturating trends. Possible causes include high dominance (space occupation) of clonal plants on poor and/or siliceous substrates and selective microenvironments (e.g. insufficient or excessive snow cover) which set a ceiling for local species richness. Snow-protected communities on calcareous soils were found to be different. A consistently positive relationship between local and regional species richness was found, as could be expected from earlier empirical work (Lawton, 1999) and theory (Rosenzweig and Ziv, 1999). In these communities there do not seem to be strong abiotic and/or biotic limitations to species coexistence (Cornell and Lawton, 1992; Zobel, 1997). Under these moderately productive situations, alpine communities appear to match the hypotheses of Huston (1999) and Pärtel et al. (2000) of a positive correlation between local and regional richness (Virtanen et al., 2002). The non-saturating relationship implies that the regional (historical) effect in European alpine communities is greatest in communities on calcareous soil with moderate snow cover.

ACKNOWLEDGEMENTS

This research was supported through grants from the Finnish Research Council of Environment and Natural Resources and the European Science Foundation (ALPNET). We also thank anonymous referees for constructive comments on the manuscript.

References

Billings WD (1973). Arctic and alpine vegetations: similarities, differences, and susceptibility to disturbance. *Bioscience*, **23**:697–704

Billings WD (1992). Phytogeographic and evolutionary potential of the arctic flora and vegetation in a changing climate. In Chapin FS III, Jefferies RL, Reynolds JF, Shaver GR and Svoboda J (eds), *Arctic ecosystems in a changing climate: an ecophysiological perspective*. Academic Press, San Diego, pp 91–109

Birks HJB (1993). Is the hypothesis of survival on glacial nunataks necessary to explain the present-day distributions of Norwegian mountain plants? *Phytocoenologia*, **23**:399–426

Birks HJB (1996). Statistical approaches to interpreting diversity patterns in the Norwegian mountain flora. *Ecography*, **19**:332–40

Blytt A (1876). Forsög til en theori om invandringen af Norges flora. *Nyt Mag Naturv*, **21**: 279–362

Chapin FS III and Körner C (eds) (1995). *Arctic and alpine biodiversity, patterns, causes and ecosystem consequences*. Ecological Studies 113. Springer, Berlin

Comes HP and Kadereit JW (1998). The effect of Quaternary climatic changes on plant distribution and evolution. *Trends in Plant Science*, 3:432–8

Conti E, Soltis DE, Hardig TM and Schneider J (1999). Phylogenetic relationships of the Silver Saxifrages (Saxifraga, Sect. Ligulatae Haworth): implications for the evolution of substrate specificity, life histories, and biogeography. *Molecular Phylogenetics and Evolution*, 13:536–55

Cornell HV and Lawton JH (1992). Species interactions, local and regional processes, and limits to the richness of ecological communities: a theoretical perspective. *Journal of Animal Ecology*, 61:1–12

Domin K (1928). The relations of the Tatra mountain vegetation to the edaphic factors of the habitat. *Acta Bot Bohem*, 6/7:133–63

DuRietz GE (1924). Studien über die Vegetation der Alpen, mit derjenigen Scandinaviens verglichen. *Veröff Geob Inst Rübel*, Zürich 1:31–138

Dullinger S, Dirnböck T and Grabherr G (2000). Reconsidering endemism in the north-eastern Limestone Alps. *Acta Botanica Croatica*, 59:55–82

Eurola S (1974). Plant ecology of Northern Kiölen: arctic or alpine? *Aquilo, Ser Bot.*, 13:10–22

Fischer AG (1960). Latitudinal variations in organic diversity. *Evolution*, 14:64–81

Geissler P and Velluti C (1996). L'ecocline subalpin-alpin: approche par les bryophythes. *Bull Murith.*, 114:171–7

Gjærevoll O (1956). The plant communities of the Scandinavian alpine snowbeds. *Kongel Nor Vidensk Selsk Skr.*, 1:1–405

Grabherr G, Gottfried M and Pauli H (1994). Climate effects on mountain plants. *Nature*, 369:448

Grabherr G, Gottfried M, Gruber A and Pauli H (1995). Patterns and current changes in alpine plant diversity. In Chapin FS III and Körner, C (eds), *Arctic and Alpine Biodiversity: patterns, causes and ecosystem consequences*. Ecological Studies 113. Springer, Berlin, pp 167–81

Grabherr G, Körner C, Nagy L and Thompson DBA (eds) (2002). *Alpine biodiversity in Europe*. Ecological Studies. Springer, Berlin

Grime JP (1979). *Plant strategies and vegetation processes*. John Wiley Sons, Chichester

Grubb PJ (1987). Global trends in species-richness in terrestrial vegetation: a view from the northern hemisphere. In Giller JHR and Giller PS (eds), *Organization of communities: past and present*. Blackwell Scientific Publications, Oxford, pp 99–118

Halvorsen R and Salvesen PH (1983). Bidrag til Vest-Hardangerviddas karplanteflora. *Blyttia*, 41:93–106

Hegi G, Merxmüller H and Reisigl H (1977). *Alpenflora*. Paul Parey, Berlin and Hamburg

Holt RD (1993). Ecology at the mesoscale: the influence of regional processes on local communities. In Ricklefs RE and Schluter D (eds), *Species Diversity in Ecological Communities*. Chicago University Press, Chicago, IL, pp 77–88

Huston MA (1999). Local processes and regional patterns: appropriate scales for understanding variation in the diversity of plants and animals. *Oikos*, 86:393–401

Jaccard P (1912). The distribution of the flora in the alpine zone. *New Phytologist*, 9:37–50

Jerosch MC (1903). *Geschichte und Herkunft der schweizerischen Alpenflora*. Verlag Wilhelm Engelmann, Leipzig

Kalliola R (1939). Pflanzensoziologische Untersuchungen in der alpinen Stufe Finnisch-Lapplands. *Ann Bot Soc Zool-Bot Fenn Vanamo*, 13:1–328

Kerner A (1863). *Pflanzenleben der Donaulaender*. Wagnerschen Univ Buchhandl, Innsbruck

Körner C (1995). Alpine plant diversity: a global survey and functional interpretations. In Chapin FS III and Körner C (eds), *Arctic and Alpine Biodiversity: patterns, causes and ecosystem consequences*, Ecological Studies 113. Springer, Berlin, pp 45–62

Körner C (1999). *Alpine plant life*. Springer, Berlin

Körner C (2000). Why are there global gradients in species richness? Mountains might hold the answer. *Trends in Ecology and Evolution*, 15:513–14

Lawton JH (1999). Are there general laws in ecology? *Oikos*, 84:177–92

Loreau M (2000). Are communities saturated? On the relationship between α, β and γ diversity. *Ecological Letters*, 3:73–6

MacArthur RH and Wilson EO (1967). *The Theory of Island Biogeography*. Princeton University Press, Princeton, NJ

Major J (1988). Endemism: a botanical perspective. In Myers AA and Giller PS (eds), *Analytical biogeography: an integrated approach to the study of plant and animal distributions*. Chapman and Hall, London, pp 117–46

Malyshev L, Nimis PL and Bolognini G (1994). Essays on the modelling of spatial floristic diversity in Europe. *Flora*, 189:79–88

Moe B (1995). Studies of the alpine flora along an east-west gradient in central Western Norway. *Nord Journal of Botany*, 15:77–89

Nordhagen R (1928). Die Vegetation und Flora des Sylenegebietes. I. Die Vegetation. Skr Nor Vidensk Akad Oslo. I. *Mat Naturv kl* 1927(1):1–612

Nordhagen R (1943). Sikilsdalen og Norges fjellbeiter. En plantesosiologisk monografi. *Bergens Mus Skr*, 22:1–607

Ochsner F (1954). Die Bedeutung der Moose in alpinen Pflanzengesellschaften. *Vegetatio*, 5–6: 279–91

Olofsson J, Kitti H, Rautiainen P, Stark S and Oksanen L (2001). Effects of summer grazing by reindeer on composition of vegetation, productivity and nitrogen cycling. *Ecography*, 24:13–24

Pärtel M, Zobel M, Liira J and Zobel K (2000). Species richness limitations in productive and oligotrophic plant communities. *Oikos*, 90:191–3

Pianka ER (1966). Latitudinal gradients in species diversity: a review of concepts. *American Naturalist*, 100:33–46

Printzen C and Lumbsch HT (2000). Molecular evidence for the diversification of extant lichens in the late Cretaceous and Tertiary. *Molecular Phylogenetics and Evolution*, 17:379–87

Ricklefs RE (1987). Community diversity: relative roles of local and regional processes. *Science*, 235:167–71

Riebesell JF (1982). Arctic–alpine plants in mountain tops: agreement with island biogeography theory. *American Naturalist*, 119:657–74

Rosenzweig ML (1995). *Species diversity in space and time*. Cambridge University Press, Cambridge

Rosenzweig ML and Ziv Y (1999). The echo pattern of species diversity: pattern and processes. *Ecography*, 22:614–28

Scharfetter R (1909). Über die Artenarmut der ostalpinen Ausläufer der Zentralalpen. *Österr bot Zeitschrift*, 1909:1–7

Stehlik I, Schneller JJ and Bachmann K (2001). Resistance or emigration: response of the high-alpine plant *Eritrichium nanum* (L.) Gaudin to the ice age within the Central Alps. *Molecular Ecology*, 10:357–70

Tilman D (1982). *Resource competition and community structure*. Princeton Univ Press, Princeton

Vestergren T (1902). Om den olikformiga snöbetäckningens inflytande på vegetationen i Sarjekfjällen. *Bot Notis*, 1902:241–68

Virtanen R, Dirnböck T, Dullinger S, Grabherr G, Pauli H, Staudinger M and Vilar L (2002). Patterns in the plant species richness of European high mountain vegetation. In Grabherr G, Körner C, Nagy L and Thompson DBA (eds), *Alpine biodiversity in Europe*. Ecological Studies. Springer, Berlin

Wahlenberg G (1813). *De Vegetatione et Climate in Helvetia Septentrionali inter flumina Rhenum et Arolam observatis et cum summi septentrionis comparatis tentamen*. Turici Helvetorum, Orell, Fuessli et Socc

Wahlenberg G (1814). *Flora Carpatorum principalium*. van den Hoeck, Göttingen

White PS and Miller RI (1988). Topographic models of vascular plant richness in the southern Appalachian high peaks. *J Ecol*, 76:192–9

Williamson M (1988). Relationship of species number to area, distance and other variables. In Myers AA and Giller PS (eds), *Analytical biogeography*. Chapman and Hall, London, pp 91–115

Wohlgemuth T (1993). Der Verbreitungsatlas der Farn- und Blutenpflanzen der Schweiz (Welten und Sutter 1982) auf EDV. Die Artenzahlen und ihre Abhängigkeit von verschiedenen Faktoren. *Bot Helvetica*, 103:55–71

Zobel M (1997). The relative role of species pools in determining plant species richness: an alternative explanation of species coexistence? *Trends in Ecology and Evolution*, 12:266–9

Environmental Determinants of Vascular Plant Species Richness in the Swiss Alpine Zone

8

Thomas Wohlgemuth

INTRODUCTION

Recently, scientific interest in alpine flora has increasingly focussed on diversity patterns (Chapin and Körner, 1995). Scientists have also began to consider the threat of species loss due to the climatically-induced upward shift of vegetation belts (Grabherr *et al.*, 1994; Gottfried *et al.*, 1999). The predicted development of vegetation in treeline ecotones and in alpine zones still has to be confirmed by long-term field observations. In order, however, to predict the future of species in alpine ecosystems, determinants of diversity need to be examined carefully. Species richness, as the most widespread indicator of diversity, was already being suggested 100 years ago, e.g. for the alpine arc in Europe. Chodat and Pampanini (1902) suggested that the high numbers of vascular plants in the Valais and in the massif of the Stilfserjoch were due to the following factors: soil diversity, dry climate, restricted glaciation during the last ice age, the proximity to glacial refugia and the lack of topographic or edaphic barriers between refugia. This enumeration mirrors what is still lively debated today: the influence of two contrasting factors, namely ecological and historical factors (Jaccard, 1902; Jerosch, 1903; Dahl, 1987; Birks, 1996; Stehlik, 2000). For a careful examination of historical factors, data on presumed nunatak areas or on genetic patterns in mountain species (Stehlik *et al.*, 2000) still need to be collected and evaluated. For the time being, it is assumed here that current species richness is predominantly caused by ecological rather than historical factors (Jaccard, 1912; Birks, 1994, 1996).

As long ago as 1908, Brockmann-Jerosch (1908) was already asking for 'good' data to explain what causes plant species richness in the Alps. Almost one century later, many more distribution data are available. Moreover, the Swiss Floristic Inventory (Welten and Sutter, 1982) resulted in data being collected systematically. These data can now be used to answer the enduring question about the causes of species richness. Based on the Swiss Floristic Inventory, the following questions were addressed: (1) How many species are there above the timberline in the Swiss Alps? (2) Which regions are rich or poor in plant species above the timberline? and (3) Which ecological factors best explain actual patterns of plant species richness above the timberline?

MATERIALS AND METHODS

Floristic data: dependent variables

The data set is drawn from the distribution atlas of pteridophytes and phanerogams of Switzerland (Welten and Sutter, 1982) and a first supplement (Welten and Sutter, 1984) to this data. The inventory area of 41 285 km^2 (extent) was divided into 593 mapping polygons (grain), with 350 lowland areas below the timberline, 215 mountain areas above timberline (Alps and Jura Mountains) and 28 lake areas (Figure 8.1). The timberline in the Swiss Alps varies between 1700 and 2300 m asl. In the present study, only the 196 mountain areas in the Swiss Alps were considered, corresponding to a total area of 9620 km^2. To avoid bringing in further variables, the mapping areas from the Jura Mountains and two

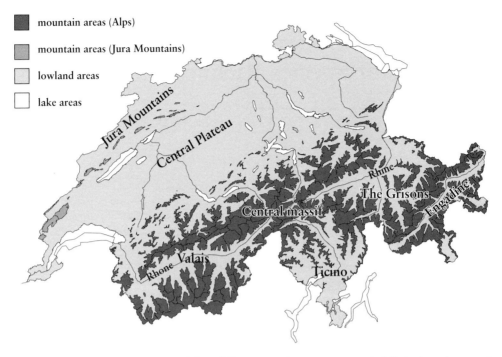

Figure 8.1 Study area, distribution of four different mapping area types and delimitation of mountain mapping polygons (according to Welten and Sutter, 1982)

mapping areas at low elevations (Napf and La Berra) have been excluded from the analyses.

The mapping areas in the Swiss Floristic Inventory were defined by topography (Welten and Sutter, 1982). It is an unusual approach aimed at producing exact distribution maps of plant species and contrasts with the quadratic mapping areas in the inventories produced in many other European countries. The Swiss Floristic Inventory is, to my knowledge, the only contiguous census of plant species worldwide which contains detailed species distribution data of exclusively alpine zones over such a large area. In inventories of other alpine countries, such as Germany (Haeupler and Schönfelder, 1989) or Austria (Niklfeld, 1976; Ellmauer, 1995), a constraint analysis of exclusively alpine zones is not possible because of the rectangular shape of the mapping areas. These areas, therefore, almost always also include low regions below the timberline. In the Swiss Floristic Inventory, defining the mapping areas according to topography was extensively discussed. The main problem is

that it results in different sized mapping areas (Welten, 1971), but it has, on the other hand, the advantage of making species distribution comparable within a unique vegetation zone.

More than 170 botanists were engaged in collecting data for the Swiss Floristic Inventory (Wohlgemuth, 1993). In total, 2573 plant species or species groups, hereafter summarised as 'species', were mapped. In the following analyses, species presence or absence values per mapping area are used. Two variables of species richness were distinguished: total species richness (SRt), and mountain species richness (SRm). Mountain species were defined as either those species which occur mainly above the timberline, i.e. alpine species, or in the subalpine forest zone above 1200–1500 m asl, i.e. subalpine species (Landolt, 1991).

Landscape data: independent environmental variables

The landscape data included ten variables that were derived from digitized factor maps. The

Table 8.1 Environmental variables used in statistical analyses of species richness

Variable	Description	Range and units	Derivation from
Geography and topography			
aR	Area size reduced by excluding unproductive lakes and glacier areas	0.1–118 km²	Geotechnical map (De Quervain *et al.*, 1963–1967)
AltMin	Minimum altitude (= actual timberline)	1700–2300 m	Digital elevation model (DEM)
Long	Longitude	549–828 km	Map of mapping areas 1:100 000
Bedrock geology			
No(R)	Number of different bedrock substrates	1–15	Geotechnical map (De Quervain *et al.*, 1963–1967)
aC	Area size of calcareous substrate	0–102 km²	Map of mapping areas 1:100 000
Ab(C)	Abundance of calcareous substrate (categorial)	0–2	Abundance of calcareous bedrock (Welten and Sutter, 1982)
Climate			
PyMin	Annual sum of precipitation; min. value	81–329 cm	Interpolated from meteorological stations using DEM
CIg^r	Continentality index of Gams (1932), corrected for temperature: CIg = atan (Alt/Py); maximum value; Alt replaced by temperature equivalent according to (Zimmermann and Kienast, 1999)	−12.5–45	(Zimmermann and Kienast, 1999): from all data points lying within a mapping area, maximum or minimum values were considered in order to express climatic extremes found at the mapping area
T7Max	Mean temperature in July; the maximum value usually corresponds to a value at the timberline	8.7–13.5 °C	
Evap6	Mean evapotranspiration in June (Turc, 1961); max. value	34–60 Cal/cm² · day	

contents of such factor maps were attached to mapping areas using geographic information system facilities (ArcInfo©). The variables represent geography and topography, bedrock geology and climate (Table 8.1). These were selected from an originally larger number of partly inter-correlated variables. For climate variables, a digital elevation model with a resolution of 250 m was used in order to interpolate meteorological data (Zimmermann and Kienast, 1999). Minimum or maximum values within mapping areas were used for the final definition of climate variables.

Species–area relationship and rarefaction analysis

The well-known 'species–area relationship' states that larger regions harbour more species than smaller regions (Gleason, 1922; Williams, 1964; Palmer and White, 1994; Rosenzweig, 1995). Because of the wide range of area sizes in the present data set, the species–area relationship was examined using correlation analysis. The rarefaction procedure was applied in order to compare increases in species richness with area sizes in different regions (Simberloff, 1979; Achtziger *et al.*, 1992). The rarefaction procedure estimates the number of species expected in a subsample of species selected at random from a larger collection. The technique can be used to standardise collections of different sample sizes, to treat them as if they were of the same size, and to make informative comparisons among them. The result of the rarefaction procedure is a hyperbolic curve that portrays the expected number of species in a given sample size. For defining floristic regions, classification methods were used with the following steps: vector transformation of presence/absence data within a mapping area, correlation coefficients to compare compositional differences among mapping areas, and minimum variance analysis to build a dendrogram (MULVA-5; Wildi and Orlóci, 1996). Vector transformation of mapping areas proved to be well-suited for comparing mapping areas according to species composition (Wohlgemuth, 1996). Plant–geographical effects of rare species with characteristic distribution areas are down-weighted with this approach.

Rarefaction curves were calculated for different floristic regions, for total species richness and for mountain species richness according to the following equation (Achtziger et al., 1992):

$$E(S_q) = \sum_{i=1}^{s} \left[\frac{Q - a_i}{q} \middle/ \frac{Q}{q} \right]$$

where $E(S_q)$ is the estimated number of species for n = 1, 2, 3...Q; q is the number of mapping areas (1 ... Q); Q is the total number of mapping areas in a floristic region; a_i is the number of mapping areas in which species i is present; and S denotes the reported number of species.

Species richness model

Correlations were determined between environmental variables and species richness. Stepwise multiple regression was performed to determine best-fitting predictor equations of species richness, with $p = 0.05$ as the selection and deselection criteria (Johnson et al., 1968; Johnson and Raven, 1970; Buckley, 1985; Wohlgemuth, 1998; Yeakley and Weishampel, 2000).

RESULTS

Species richness above the timberline in the mapping areas

A total of 1283 plant species, of which 567 were mountain species, have been found above the

timberline in the Swiss Alps. Only 22 species, or 1% of the total number of species, were restricted to the alpine zone. Hence, from the 2573 plant species reported for the whole of Switzerland, 2551 species, or 99%, were found growing below the timberline. The number for total species richness (SRt) reported from the 196 mapping areas ranged from 135 to 549 species (mean = 357), while those for mountain species richness (SRm) ranged from only 64 to 339 (mean = 243; Figure 8.2). The 1283 species occurring above the timberline correspond to about 50% of the flora of Switzerland. The eastern parts of the Swiss Alps has a larger SRt as well as SRm (Figure 8.2). Mapping areas in the Grisons, which roughly correspond to the basin of the river Rhine, were characterised by especially high species richness. Mapping areas with reduced species richness were generally located at the two front ranges of the alpine arc. If only mountain species are considered, differences between species-rich and species-poor mapping areas were less extreme. The mapping areas in Ticino, which were at comparatively low elevations, appeared to be the least species rich.

The topographical delimitation of the mapping areas resulted in differently-sized mapping areas, ranging from 0.1 to 139 km². If these areas are reduced by excluding unproductive lake and glacier areas (aR), the range is from 0.1 to 118 km². Species richness correlated well with area size, aR, for small mapping areas with area sizes below 20 km² (Figure 8.3; SRt: $r = 0.60$, $p < 0.001$; SRm: $r = 0.73$, $p < 0.001$). In contrast, the correlation coefficients between aR and species richness for mapping areas larger than 20 km² were small (SRt: $r = 0.20$, $p = 0.016$; SRm: $r = 0.25$, $p = 0.002$). For the whole range of mapping areas, a logarithmic smoothing, E(y) = a + bln(x), was also performed. By taking the correlation coefficients after logarithmic transformation for area size, better fits for species–area relationship, indicated by $r = 0.51$ for SRt ($p < 0.001$), $r = 0.73$ for SRm ($p < 0.001$), could be obtained along the whole range of mapping area sizes.

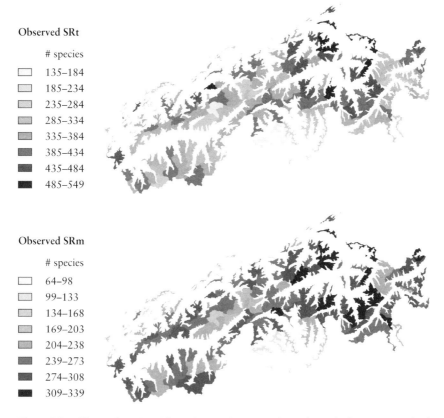

Figure 8.2   Observed species richness in mapping areas above the timberline in Switzerland for all species (SRt) and for mountain species (SRm)

Species-rich and species-poor regions

In order to group mapping areas according to compositional similarity, only mountain species were considered. The resulting dendrogram was limited to six groups. Mapping areas within a group had similar species composition as well as similar species richness. Groups differed most in average area sizes, substrate and average species richness (both SRt and SRm, see Table 8.2). Grouping mapping areas meant that more or less contiguous regions could be formed (Figure 8.4).

Rarefaction curves for the mapping areas of the six groups revealed distinct differences with respect to slope and the two types of species richness, SRt and SRm (Figure 8.5). For SRt, the slope for the lower Pre-Alps was very steep. The northern calcareous Alps showed the second steepest slope. Among the small areas,

the lower Engadine calcareous Alps also began with a steep slope, indicating that there is a sharp gradient for species richness within these small areas. The Ticino, the Valais and the Central massif are on silicate bedrock and had more moderate slopes.

Rarefaction curves for SRm showed some distinct differences. The three groups on calcareous bedrock (groups 1, 2, 3) were characterised by almost identical slopes. Species richness again increased moderately within the three groups of the silicate Alps. Ticino had the smallest increase in species richness.

Ecological factors in species richness models

Bedrock geology variables turned out to be the major explanatory variables in stepwise

107

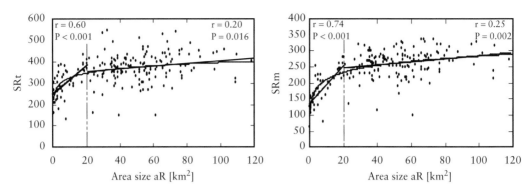

Figure 8.3 Plant species–area relationship of mapping areas above the timberline in Switzerland (all species, SRt, alpine species, SRm). Original area sizes are reduced by excluding lake and glacier areas (aR; see Table 8.1). Regression curves are distinguished for small (n = 46) and large (n = 150) area sizes. Logarithmic smoothing gives a better estimate along the whole range of area sizes. Regressions SRt: y(small) = 6.77x + 251.4, df = 25.3; y(large) = 0.66x + 340.2, df = 5.95; Regressions SRm: y(small) = 6.01x + 137.6, df = 51.6; y(large) = 0.50x + 236.7, df = 9.70

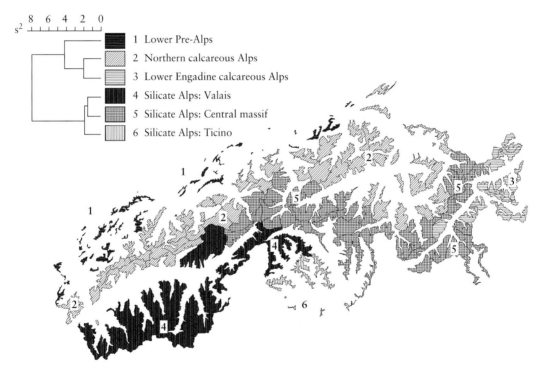

Figure 8.4 Regionalization of mapping areas above the timberline in Switzerland and dendrogram. Results are based on correlation coefficients of standardised mountain species composition in mapping areas and on minimum variance analysis (s^2) of correlation coefficients

multiple regression analyses (Table 8.3). As a single variable, the logarithmic area size of calcareous bedrock, ln(aC), explained 50% of the variance for SRt and 61% for SRm. The variable ln(aC) could be replaced by a combination of two variables (a) logarithmic area without glaciers and lakes, ln(aR), and (b) abundance of calcareous substrate, Ab(C).

Table 8.2 Averaged area sizes, maximum elevations and numbers of species for groups of mapping areas derived from a classification analysis (see text for details). Standard deviations are given in brackets

No	Description	Main substrate	Location in alpine arc	n	Average area size km²	Average maximum elevations	Average SRt	Average SRm
1	Lower Pre-Alps	Calcareous, Molasse	Front range	39	7.65 (7.5)	2109 (203)	310 (65)	181 (44)
2	Northern calcareous Alps	Calcareous, Bündner Schist	Front range/ Central range	53	57.04 (27.1)	3070 (452)	416 (56)	281 (29)
3	Lower Engadine calcareous Alps	Calcareous	Central range	16	39.66 (16.1)	3183 (127)	365 (51)	277 (27)
4	Silicate Alps: Valais	Silicate	Central range	43	69.32 (24.8)	3793 (494)	355 (58)	256 (40)
5	Silicate Alps: Central massif	Silicate	Central range	37	67.60 (21.8)	3270 (299)	357 (56)	251 (32)
6	Silicate Alps: Ticino	Silicate	Front range	8	23.18 (23.9)	2460 (298)	203 (36)	131 (40)

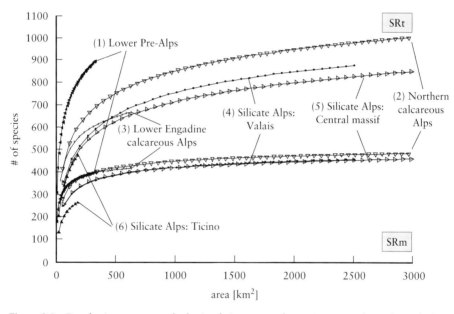

Figure 8.5 Rarefaction curves on the basis of six groups of mapping areas above the timberline. Upper curves show results for all species (SRt), lower curves show results for mountain species (SRm)

The best combinations for explaining variance almost all included a climate variable. Models for SRm could explain more of the variance than models for SRt. The residuals of models listed in Table 8.3 were normally distributed (tested by probability plots of residuals).

Most of the correlation coefficients between all the above variables were significant (Table 8.4), indicating the complex interrelationships of geography, geology and climate in the Swiss Alps. For SRt, the largest correlates were found with variables ln(aC), ln(aR), No(R). For SRm, the same variables had the largest coefficients, generally showing larger values. T7Max correlated fairly well, albeit negatively, with SRm, indicating that species richness was greater in those mapping areas at lower temperatures at the timberline. Longitude seemed to influence

Table 8.3 Multiple stepwise regression results for total species richness (SRt) and mountain species richness (SRm) in mapping areas above the timberline in the Swiss Alps. Negative influences are indicated. All independent variables shown below were significant in partial F-tests ($p < 0.001$) (For abbreviations, see Table 8.1)

Dependent variable	Independent variable	F-ratio	R^2
SRt	ln(aR)	66.7	0.256
	No(R)	69.4	0.263
	ln(aC)	196.5	0.503
	No(R), ln(aC)	116.1	0.528
	ln(aC), −Evap6	107.8	0.531
	Ab(C), ln(aC)	109.3	0.546
	Ab(C), ln(aC), ln(aR)	82.9	0.564
	Ab(C), ln(aC), −Evap6	83.6	0.566
	No(R), Ab(C), ln(aC)	86.4	0.574
	−CIg^t, Nr(R), ln(aC), −Evap6	67.0	0.584
	−AltMin, Ab(C), ln(aR),	68.0	0.587
	−T7Max	69.4	0.592
	No(R), Ab(C), ln(aC), Evap6	70.9	0.598
	−AltMin, No(R), −T7Max, ln(aC)		
SRm	CIg^t	139.3	0.418
	No(R)	171.2	0.469
	ln(aC)	302.8	0.610
	ln(aC), −T7Max	202.1	0.677
	No(R), ln(aC)	234.3	0.708
	Ab(C), ln(aR)	254.4	0.725
	Ab(C), CIg^t, ln(aR)	187.1	0.745
	Long, No(R), ln(aC)	195.9	0.754
	Ab(C), ln(aR), −T7Max	201.9	0.759
	Ab(C), ln(aR), −T7Max, PyMin	153.1	0.763
	No(R), Ab(C), ln(aR), −Evap6	155.7	0.765
	Long, No(R), ln(aC), −T7Max	157.8	0.768
	No(R), Ab(C), ln(aR), −T7Max	160.0	0.770

Table 8.4 Pearson correlation coefficient matrix for the dependent variables total species richness (SRt), mountain species richness (SRm) and independent environmental variables. Significant correlations ($p \leq 0.05$) are shown in bold (For abbreviations, see Table 8.1)

	SRm	ln(aR)	AltMin	Long	No(R)	ln(aC)	Ab(C)	PyMin	CIg^t	T7Max	Evap6
SRt	0.89	0.51	0.19	0.18	0.51	0.71	0.45	−0.30	0.34	−0.40	−0.39
SRm		0.73	0.50	0.37	0.69	0.78	0.35	−0.52	0.65	−0.57	−0.41
ln(aR)			0.70	0.33	0.67	0.73	−0.12	−0.47	0.78	−0.46	−0.21
AltMin				0.32	0.46	0.25	−0.16	−0.77	0.81	−0.70	−0.18
Long					0.13	0.18	0.02	−0.41	0.43	−0.34	−0.41
No(R)						0.54	0.15	−0.43	0.64	−0.45	−0.21
ln(aC)							0.36	−0.44	0.51	−0.43	−0.32
Ab(C)								−0.05	−0.12	−0.12	−0.23
PyMin									−0.75	0.63	0.30
CIg^t										−0.73	−0.37
T7Max											0.76

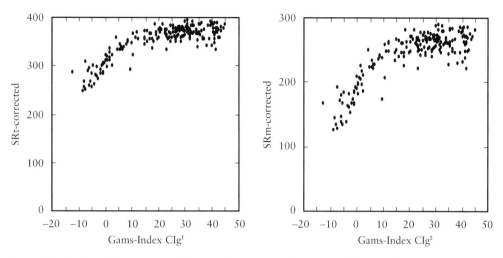

Figure 8.6 Species richness corrected for mapping area size, ln(aR), as a function of Gams' continentality index. Corrected values for all species (SRt corrected) and mountain species (SRm corrected)

SRm only slightly, whereas CIg^t had a good positive coefficient for SRm. In order to consider the complex relation between the size of mapping areas and their location within the Alpine arc, SRt and SRm were corrected by the effect of area size using ln(aR), with a regression for SRt: $y = 84.9x + 231.5$, $R^2 = 0.27$, df = 71.4, and regression for SRm: $y = 87.8x + 113.4$, $R^2 = 0.53$, df = 216.2. The resulting SRt-corrected and SRm-corrected were displayed as a function of CIg^t (Figure 8.6).

DISCUSSION

The importance of area size and substrate

The species–area relationship (Figure 8.3) indicated that species richness increased with increasing area size, but regression curves for small and large mapping areas differed considerably in their slopes. One explanation is that small mapping areas generally correspond with isolated summits at the front ranges of the Alps, in the north as well as in the south. With increasing proximity to the central ranges of the Alps, summits are higher in altitude and alpine zones are contiguous over large areas. It therefore made sense, when designing the Swiss

Floristic Inventory, to have larger mapping areas in the more central ranges. With increasing proximity to the central ranges of the Alps, the SRm also increases (Lüdi, 1928). Mapping areas in the central ranges had generally higher mean altitudes, encompassing the highest summits in the Alps (Table 8.2). But species richness increased only slightly in the central ranges with increasing area size (Figure 8.3). This is because plant life is limited by the harsh conditions prevailing at higher altitudes (Wohlgemuth, 1993). A better estimate of the relation between area size and species richness was found with logarithmic smoothing (Figure 8.3).

There is a distinct though modest relationship between species richness and area size, as demonstrated above. But the regionalization of the mapping areas and the corresponding rarefaction curves (Figures 8.4 and 8.5), showed that bedrock geology is probably the most important determinant of species composition and species richness in the Alps. The silicate Alps (groups 4, 5, 6) consist mainly of granite and gneiss, resulting in a largely acidic substrate. The rough distinction between the calcareous and silicate groups mirrors the geological feature within the Swiss Alps (Landolt, 1971). The figures for average species richness of the mapping areas of the silicate Alps Valais (group 4) and the Central massif (group 5) were lower than those for the northern calcareous Alps (group 2) and for the lower Engadine calcareous Alps (group 3). The mapping areas in the front ranges had lower average species richness, mainly because of the smaller average sizes of mapping areas. Again, the average species richness of the silicate mapping areas of the Ticino were considerably lower than those in the lower Pre-Alps, although the average area size of the Ticino areas was three times larger than the average area sizes of the lower Pre-Alps. There was no difference between the groups Valais (4) and Central massif (5) of the silicate Alps with respect to species richness. The reasons for distinguishing the central ranges of the silicate Alps are largely ecological and historical. Ozenda (1985) identified quite similar regions in the Alps by means of continentality. Many western alpine plant species have their most eastern sightings in the Valais. In contrast, many eastern alpine plant species re-immigrated into the Alps no further than to the Central massif (Merxmüller, 1952–54; Wagner, 1966; Becherer, 1972).

The very steep slope for the lower Pre-Alps in the rarefaction curves, eventually resulting in a large species richness (SRt and SRm) at about 300 km^2, contrasted with the moderate slope and the small total of species (SRt and SRm) in the Ticino, at about 185 km^2. Obviously, the three regions on calcareous bedrock (groups 1, 2, 3) had the steepest slopes. These results confirm for larger scales what is widely known from the field on smaller scales: many species are found in relevés on calcareous substrates, in contrast to the often disappointingly few species found in relevés on acidic substrates (Grime, 1979; Keller et al., 1998).

There are a multitude of reasons why there are more species on calcareous than on acidic substrates. An ecological explanation may be the fact that, for plant species, a calcareous substrate generally means a chemically and physically special site with extreme conditions regarding water and nutrient availability (Ellenberg, 1958; Walter, 1960; Gigon, 1971).

No calcicoles at all grow on acidic substrates, mainly because of competition for nutrients. In contrast, on calcareous substrates, site diversity is greater, with many places for calcifuges to grow. For example, they can grow on acidic humus layers that have successively accumulated through precipitation and leaching. This explains the wider distribution areas of calcifuges in contrast with the more restricted distribution areas of calcicoles. For example, the distribution ranges of the two species of the Alpenrose strongly differ: the calcifuge *Rhododendron ferrugineum* has a wide distribution range in the Swiss Alps whereas the range of the calcicole *R. hirsutum* is restricted to calcareous substrates.

The reservoir of calcicoles in Switzerland is considerably larger than that of calcifuges (Landolt, 1977; Wohlgemuth, 1998), indicated by the ratio of calcicoles to calcifuges of 1.77. Why are there more calcicoles than calcifuges? Historical data may well help to answer this question. In Britain, where the reservoir of calcicoles is also larger, Grime (1979) hypothesised that calcicoles evolved mainly at lower latitudes where the effect of low precipitation and high evapotranspiration maintained a high base status in soils. This explanation may, to some extent, also explain the richness of lowland species on calcareous bedrock in Switzerland. The larger reservoir of alpine calcicoles (274 species in the Swiss Alps, in contrast to 189 calcifuges; ratio = 1.45), however, may have developed while large areas of calcareous bedrock were forming during the evolution of the Alps (Merxmüller, 1952–54). Angiosperms (actual calcicole/calcifuge ratio 1.58) evolved in the late Cretaceous, around 100 million years ago (Anonymous 1991), at a time when calcareous bedrocks (e.g. clay and chalk) were forming worldwide, and schists evolved locally in the Alps. Surprisingly, one of the hotspots of mountain species richness corresponded with the Bündner schist region in the basin of the River Rhine. The ratio of calcicoles to calcifuges for grasses which evolved in the Oligocene (Tertiary) around 25 million years ago, is smaller, with a current ratio in Switzerland of 1.18, and a ratio of 1.45 for alpine grasses only. Grasses evolved after the evolution of the Alps had been completed.

To further explain substrate-related differences in species pools, glaciation and the special array of bedrock geology in the Alps need also to be considered. The higher elevated central ranges on mainly silicate bedrock were much more glaciated than the lower elevated and calcareous front range, eventually resulting in a larger number of nunataks at the edge of the Alpine arc (Merxmüller, 1952–54; Landolt, 1992; Stehlik, 2000). More calcicoles may have survived the glaciation period at these refugia than did calcifuges in the central ranges. This point of view is supported by the lack of endemic calcifuges in the Alps, which contrasts with the variety of regional endemic calcicoles occurring, especially in the southern calcareous front range in the Italian Bergamask (Landolt, 1992).

The importance of climate

Climate variables improved the models for both SRt and SRm. The corresponding variables used in the models expressed conditions at the timberline where the lowest or highest values are usually found. Surprisingly, species richness was increased if the mean temperature in July (T7Max) is low. The negative correlation coefficients of T7Max with SRt and SRa may be, at first sight, counter-intuitive. Generally, temperature is considered a limiting factor on plant life and species richness is reported to decrease with increasing altitudes (e.g. Grabherr et al., 1995; Odland and Birks, 1999). A reason for this result may be the complex effect of continentality on plant life. Both SRt and SRm are positively correlated with CIg^t, i.e. species richness generally increases in the central, more continental ranges where annual sums of precipitation (e.g. PyMin) are smaller. Not only are mapping areas in the central ranges larger in size, they are also characterised by a broader range of climatic conditions. The influence of continentality is particularly marked when area-corrected species richness is related to continentality CIg^t (Figure 8.6): there is a hyperbolic increase in

species richness with increasing values of continentality. Similarly, Zimmermann and Kienast (1999) used continentality for the modelling of alpine grassland vegetation in the Swiss Alps.

The possible influences of climate change on species richness above the timberline is not the subject of this paper. However, I consider the results suitable to evaluate possible variables for risk assessments based on the Swiss Floristic Inventory. With respect to such an evaluation, both the effects of continentality on plant distribution and the correspondences of continentality with other geo-factors in the Alps, are of special interest.

Observed and modelled species-rich and species-poor mapping areas

As can be seen from the observed SRt and SRm in Figure 8.2, hot spots of species richness were mostly located in the eastern parts of the Swiss Alps, in the Engadine as well as in the Rhine basin. Marked features were more conspicuous for SRt than for SRm, indicating that the distribution of mountain species is more even than that of all plant species. The silicate Alps of the Ticino exhibited especially low species richness. Variables used in the model overestimated SRt as well as SRm in this region. Jaccard (1902) provided a plausible explanation for this low species richness: he believed that the lack of variation in the bedrock led to such floristic poverty. Whereas the bedrock geology in most alpine regions is usually quite varied, in the gneiss zone of the Ticino it is uniform over larger areas.

The determinants found to statistically best explain species richness above the timberline in Switzerland corresponded with the factors 'dry climate', correlated to T7max, Evap6, and CIg^t for continentality, and 'soil diversity', i.e. ln(aC), ln(aR), No(R) for substrate. These had already been suggested by Chodat and Pampanini in 1902. Longitude is another variable that moderately but positively correlates with mountain species richness. This indicates that the numbers of mountain species increase from the western to the eastern parts of the

Alps. The mapping areas in the Rhine basin were generally richer than the mapping areas in, e.g. the Valais. To some extent, this pattern corresponds to the more general pattern that the eastern Alps are richer in alpine plants than the western Alps (Favarger, 1972). As a historical explanation, Lang (1994) suggests that mountain plant species re-immigrated after glaciation from refugia in the southeast of Europe. Merxmüller (1952–54), however, did not make such a general claim, but rather carefully distinguished between several historical explanations.

CONCLUSIONS

The type of substrate is the predominant factor in the Swiss Alps that affects species richness of vascular plants. A calcareous substrate is favoured by more species than a silicate substrate – a fact that leads to the basic question of why there are differences in species pools for calcicoles and califuges. Hypotheses about their origin need to be tested.

Continentality is the other important factor influencing species richness. Continental regions are more species rich than oceanic regions. Further analyses are needed to better understand the effects of the variable continentality on species richness, since it is a complex variable.

This analysis of detailed distribution data of Swiss plant species above the timberline shows that questions about the composition, the origin and the diversity of the mountain flora, which have been put forward as early as at the beginning of the twentieth century, now can be answered more accurately. An evaluation of the factors determining species richness is important in assessing the possible effects of climate change in alpine regions.

ACKNOWLEDGEMENTS

I would like to thank R Holderegger and W Keller for greatly helping to improve the earlier drafts of this chapter with their feedback; E Landolt for providing me with valuable information; Ch Körner and an anonymous referee for helpful comments; and S Dingwall for improving the English.

References

Achtziger R, Nigmann U and Zwölfer H (1992). Rarefaction-Methoden und ihre Einsatzmöglichkeiten bei der zooökologischen Zustandsanalyse und Bewertung von Biotopen. *Z Ökol Naturschutz*, 1:89–105

Anonymous (1991). Geochronology: the interpretation and dating of the geological record. In *The New Encyclopædia Britannica* vol. 19 (Macropædia), Encyclopædia Britannica Inc, Auckland, pp 148–76

Becherer A (1972). *Führer durch die Flora der Schweiz*. Schwabe, Basel und Stuttgart

Birks HH (1994). Plant macrofossils and the nunatak theory of per-glacial survival. *Diss Botanicae*, 234:129–43

Birks HJB (1996). Statistical approaches to interpreting diversity patterns in the Norwegian mountain flora. *Ecography*, 19:332–40

Brockmann-Jerosch MC (1908). Die Geschichte der schweizerischen Alpenflora. In Schröter C (ed.) *Das Pflanzenleben der Alpen*. Raustein, Zürich, pp 743–77

Buckley RC (1985). Distinguishing the effects of area and habitat type on island plant species richness by separating floristic elements and substrate types and controlling for island isolation. *Journal of Biogeography*, 12:527–35

Chapin FS III and Körner C (1995). *Arctic and alpine biodiversity: Patterns, causes and ecosystem consequences*. Ecological Studies, vol. 113. Springer, Berlin, Heidelberg, New York

Chodat R and Pampanini R (1902). Sur la distribution des plantes des Alpes austro-orientales et plus particulièrement d'un choix de plantes des Alpes cadoriques et venitiennes. *Le Globe*, 41:63–132

Dahl E (1987). The nunatak theory reconsidered. *Ecological Bulletin*, 38:77–94

De Quervain F, Hofmänner F, Jenny V, Köppel V and Frey D (1963–1967). *Geotechnische Karte der Schweiz*. 1:200 000 (2nd ed). Kümmerly & Frey, Bern

Ellenberg H (1958). Bodenreaktion (einschliesslich Kalkfrage). In Ruhland W (ed.), *Handbuch der Pflanzenphysiologie*, vol. 4, Die mineralische

Ernährung der Pflanze. Springer, Berlin, pp 638–708

Ellmauer T (1995). Biodiversity hot-spots in Österreich – eine erste Annäherung. *Z Ökol Naturschutz*, 3:271–9

Favarger C (1972). Endemism in the montane floras of Europe. In Valentine DH (ed.), *Taxonomy, phytogeography and evolution*. Academic Press, London and New York, pp 191–204

Gams H (1932). Die klimatische Begrenzung von Pflanzenarealen und die Verteilung der hygrischen Kontinentalität in den Alpen, mit Vegetationskarte: Verteilung der hygrischen Kontinentalität. *Z Ges Erdkunde*, 52–68:178–98

Gigon A (1971). Vergleich alpiner Rasen auf Silikat- und Karbonatboden. Konkurrenz- und Stickstofformenversuche sowie standortskundliche Untersuchungen im Nardetum und im Seslerietum bei Davos. *Veröff Geobot Inst ETH*, 48:1–159

Gleason HA (1922). On the relation between species and area. *Ecology*, 3:158–62

Gottfried M, Pauli H, Reiter K and Grabherr G (1999). A fine-scaled predictive model for change in species distribution patterns of high mountain plants induced by climate warming. *Diversity and Distribution*, 5: 241–51

Grabherr G, Gottfried M, Grubler A and Pauli H (1995). Patterns and current changes in Alpine plant diversity. In Chapin FS III and Körner C (eds), *Arctic and alpine biodiversity: patterns, causes and ecosystem consequences*. Ecological Studies, vol 113. Springer, Berlin, Heidelberg, New York, pp 167–81

Grabherr G, Gottfried M and Pauli H (1994). Climate effects on mountain plants. Nature, 269:448

Grime JP (1979). *Plant strategies and vegetation processes*. Wiley, Chichester

Haeupler H and Schönfelder P (1989). *Atlas der Farn- und Blütenpflanzen der Bundesrepublik Deutschland*. (2nd ed). Ulmer, Stuttgart

Jaccard P (1902). Gesetze der Pflanzenverteilung in der alpinen Region. *Flora*, 90:349–77

Jaccard P (1912). The distribution of the flora in the alpine zone. *New Phytologist*, 11:37–50

Jerosch MC (1903). *Geschichte und Herkunft der schweizerischen Alpenflora. Eine Übersicht über den gegenwärtigen Stand der Frage*. Engelmann, Leipzig

Johnson MP, Mason LG and Raven PH (1968). Ecological parameters and plant species diversity. *American Naturalist*, 102:297–306

Johnson MP and Raven PH (1970). Natural regulation of plant species diversity. *Evolutionary Biology*, 4:127–62

Keller W, Wohlgemuth T, Kuhn N, Schütz M and Wildi O (1998). Waldgesellschaften der Schweiz auf floristischer Grundlage. Statistisch überarbeitete Fassung der 'Waldgesellschaften und Waldstandorte der Schweiz' von Heinz Ellenberg und Frank Klötzli 1972. *Mitt Eidg Forschungsanst Wald Schnee Landschaft*, 73:91–355

Landolt E (1971). Ökologische Differenzierungsmuster bei Artengruppen im Gebiet der Schweizerflora. *Boissiera*, 19:129–48

Landolt E (1977). Ökologische Zeigerwerte zur Schweizer Flora. *Veröff Geobot Inst ETH*, 48:1–208

Landolt E (1991). *Gefährdung der Farn- und Blütenpflanzen in der Schweiz mit gesamtschweizerischen und regionalen roten Listen*. EDMZ, Bern

Landolt E (1992). *Unsere Alpenflora*. Fischer, Stuttgart Jena

Lang G (1994). *Quartäre Vegetationsgeschichte Europas: Methoden und Ergebnisse*. Gustav Fischer, Jena

Lüdi W (1928). Die Alpenpflanzenkolonien des Napfgebietes und die Geschichte ihrer Entstehung. *Mitt Natf Ges Bern*, 1927/1928:195–265

Merxmüller H (1952–1954). Untersuchung zur Sippengliederung und Arealbildung in den Alpen. *Jahrb Verein Schutz Alpenpflanz Tiere*, 17:96–133, 18:135–58, 19:97–139

Niklfeld H (1976). Der Stand der Kartierung der Flora Mitteleuropas in Österreich und Liechtenstein zu Beginn der Vegetationsperiode 1976. *Nachr Florist Kartierung*, 4:1–10

Odland A and Birks HJB (1999). The altitudinal gradient of vascular plant richness in Aurland, western Norway. *Ecography*, 22:548–66

Ozenda P (1985). *La végétation de la chaîne alpine dans l'espace montagnard européen*. Masson, Paris

Palmer MW and White PS (1994). Scale dependence and the species-area relationship. *American Naturalist*, 144:717–40

Rosenzweig ML (1995). *Species diversity in space and time*. Cambridge University Press, Cambridge

Simberloff D (1979). Rarefaction as a distribution-free method of expressing and estimating diversity. In Grassle JF, Pati GP, Smith W and Taillie C (eds), *Ecological diversity in theory and practice*. International Cooperative Publishing House, Fairland, pp 159–76

Stehlik I (2000). Nunataks and peripheral refugia for alpine plants during quarternary glaciation in the middle part of the Alps. *Botanica Helvetica*, 110:25–30

Stehlik I, Holderegger R, Schneller JJ, Babbott RJ and Bachmann K (2000). Molecular biogeography and population genetics of alpine plant species. *Bull Geobot Inst ETH*, 66:47–59

Turc L (1961). Evaluation des besoins en eau d'irrigation, évapotranspiration potentielle, formule simplifié e mise à jour. *Ann Agron*, 12:13–49

Wagner H (1966). Ost- und Westalpen, ein pflanzen-geographischer *Vergleich*. *Angew Pflanzensoz*, **21**: 265–78

Walter H (1960). *Einführung in die Phytologie*, vol 3(1) Grundlagen der Pflanzenverbreitung, Standortslehre. (2nd ed). Ulmer, Stuttgart

Welten M (1971). Die Kartierung der Schweizer Flora. *Boissiera*, 19:97–105

Welten M and Sutter R (1982). *Verbreitungsatlas der Farn- und Blütenpflanzen der Schweiz*, vol. 1, 2. Birkhäuser, Basel

Welten M and Sutter W (1984). *Erste Nachträge und Ergänzungen zum 'Verbreitungsatlas der Farn- und Blütenpflanzen der Schweiz'*. Zentralstelle der floristischen Kartierung der Schweiz, Bern.

Wildi O and Orlóci L (1996). *Numerical exploration of community patterns*. (2nd ed). SPB Academic Publishing, The Hague

Williams CB (1964). *Patterns in the balance of nature*. Academic Press, New York

Wohlgemuth T (1993). Der Verbreitungsatlas der Farn- und Blütenpflanzen der Schweiz (Welten und Sutter, 1982) auf EDV. Die Artenzahlen und ihre Abhängigkeit von verschiedenen Faktoren. *Botanica Helvetica*, **103**:55–71

Wohlgemuth T (1996). Ein floristischer Ansatz zur biogeographischen Gliederung der Schweiz. *Botanica Helvetica*, **106**:227–60

Wohlgemuth T (1998). Modelling floristic species richness on a regional scale: a case study in Switzerland. *Biodiversity and Conservation*, 7: 159–77

Yeakley JA and Weishampel JF (2000). Multiple source pools and dispersal barriers for Galápagos plant species distributions. *Ecology*, 81:893–8

Zimmermann N and Kienast F (1999). Predictive mapping of alpine grasslands in Switzerland: species versus community approach. *Journal of Vegetation Science*, 10:469–82

Plant Diversity and Endemism in High Mountains of Central Asia, the Caucasus and Siberia

9

Okmir Agakhanjanz and Siegmar-W. Breckle

INTRODUCTION

Biodiversity should always be regarded as the result of evolution within given biota. The three-dimensional structure of mountain regions and the relevant ecological situation govern this process. Thus, regional differences may be great (Walter and Breckle, 1999). The evolution of distinct mountain flora is influenced by the following factors:

(1) Character of orogenesis (volcanism, bloc tectonics) which influences the floristic migration between plains and the developing mountains;

(2) Speed of orogenesis (thermal conditions, erosion speed);

(3) Climatic situation within planetarian belts (belts parallel to geographical latitudes) and wind systems (oceanic, continental);

(4) Situation of the forest belts (on all exposure levels, only on distinct favourable exposures, or lacking on exposures); and

(5) Glaciation (number of glaciations and extent).

The oreophilization of the flora has mostly been derived from floristic ancestors in the plains and adjacent regions of lower hill areas or older mountains thus leading to a new flora. This has developed by specific migrations of species, by evolution of new species and by extinctions of existing species. Comparison of various mountains, isolated or larger mountain chains, can be used to detect migrations and the degree of isolation of the mountain flora. This is the basis for a quantitative geographical scheme of the mountain florogenesis (Agakhanjanz and Breckle, 1995).

GEOGRAPHICAL SITUATION

Floristic investigation of the major regions in Eurasia has been intensive and has led to a good knowledge of the flora of the relevant mountains. Nevertheless, each new expedition still finds new species for these regions, or even species new to science. Here, three complex mountain systems and their elevational belts are compared. We will consider Central Asia, including the endorrheic basins of the Caspian and the Aral Sea, with the Kopetdag and Balkan mountains, the Pamir-Alai, the Tien Shan and the Hindu Kush Mountains. These are mountains from the Tertiary period that have developed by complicated tectonics, with bloc orogenesis and folded systems. Second, we compare data from the Caucasus (Great and Small Caucasus), having a Mesozoic-Kenozoic volcanic character and where strong bloc orogenesis took place later. Finally, we will examine the southern Siberian mountains (Altai, Sajany) that are volcanic in origin, but again have undergone later recent bloc tectonics.

All three mountain systems range more or less latitudinally from west to east, however, the mountain knots (Pamir-Alai, Tien Shan, Hindu Kush, Karakoram, etc.) are much more complex and also exhibit meridional chains (Safidihyrs, Ishkashim, Kashgar, Fergana chains, etc.). The latitudinal chains are somewhat symmetric in their floral belts, the temperature differences between northern and southern exposures are only striking in more northern mountains. The meridional chains are orographically rather asymmetric: precipitation of the luv and lee sides differs considerably (e.g. in the Pamirs between 2000 mm a^{-1} in the

Table 9.1 Phytodiversity of Eurasian mountain regions, part 1: Caucasus

Region	Families	Genera	Species	Endemism %	Species/genus	Source
Caucasus	155	1286	6350	25.2	4.93	Gadjiyev & Sakhokia 1966
Armenia (28 800 km²)	111	770	3000		3.89	Gadjiyev & Sakhokia 1966
– Ararat region		542	1452		2.67	Takhtadjan & Fedorov 1972
– Ararat basin	97		1026			Takhtadjan & Fedorov 1972
Aserbeidjan (86 600 km²)	125	930	4110		4.42	Gadjiyev & Sakhokia 1966
– Nakhichevan region			2500	11.0		Ibrahimov 1974
– oreophytic belt	66	366	892		2.43	Ibrahimov 1974
– Babadag	80	371	1014	27.2	2.73	Effendijev 1974
– Talysch region, oreophytic belt	70	328	708	12.2	2.19	Gadjiyev 1979
– Aserbeidjan, subalpine belt	80	324	860	19.7	2.65	Gadjiyev 1979
– Aserbeidjan, alpine belt	43	224	309		1.37	Gadjiyev 1979
– Bozdag Alasan region	84	459	956		2.08	Gadjiyev 1979
Georgia (69 700 km²)	138	881	4028		4.57	Gadjiyev & Sakhokia 1966
N Caucasus	154	904	3800	22.8	4.20	Seredin 1977
– Endemic flora, foothills			87	2.2		Seredin 1977
– Endemic flora, forest belt			232	6.1		Seredin 1977
– Endemic flora, oreophytic belt			554	14.5		Seredin 1977
– NE Caucasus, subnivale belt	25	81	181	15.4	2.23	Prima 1976
– Stavropol region	128	752	2750		3.47	Guide 1977
Central Caucasus	115	640	2259	5.4	3.52	Galuschko 1976
– mountains in Daghestan			1600	12.5		Grossheim 1948, Elenevski 1966
– Botlich, oreophytic belt			119	29.4		Lvov 1971
Smaller Caucasus, subnivale belt		65	106		1.63	Prima 1976

west and 60–200 mm a^{-1} in the east), thus humid and arid sides of the mountains have led also to differing glaciation during the glacial periods. Most mountains in Central Asia are moderately dry, with several arid months per year. However, there is a great variability in climatic conditions and the geological and geomorphological situation. These are generally characterised by a very large relief energy that causes landslides and erosion, and by huge flat areas in which fine colluvial and alluvial material accumulates. This material resembles catenas, with typical particle size distribution from coarse in the upper to very fine particles in the lowest parts. All these factors have influenced the development of distinct floras in each part of the various mountains studied.

SPECIES NUMBERS

The species numbers can be obtained by reference to the various floras of each region (Tables 9.1 to 9.3). However, in most cases, it is not possible to provide relevant data for distinct mountain regions or altitudinal belts. Most of the published floras cover areas according to administrative borders (states, provinces) and/or rare to natural regions (valleys, foothills, vegetation belts). Thus, the relevant data from different regions are rarely comparable. In those frequent cases where the data were doubtful, outdated or too hypothetical, they were rejected.

Tables 9.1 to 9.3 contain data on the number of families, genera and species, as well

Table 9.2 Phytodiversity of Eurasian mountain regions, part 2: Central and middle Asia (within the border of the former USSR)

Region	Families	Genera	Species	Endemism %	Species/genus	Source
Middle Asia						
(Turan, former USSR)			7500			Kamelin 1973
Orobiomes			5500			Kamelin 1973
– Seravschan mountains			1650			Zakirov 1955
Turanian Plains			2000			Kamelin 1973
– Syrdarjinian Karatau	94	573	1666	9.2	2.9	Kamelin 1990
– Turanian sand deserts	40		536	30		Bykov 1979
– Seravschan basin			2588	7.5		Zakirov 1955
– Seravschan valley			938			Zakirov 1955
Turan, alpine	41	201	668	48.5	3.32	Kamelin 1973
Turan, subalpine			1300			Kamelin 1973
W Tien Shan	191	1323	2812	17.0	2.12	Pavlov 1974
N Tien Shan			2230			Rubtsov 1956
Fergana foothills	39	169	246	12.1	1.45	Rakhimova 1979
Turkmenia (488100 km^2)		683	2200	14.7	3.22	Cherepanov 1981
– Kopet Dagh, in general		620	1170	18.0	2.85	Kamelin 1970
– Kopet Dagh, oreophytic belt	60	318	657	11.4	2.06	Kurbandurdjev 1975
– Dsungarian Alatau	112	622	2168	3.5	3.48	Goloskokov 1984
– Great Balkan mountains			457			Proskurjakova 1964/67
– Kaplankyr plateau	39	179	341		1.90	Sarybajev 1985
– Repetek nature reserve	32	95	132	31.8	1.38	Ischankuliev 1982
– Badghys			1050			Botshantsev 1982

Table 9.3 Phytodiversity of Eurasian mountain regions, part 3: Central and middle Asia continued (within the border of the former USSR)

Region	Families	Genera	Species	Endemism %	Species/genus	Source
Middle Asia continued						
(Turan of the former USSR)						
Tadjikistan with Pamir (143 100 km^2)	116	994	4513	14.2	4.54	Flora Tadjik 1957ff.
– E Pamir	63	267	738	4.6	2.76	Ikonnikov 1991
– E Pamir, subnivale belt	14	34	50		1.47	Ukhacheva 1983
– W Pamir	108	548	1667	9.7	3.04	Ikonnikov 1991
– W Pamir + Darwaz	101	568	1952		3.43	Kassatsch & Dengubenko 1990
– Darwaz region	104	631	1786		2.83	dto.
– Surkho mountains	82	438	886	0.2	2.02	Chukavina 1983
– Schachristan region	64	335	838	15.0	2.50	Konnov 1974
– Peter I Chain, oreophytic belt	57	241	648	23.4	2.69	Strizhova 1974
– Warzob basin			1671	1.7		Kamelin 1990
Uzbekistan (447 400 km^2)		880	3663	9.9	4.16	Cherepanov 1981
Kyrghisia (198 500 km^2)		831	3276	9.8	3.94	Cherepanov 1981
– Central Tien Shan	86		1870			Golovkova 1959
Kazakhstan (2 717 300 km^2)		1022	4759	11.5	4.65	Cherepanov 1981
– Mangyschlak region	65		616			Safronova 1993
Afghanistan (650 000 km^2)			3500	35.0		Groombridge 1992
Afghan Wakhan	61	247	690		2.79	Podlech & Anders 1977
Iran (1 640 000 km^2)			6500	35.0		Groombridge 1992
Turkey (767 130 km^2)			8472	30.9		Groombridge 1992
Altai–Sajan mountains	123		3020	13.0		Revjakina 1996
Mongolia	128	662	2823		4.26	Gubanov 1996

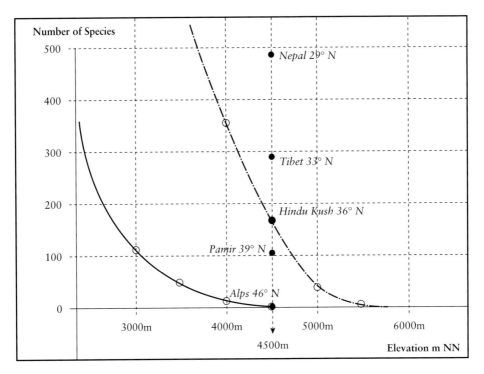

Figure 9.1 The 'floristic drainage'; in mountains. With increasing altitude, the number of species decreases. This is shown for the European Alps and the Afghan Hindu Kush. Data for some other mountain regions at 4500 m asl are given (Breckle, 1981)

as the percentage of endemism of the relevant floristic region. When possible, the relation between genera and species (coefficient species per genus) was calculated.

It was not possible to give an idea on the relationship of the area dependant species number in most regions. This is important since the Arrhenius equation (1920) is exponential, where c and z vary from region to region:

$$\log S = c + z \log A$$

where S is the number of species, A is the relevant area of the region, and c and z are typical constants for distinct regions or taxa.

In the biodiversity index map compiled by Barthlott *et al.* (1996) the species numbers was calculated for 10 000-km² grids, but this is often not applicable in mountain regions.

Phytosociological methods used by Braun-Blanquet (1964) and Dierschke (1994) depend on relevés of 10–1000 m², whereas transects used by Russian geobotanists such as Tolmachev (1974) describe a 'Konkret Flora' (a realistic portion of the flora) for an area of 100 km². The GMBA and Gloria Network intend to work on small representative sites (1–20 km²) for each mountain region and might be more applicable to such regions.

The so-called floristic 'drainage' with increasing altitude is shown in Figure 9.1 for the Alps and the Hindu Kush mountains. At 4500 m asl, latitude figures are included for other mountain ranges. This demonstrates the increasing species number at the same altitude but with differing latitude from north to south. In most mountain ranges there are species which exhibit a very broad amplitude according to altitude. These belts can stretch more than 3000 m and are exemplified by species

Table 9.4 Examples of plant species with their altitudinal ranges (m) within the Hindu Kush and Wakhan Mountains (Afghanistan, partly from Agakhanjanz and Breckle, 1995)

part 1: creek, valleys and river meadows in the Pamir-Alai (mNN)

Species	Range
Carex divisa	400–3000
Carex diluta	400–3600
Carex soongorica	400–4000
Salix wilhelmsiana	600–3300
Agrostis alba	600–4000
Calamagrostis pseudophragmites	600–4100
Poa supina	800–3400
Poa pratensis	800–4600
Hordeum brevisubulatum	900–4100
Salix turanica	1000–3800
Blysmus compressus	1000–4100
Salix pycnostachya	1100–4100
Carex orbicularis	1400–4700
Trichophorum pumilum	1700–4200
Carex stenocarpa	1700–4400
Roegneria schugnanica	1900–4499
Kobresia stenocarpa	2200–4200
Salix schugnanica	2300–4200
Oxytropis hirsutiuscula	2300–4300
Elymus dasystachys	2300–4300
Carex pseudofoetida	2400–4200
Carex microglochin	2700–4300
Carex nivalis	2700–3900
Oxytropis gorbunowii	2700–4400
Kobresia persica	2700–4300
Kobresia capillifolia	2700–4600
Carex melanantha	2700–4600
Kobresia pamiroalaica	2900–4500
Calamagrostis tianschanica	3200–4500

part 2: dry stands in the Pamirs (brackets: whole elevational amplitude), main figures: dominant occurrence (mNN)

Species	(lower)	Dominant	(upper)
Salsola arbuscula		400–2500	(2900)
Salsola orientalis		400–3100	(3400)
Salsola gemmascens		400–3100	(3500)
Carex pachystylis		400–2500	(3400)
Poa bulbosa		400–3500	(3700)
Astragalus bakaliensis	(400)	700–600	(3800)
Halogeton glomeratus		400–3600	(4100)
Climacoptera lanata	(400)	800–3800	(4100)
Kochia prostrata		2000–4300	
Salsola paulsenii	(400)	600–3600	(4000)
Salsola montana	(400)	700–3800	(4100)
Eurotia ceratoides	(400)	1000–4700	
Carex stenophylloides	(500)	1300–4400	(4700)
Salsola laricifolia	(600)	900–3300	
Carex bucharica		600–3200	(3700)
Carex turkestanica		800–3200	(3600)
Piptatherum vicarium	(800)	1000–2600	(2900)
Stipa szowitsiana	(800)	1100–2700	(3100)
Allium barszczewskyi	(800)	1600–3300	(3600)
Poa bactriana	(800)	1200–3200	(3400)
Stipa kirghisorum	(1100)	1500–3300	(3800)

(Continued)

121

Table 9.4 *(Continued)*

Stipa caucasica	(1100)	1500–4000	(4400)
Poa relaxa	(1100)	1500–4400	
Astragalus andaulgensis	(1400)	2200–4550	
Carex stenocarpa	(1400)	1900–4600	
Piptatherum laterale	(1800)	2200–4000	(4700)

part 3: Examples of plant species from the Hindu Kush and Afghan Wakhan region, with very narrow or very broad elevational amplitude (T = dry stands; A = river meadows and creek margins) mNN

Stipa szowitsiana (= S. barbata)	T	500–3300
Poa bulbosa	T	500–3800
Lepirodyclis holosteoides	T	600–3300
Calamagrostis pseudophragmites	A	800–3600
Salix wilhelmsiana	A	1100–2200
Acanthophyllum macrodon	T	1100–3200
Dionysia tapetodes	T	1200–3200
Primula edelbergii	A	1400–1600
Salix pycnostachya	A	1500–3800
Poa bactriana	T	1600–2700
Agrostis alba (A. stolonifera)	A	1700–2900
Stipa caucasica	T	2000–2700
Poa supina	A	2000–3500
Cerastium cerastioides	T/A	2000–5100
Juncus membranaceus	A	2100–4600
Psathyrostachys caduca	T	2200–3900
Cortusa brotheri	T	2200–4000
Androsace himalayica	T	2300–3600
Elymus dasystachs	A	2400–3000
Oxytropis hirsutiuscula	A	2600–3800
Oxytropis gorbunowii	A	2700–3800
Salix schugnanica	A	2900–3800
Silene luciliae	A	2900–4700
Androsace villosa	T	2900–5200
Hordeum brevisubulatum	A	3000–3200
Primula pamirica	T/A	3000–4500
Chorispora macropoda	T	3100–4500
Oxytropis oreophila	T	3100–4900
Primula macrophylla	T/A	3400–5500
Phaeonychium surculosum	T	3700–5100
Ermania flabellata	T	3800–5000
Sibbaldia procumbens	T	3900–5500
Stellaria decumbens	T	3900–4400
Holosteum kobresioides	A	4500–4700

such as *Poa bulbosa* or *Cerastium cerastioides* (Table 9.4), while other species are restricted to a rather narrow elevational belt (e.g. *Holosteum kobresioides*, Table 9.4).

Some typical 'alpine belt' species start above the timberline or in the elfinwood line (in arid mountains) and reach the nival belt. An example of this is *Primula macrophylla* which holds the record in the Hindu Kush at 5500 m asl, together with *Sibbaldia procumbens*. This is normally valid for species restricted to typical azonal wet sites as well as those on zonal scree or other substrates.

LIFE FORMS

The biological characteristics of the high mountain flora (above 4000 m asl in the Hindu Kush) is demonstrated by the life

Table 9.5 Life form spectra of the alpine flora of the Hindu Kush above 4000 m asl (Breckle, 1988)

Life form	Phanerophytes	Nano-Phanerophytes	Chamaephytes	Hemicryptophytes	Biennials	Annuals (Therophytes)	Geophytes	Parasites	Hydrophytes
Altitudinal range:									
> 5400 m asl	–	–	10	70	10	–	10	–	–
5200–5400 m	–	–	16	74	5	–	5	–	–
5000–5200 m	–	–	12	71	8	6	4	–	–
4800–5000 m	–	–	12	72	6	5	4	–	–
4500–4800 m	–	1.6	11.7	69.2	4.0	6.5	6.5	–	0.4
4000–4500 m	0.1	2.8	17.1	61.2	2.3	9.1	7.2	0.1	0.1

Table 9.6 Percentage values of chorological groups from the alpine and nivale flora of the Hindu Kush above 4000 m asl (Breckle, 1988)

Chorological group	Cosmopolitan	Northern hemisphere Arctic	Central and middle Asian	Central and middle Asian Himalaya	Himalayan	Endemic
Altitudinal Range:						
> 5400 m asl	–	10	20	40	30	–
5200–5400 m	5	15	15	45	20	–
5000–5200 m	6	16	14	38	24	2
4800–5000 m	4	14	14	36	28	4
4500–4800 m	3	10	14	32	28	13
4000–4500 m	2	10	11	28	27	22

form pattern (Table 9.5). As expected, hemicryptophytes dominate in the upper belt. Chamaephytes, geophytes and biennial species also reach the highest parts of mountain. The therophytes are rare, with one remarkable exception, the small plant species *Koenigia islandica* which grows during a very short vegetation period of a few weeks in melting snow water channels in the Afghan Wakhan and the eastern Hindu Kush, this being one of the southernmost stands of this arctic annual plant species. With decreasing altitude, the percentage of woody plants increases and reaches an optimum between 1600 and 2300 m asl in

both abundance and dominance. Also, the percentage of the annuals increases with decreasing altitude, but their importance is greatest in the arid lowland semideserts and deserts below 1000 m asl.

CHOROLOGY AND ENDEMISM

The data on the flora and species distribution can be analysed according to chorological types as well as life form spectra. Table 9.6 gives an example from the upper mountain parts of the Hindu Kush. The predominance of Central Asian and Himalayan chorological elements

Table 9.7 Vertical distribution of endemic plants in Asian mountains, given as percentages of the relevant belt. State of forest belt: *yes* closed forest belt, *ex* forests on some exposures only, *no* forest belt lacking

Data Mountain region	No. of species	Above forest line (oreophytic belt)	Forest belt	Below forest belt	Total mountain region %	Source
Caucasian mountains						
NW Caucasus	2259	–	yes	–	18.5	Galuschko 1976
Aserbaydjan	4110	–	ex	–	19.7	Gadjiyev 1966
NE Caucasus	181	–	ex	–	15.4	Prima 1976
NW Caucasus	819	14.4	yes	–	35.0	Altuchov 1974
Nakhichewan	2500	2.4	ex	–	11.0	Gadjiyev 1979
Talysch chain	708	12.2	yes	–	–	Gadjiyev 1979
Babadag region	1014	–	ex	–	27.2	Effendijev 1974
Daghestan	1600	7.4	ex	–	12.6	Elenevski 1966
N Caucasus	3800	> 14.5	yes	2.2	–	Seredin 1977
Siberian mountains						
Altai Sajan	3020	5.2	yes	–	13.0	Revjakina 1996
W Sajan, alpine	601	9.2	yes	–	–	Krasnoborov 1976
Mountain regions of middle Asia						
High mountains in general[1]	1643	19.6	ex	–	–	Kamelin 1973
Tadjikistan[2]	4513	–	ex	–	14.2	Flora Tadsh. 57 f.
W Tien Shan	2812	–	yes	–	17.0	Pavlov 1974
Afghan Hindu Kush	450	13.7	ex	–	–	Breckle 1988
E Pamir	738	4.6	no	–	4.6	Ikonnikov 1991
W Pamir	1667	–	no	–	9.7	Ikonnikov 1991
Serawshan basin	2588	–	ex	–	7.5	Zakirov 1955
Peter I chain	648	–	yes	–	23.4	Strizhova 1974
Shakhristan region	838	–	ex	–	15.0	Konnov 1974
Kopetdagh, Turkmenia	1770	11.4	ex	–	18.0	Kamelin 1970, Kurbandurdyev 1975
Warsob valley	1679	–	ex	–	1.7	Kamelin 1970, 1990
Turkmenia	2200	–	ex	–	14.7	Cherepanov 1981
Repetek region	132	–	no	–	31.8	Ischankuliev 1982
Fergana region	246	–	no	12.1	12.1	Rakhimova 1979
Afghanistan	3500	–	ex	–	35.0	Groombridge 1992
Iran	6500	–	ex	–	35.0	Groombridge 1992
Turkey	8472	–	ex	–	30.9	Groombridge 1992

[1]Pamir-Region excluded
[2]Pamir-Region included

corresponds to the orographic contacts of the mountain ranges. Cosmopolitan and widespread floristic elements from the boreal and arctic are also present, especially in the very high mountain belts (Breckle, 1988), whereas endemics are almost absent in the nival belt, but reach a high proportion in the subalpine zone in the Hindu Kush, which is characterised by a thorny cushion plant belt (Tragacanth Belt), generally between 3000 and 3600 m asl.

Despite the fact that there are many floras and vegetation monographs available, we still lack data on the distribution of species within the various elevational belts in mountain regions. Normally, the exposure of mountain ridges plays a role in floristics. The deep shady

Table 9.8 General trends in endemism percentages in mountains of Asia according to climate and status of forest belts

Mountain belts	Continuous forest belt: N Caucasus, Siberian mts	Forest belt only on favourable exposures: middle Asia mts. (Hindu Kush)	Lacking forest belt: Pamir, Tibet, N Karakoram (Kopet-Dagh, N exposure)
Above forest line (oreophytic belt)	**maximum:** **2.4–14.5%**	minimum: 0–0.7 % (ca. 10%)	minimum: 2.0–4.6% (11.4%)
Forest belt		**maximum:** **1.7–15.0%** (> 20% ?)	(18.0%)
Below forest belt	minimum: 0–2.0%	minimum: 0–1.6% (ca. 5%)	**maximum:** **5.0–9.7%** (31.8%)

valleys sometimes provide refugia for alpine and nival plants. In arid mountain regions the vegetation belts gradually change with altitude, and a distinct borderline between belts is often difficult to draw. The situation is different in mountains with forest belts, but still data on percentages of plant species below, within or above forest belts are scarce. Table 9.7 is an attempt to survey some of the available data for endemic plants from Asian mountains, indicating the types of mountain forest belts in this region.

CONCLUSIONS

It is not easy to draw generally valid conclusions on the florogenesis of Central Asian mountains, but it seems that the endemism in such mountains is strongly influenced by climatic conditions which govern the existence or absence of forest belts. Table 9.8 provides a general scheme on the mean percentages of endemics in the various mountain types in this region. In mountains with a continuous forest belt (often deciduous trees as a lower belt, conifers as an upper belt) the maximum percentage of endemics is above the treeline in the oreophytic belt. In contrast, in arid mountains lacking any forest belt or forest patches, the maximum percentage of endemism occurs in the foothills of the mountains. In mountains with forest patches those, together with the rich vegetation mosaic of this belt, are the main sites rich in endemism. Thus, we can summarise:

(1) Endemism is related to degree of isolation, the use of reference areas, and to ecological conditions past and present. Thus, semi-humid mountains with distinct altitudinal belts (steppe, forest, oreophytic mosaic) differ considerably from arid mountains with continuous, gradually changing formations of vegetation types (desert, semidesert, oreophytic steppe).

(2) The biodiversity indices (related to 10 000 km^2 and derived from the Arrhenius equation) for rich mountain regions with their small-scale mosaic are too generalised. The Caucasus Mountains seem to be richer in species than the Middle and Central Asian Mountains, when related to surface area.

(3) A biodiversity index related to ecological resources (precipitation, available area in an elevational restricted belt) might reveal totally different relations (Breckle, 2000). A new methodological approach should be considered. Small permanent plots and transect studies along all elevational gradients may help in future research.

(4) Biodiversity in Asian mountains is not sufficiently known, despite an immense data set in floras and monographs. More standardised investigations are needed in the future.

ACKNOWLEDGEMENTS

The help of many people, especially the hospitality of the inhabitants in the mountain areas, is greatly acknowledged. Thanks are due also to the University of Bonn (affiliation with the University of Kabul), to the Schimper Foundation (Stuttgart Hohenheim) and other institutions which were helpful for mountain trips of S-W Breckle in Iran and Afghanistan.

Thanks are due also to many institutions which helped O Agakhanjanz during his visit to Afghanistan and during his long and intensive mapping work in the Pamirs and other Asian mountains.

References

Agakhanjanz O and Breckle S-W (1995). Origin and evolution of the mountain flora in middle and central Asia. In Chapin III FS and Körner C (eds), Arctic and alpine biodiversity: patterns, causes and ecosystem consequences. *Ecological Studies*, **113**:63–80

Altuchov MD (1974). Zur Charakteristik der Hochgebirgsflora des nordwestlichen Kaukasus. *Probl Bot*, **XII**:9–14

Arrhenius O (1920). Distribution of the species over the area. *Meddeland Vetenskaps Akad Nobelinstit*, **4**:1–6

Barthlott W, Lauer W and Placke A (1996). Global distribution of species diversity in vascular plants: Towards a world map of phytodiversity. *Erdkunde*, 50:317–27

Botshantsev VP (1968). Weitere Funde zur Flora des südlichen Mittelasiens. *Botanical Journal*, **40**:308–09

Botshantsev VP (1982). Übersicht über die Wüstenflora von Zentral-Asian. *Neues Zur Systematik der Höheran Pflanzen* (Leningrad), **19**:19–37

Braun-Blanquet J (1964). Pflanzensoziologie. *Grundzüge der Vegetationskunde*. 3.Aufl, Springer, Berlin

Breckle S-W (1981). Zum Stand der Erforschung von Flora und Vegetation Afghanistans. In Rathjens C (ed.), *Neue Forschungen in Afghanistan*. Leske, Opladen, pp 87–104

Breckle S-W (1988). Vegetation und Flora der nivalen Stufe im Hindukusch. In Grötzbach E (ed.), Neue Beiträge zur Afghanistanforschung. *Schriftenreihe der Stiftung Bibliotheca Afghanica, Liestal*, **6**:133–48

Breckle S-W (2000). Biodiversität in Wüsten und Halbwüsten. *Ber Reinh Tüxen-Ges.*, **12**: 207–22

Bykov BA (1979). *Abriß der Geschichte der Pflanzenwelt Kasakhstans und Mittelasiens*. Alma-Ata

Cherepanov SK (1981). *Die Gefäßpflanzen der UdSSR*. Nauka, Leningrad

Chukavina AG (1983). *Die Vegetationsdecke der Surkho-Kette*. Dissertation, Dushanbe, 20 pp

Dierschke H (1994). *Pflanzensoziologie*. Ulmer, Stuttgart

Effendijev PM (1974). Die Analyse der Flora der Babadag-Region. In *Thesen der Vorträge des VI Symposium zur Erforschung und Nutzung der Flora und Vegetation der Hochgebirge* (TVVIS-FVH), Stavropol, pp 202–03

Elenevski AG (1966). Über einige interessante Besonderheiten der Flora der inneren Gebirge Daghestans. *Bull Moskauer Ges für Naturprüfung, Ser Biol*, **21**:302–12

Fedorov (ed.) (1957–1991). Flora of *Tadjikistan*. vol. 1–10. Nauka, Leningrad

Flora von Tadshikistan. (1957ff.) (ed). Ovchinnikov P. Moskau-Leningrad, Band 1. 548 pp

Gadjijev VD and Sakhokhia MF (1966). Flora der subalpinen Gürtel des Großen Kaukasus in den Grenzen von Aserbidshan. *Problem Botan*, **VIII**:18–27. Nauka, Moskau-Leningrad

Gadjiyev VD (1979). *Flora und Vegetation des Talysch-Hochgebirges*. Baku

Galuschko AI (1976). Analyse der Flora des westlichen Teils des Zentralen Kaukasus. In *Flora des Nordkaukasus und seine Geschichte*. Stavropol, pp 5–130

Galuschko E (ed.) (1977). *Guide of the wild growing plants of the Stavropol region*. Rostov/Don 96 pp

Goloskokov VP (1984). *Flora des Djungarischen Alatau*. Nauka, Almaty

Golovkova AG (1959). *Vegetation des Zentralen Tien Shan*. Frunse (=Bischkek)

Groombridge B (ed.) (1992). *Global Biodiversity*. Chapman & Hall, London

Grossheim AA (1948). *Vegetation des Kaukasus*. Moskwa

Gubanov IA (1996). *Konspekt der Flora der Inneren Mongolei*. Walang Verlag, Moskau

Heywood VH and Watson RT (eds) (1995). *Global Biodiversity Assessment*. UNEP, Cambridge Univ Press, Cambridge

Ibrahimov AS (1974). *Neue Angaben zur Flora der Hochgebirge der Nakhichevan-Republik*. TVVIS-FVH, Stavropol pp 168–70

Ikonnikov SS (1991). *Flora Badakhshans und des Pamir* (Bestand, vergleichende Analyse, botanisch-geographische regionale Verbreitung). Dissertation, Saint Petersburg

Ischankulijev M (1982). *Flora des Repetek-Reservats*. Manuskript (Dissertation), Aschchabad

Kamelin RW (1970). Botanisch-geographische Besonderheiten der Flora des Kopetdag. *Botanical Journal*, **55**:377–83

Kamelin RW (1973). *Florengenetische Analyse der natürlichen Flora der Gebirge Mittelasiens*. Nauka, Leningrad

Kamelin RW (1990). *Flora des Syrdarjinischen Karatau*. Nauka, Saint Petersburg

Kassatsch A and Dengubenko AV (1990). *Floristisches Verzeichnis der Flora der Gebirge des badachschanischen autonomen Gebietes Tadjikistans*. Manuscript, Khorog

Konnov AA (1974). Über die Flora der Wacholderwälder Schachristans. *Probl Bot Leningrad*, **12**:52–4

Krasnoborov IM (1976). *Hochgebirgsflora des West-Sajangebirges*. Nauka, Novosibirsk

Kurbandurdyjev R (1975). Flora des Oreophytengürtels des Kopetdags. *Problem Wüstennutzung, Aschchabad*, 7:48–50

Lvov PL (1971). *Flora von Daghestan*. Machachkala

Pavlov VN (1974). Die Besonderheiten der Flora des westlichen Tienshan. *Problem Botanik, Leningrad*, **12**:63–70

Podlech D and Anders O (1977). *Florula des Wakhan (NE Afghanistan)*. Mitt Bot München pp 361–502

Prima VM (1976). Einige Besonderheiten der oberalpinen Flora des NE-Kaukasus. In *Flora des Nordkaukasus und seine Geschichte*. University of Stavropol pp 163–6

Proskurjakova GM (1964–1967). *Konspekt der Flora des Großen Balkhan*. part 1–4, Wiss Vorträge der Hochschule, Moskau

Rakhimova T (1979). Zur Flora des Chustpap-Adyren des Fergana-Tals. *Nachrichten der Akad Wiss Uzbekistans*, **8**:64–5

Revjakina NV (1996). *Die rezente periglaziale Flora des Altai-Sajan-Gebirges*. Barnaul

Rubtsov NI (1956). Flora des nördlichen Tienshan und seine geographischen Verbindungen. *Botanical Journal*, **41**:11–19

Safronova IN (1993). Über die Flora des Mangyschlak. *Botanical Journal*, **78**:60–7

Sakhokia MF (1965). *Konspekt der Flora des Kaukasus*. Manuskript. Tbilisi

Sarybajev B (1985). Der floristische Bestand im Kaplankyr-Plateau. *Vestnik Karakalpak-Filiale der Akad der Wiss Uzbekistans*, 3:12–16

Seredin RI (1977). Endemismus in der Flora des Nordkaukasus. In *Die Fragen der Erforschung und Nutzung der Flora und Vegetation der Hochgebirge*. Novosibirsk, pp 40–2

Strizhova TG (1974). Die Hochgebirgsflora der Peter I.-Kette. *TVVISFVH* Stavropol pp 190–1

Takhtadjan AL and Fedorov AA (1972). *Flora von Erivan* (Guide to the wild flowers of the Ararat-Kessel). Nauka, Leningrad

Tolmachev AI (1974). *Einführung in die Pflanzengeographie*. Nauka, Leningrad

Ukhacheva VN (1983). Flora des Nivalgürtels des Pamir. *Vestnik Univ Leningrad*, **21**:97–9

Walter H, Breckle S-W (1999). *Vegetation und Klimazonen*. 7. Aufl; Ulmer, Stuttgart

Zakirov K (1955). *Flora und Vegetation des Seravshan-Basin*. Akad. Wiss. Uzbek., Tashkent

Cold Spots in the Highest Mountains of the World – Diversity Patterns and Gradients in the Flora of the Karakorum

10

W. Bernhard Dickoré and Georg Miehe

HOT SPOTS AND COLD SPOTS IN HIGH ASIA: INTRODUCTION AND OBJECTIVES

The issue of mountain biodiversity gains more and more importance in the scientific community as well as in international organisations (Spehn, 2001). From a global perspective, mountain areas accommodate all six major diversity centres that surpass an estimated 5000 species of vascular plants per 10 000 km^2 (Barthlott *et al.*, 1996). Of these, the East Himalayan–Yunnan 'hot spot' to the SE of the Tibetan Plateau is the only one to extend considerably into extratropical latitudes. On the opposite edge of the world's highest and largest plateau (occupying 2.5 million km^2 at around 4000–5000 m asl), the 'cold spots' of northwest Tibet, the Karakorum Mountains and the Takla Makan Desert rank with the poorest floras of the world.

The term 'cold spot' is employed here for distinct low diversity areas, as opposed to the more familiar 'hot spots' of biodiversity. With this terminology originally used in evolution biology, we intend to emphasise congruence in definition. Accordingly, we would also expect complex mosaics of populations being strongly shaped by the geographic distribution of 'hot spots' and 'cold spots', due to asymmetries in local selection and gene flow (Gomulkiewicz *et al.*, 2000).

Environmental and biosystematic gradients across the diverse and contrasting high mountains of Central Asia received much attention with regard to their paleoecological evolution (e.g. Agakhanjanc, 1980; Hsü Jen, 1982, 1984; Zhang, 1983; Axelrod *et al.*, 1998; Kubitzki and Krutzsch, 1998). With the noteworthy exception of Zheng Du (1983) for Tibet (Xizang), chorological classifications and analyses of diversity gradients are almost absent. Local studies dealing with altitudinal gradients in the rich warm-temperate relict floras relating to the Yunnan hot spot were published by Zhong *et al.* (1999) for Minya Gongka and Tang and Ohsawa (1997) for Emei Shan in Sichuan. Local information for the tropical margin was provided for tropical Yunnan by Zhu (1997) and for Fan Si Pan Mountain in north Vietnam by Thin and Harder (1996). Rare examples of biodiversity studies in the Himalayas are confined to small plots, selected habitats or small sections of the altitudinal gradient (Adhikari *et al.*, 1992; Samant *et al.*, 1998; Vetaas, 1997), and mainly focussed on montane forests. Negi and Upreti (2000) investigated the distribution of lichens over an extensive altitudinal gradient in Ladakh while concentrating on a very limited number of species.

The present study aims at analysing altitudinal and latitudinal diversity gradients of the Karakorum Divide, based on a large set of original data on vascular plant distribution. Main objectives of the present study are: 1) to provide baseline data of plant diversity for the Karakorum Divide, 2) to identify patterns in large-scale altitudinal and periphery–central diversity gradients, and 3) to discuss their significance with respect to biogeography, environmental evolution and human impact.

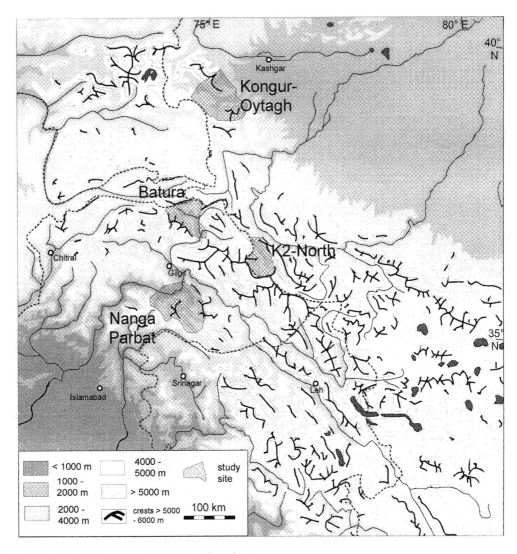

Figure 10.1 Map of the western High Asia with study sites

STUDY AREA, MATERIAL AND METHODS

Transect study sites

The Karakorum Mountains, comprising the inner section of the ranges connecting the plateaus of Pamir and Tibet, occupy approximately 180 000 km², with many peaks exceeding altitudes of 7000–8000 m. The high mountain ranges of Kunlun Shan and Himalaya frame the Tibetan Plateau to the north and to the south. To the northwest of Tibet, all three major mountain ranges link up to form a series of roughly parallel west–northwest to east–southeast striking high mountain ridges. The Kunlun–Karakorum–Himalaya divide, here briefly termed the Karakorum Divide, separates the Tarim Basin in Central Asia from the Punjab Plains of northwest India.

The Karakorum Divide delimits the fundamentally different Irano–Turanian (winter rain), Sino-Himalayan (summer rain), and Central Asiatic (permanently arid) floristic regions (Grubov, 1963; Walter and Breckle, 1984).

Table 10.1 Study site properties

Study site	Kongur-Oytagh	K2-North	Batura-Hunza	Nanga Parbat
Region	NW Kunlun Shan	Central N Karakorum	Central S Karakorum	NW Himalaya
Coordinates (NW)	38° 53′ N 74° 20′ E	36° 03′ N 76° 04′ E	36° 07′ N 74° 00′ E	35° 14′ N 73° 58′ E
Approximate size [km²]	6100	2290	4010	7070
Altitudinal range [m]	1700–3900	3800–5100	2400–4900	1100–5200
Number of records	2724	937	6506	4668
Total species number	416	134	420	871
Area investigated/ study site area(%)	30	20	60	70
Proportion of species encountered (%)	80	85	90	85

Precise delineations of these phytogeographical regions do not exist, although overall climatic and topographical provisions are obvious. Chitral, Gilgit, and Baltistan in north Pakistan are transitional and difficult to classify.

Along an idealised north–south transect, from the Tarim Basin (1300–1400 m) to the Punjab Plains (800–1000 m), the Karakorum Divide culminates in K2 (8611 m), while stretching over only ca. 700 km. It thus comprises the highest bi–directional altitude gradient on Earth.

Four artificially delimited study sites (Figure 10.1) were chosen along the Karakorum transect. These are: Kongur–Oytagh (west Kunlun), K2-North (northern central Karakorum), Batura–Hunza (west Karakorum/ south of the continental watershed), and Nanga Parbat (inner west Himalaya). The study sites range in size between ca. 2300 and 7100 km² (Table 10.1). They are considered relatively homogeneous and representative with respect to latitudinal gradients (for Nanga Parbat: see Dickoré and Nüsser, 2000).

Methodology for mapping species richness of the Karakorum Divide beyond these detail study sites followed Barthlott *et al.* (1996) and Barthlott *et al.* (1999). 'Isotaxas' (zonation boundaries with equal taxa density) derived from original specimen and weighted literature presence–absence data of a preliminary database containing approximately 92 000 records, were superimposed on the respective section of the *World map of phytodiversity* (Barthlott *et al.*, 1996).

Climate gradients, vegetation belts and phytogeographical zones

The Karakorum Divide is a cool, arid continental area with moderate monsoonal summer rains in the southwest, and some winter rain in the northwest. The most prominent climate gradient is a marked decrease of precipitation from southwest to northeast. Climate stations representing much of the variation of the transect are Murree (2168 m, mean annual temperature 12.7 °C, precipitation 1724 mm/a) in the outer northwest Himalaya and Tianshuihai (4880 m, mean annual temperature –7.8 °C, precipitation 24 mm/a) on the east Karakorum–Tibetan border (Miehe *et al.*, 2001). Steep vertical moisture and temperature gradients prevail throughout the Karakorum Divide. Humidity increases markedly from the arid valley bottoms to the highest summits (Weiers, 1995).

By virtue of vegetation cover and species composition, the arid mountain ranges to the west of Tibet exhibit conspicuous altitudinal belts (Thomson, 1852; Troll, 1939; Dickoré, 1991). Synoptic studies of (parts of) the Karakorum transect were provided by Paffen *et al.* (1959), Zheng Du (1983), Dickoré (1991), Miehe *et al.* (1996), Braun (1996), and Richter *et al.* (1999). Based on humidity gradients, Miehe *et al.* (1996) and Miehe and Miehe (1998) classified altitudinal sequences of vegetation belts of the Karakorum Mountains, which range from sub-humid to eu-arid regimes. Accordingly, common vegetation types of the Karakorum Divide range from forests to widespread deserts.

Montane, 'boreal' coniferous forests do occur on the outer slopes of west Kunlun (Kongur–Oytagh: *Picea schrenkiana*, very local; Skrine, 1925) and from southwest Karakorum through west Himalaya (Nanga Parbat: *Picea smithiana, Pinus wallichana, Abies pindrow*; Troll, 1939; Schickhoff, 1996; Dickoré and Nüsser, 2000). These forests form conspicuous bands or pockets that are sharply delimited on their lower and upper edges, obviously due to moisture and temperature constraints. Except for local juniper groves (Batura–Hunza: *Juniperus semiglobosa, J. turkestanica*), the inner ranges of the Karakorum are devoid of a forest belt.

Aspect differences are considerable in relation to vegetation cover, most pronouncedly so at middle altitudes (Miehe and Miehe, 1998; Dickoré and Nüsser, 2000). Where present, forests are almost completely confined to northern exposures (Skrine, 1925; Troll, 1939). Southerly exposures, even at high altitudes, are widely occupied by steppe and desert vegetation. East–west divergences, i.e. different diurnal energy and evaporation budgets of opposite slopes and intricate spatial interference of winter and summer rain regimes further modify the distribution of species and vegetation types.

From the outer (Kunlun, Himalaya) to the inner mountain ranges (Karakorum), a general decrease of vegetation diversity and species richness correlates to increasing aridity. Accordingly, a huge temperature effect of mass elevation corresponds to a substantial altitudinal shift of vegetation belts and individual species' distribution ranges (Dickoré, 1991, 1995). Generalised vegetation belts based on vegetation physiognomy and indicator species are summarised in Table 10.2.

Taxonomic basis and species concepts

Consistent data on flora and vegetation relating to parts of the transect are rare (Pampanini, 1930, 1934; Persson, 1938; Troll, 1939; Hartmann, 1966, 1968, 1972; Schickhoff, 1993; Dickoré, 1991, 1995; Dickoré and Nüsser, 2000). Floras covering parts of the area under consideration are *Flora of Pakistan* (Nasir and Ali, 1970–), *Flora of China* (Wu and Raven, 1994–), *Plantae Asiae Centralis* (Grubov, 1963–), *Flora Tadzhikistana* (Ovczinnikov, 1957–), and *Flora Xizangica* (Wu, 1980–1986). These floras are mostly monumental, largely controversial and, except for the last, not yet completed. Stewart's (1972) checklist, referring to Pakistan and Kashmir, is largely outdated. Accordingly, there is no coherent nomenclature checklist for the complete area. New species are constantly described from the Karakorum and adjacent mountain ranges, but few of these seem well founded regarding genetic, morphological or chorological isolation.

Sound species concepts were considered a necessary prerequisite to the present study, also implying interpretation and taxonomic 'pruning' of literature records. The *Flora Karakorumensis* (Dickoré, 1995) is an endeavour to update *La flora del Caracorùm* (Pampanini, 1930, 1934) and to amalgamate taxonomic concepts over adjacent areas. In the course of our on-going studies, a large specimen-based database was assembled, including numerous unpublished collections and taxonomic interpretations of historical records. This served as a synonymized checklist for the study sites, and for preliminary analyses of species richness of adjacent areas.

Based on ample material and largely coherent revisions, a rather rigorous ('lumping') morphological–chorological species concept was thought to best fit the regional conditions. Distribution areas of vascular plant species present in the Karakorum Mountains are usually large and contiguous over wide areas. Rather complex three-dimensional vicariance patterns occasionally blur individual species' definitions, but do not seem to substantially contribute to taxonomic proliferation. Spurious taxonomic groups were treated under a limited number of morphologically defined entities in the present study. Except for *Taraxacum* and possibly *Poa*, apomictic taxa do not seem to be important in the flora of the Karakorum in terms of absolute numbers.

Table 10.2 Approximate altitudinal vegetation belts of study sites

	Colline	Submontane	Montane	Subalpine	Alpine	Subnival
Kongur-Oytagh						
Altitude	1700–2000	2000–2700	2700–3300	3300–3700	3700–3900	–
Climate	Warm temperate–arid	Temperate–subarid	Temperate–subhumid	Cool–subhumid	Cold–humid	–
Vegetation	Desert	Desert, semi-desert, scrub	Coniferous forest, scrub, steppe	Scrub, meadow, steppe	Dwarf scrub, turf	–
Representative species	Zygophyllum xanthoxylum, Calligonum mongolicum	Sympegma regelii, Reaumuria songorica, Kalidium cuspidatum	Picea schrenkiana, Juniperus semiglobosa, Artemisia brevifolia	Juniperus pseudosabina, Festuca olgae, Krascheninnikovia ceratoides	Carex alexeenkoana, C. stenocarpa, Kobresia karakorumensis	–
K2-North						
Altitude	–	–	–	3800–4400	4400–4800	4800–5100
Climate	–	–	–	Cool–arid	Cold–subarid	Frigid–subhumid
Vegetation	–	–	–	Desert, dwarf scrub, scrub	Desert, dwarf scrub, forbs, turf	Scree and rock vegetation
Representative species	–	–	–	Krascheninnikovia ceratoides, Tanacetum fruticulosum, Stipa caucasica ssp. glareosa	Kobresia karakorumensis, Carex borii, Tanacetum tibeticum	Saussurea gnaphalodes, Sibbaldia tetrandra, Thylacospermum caespitosum
Batura–Hunza						
Altitude	–	2500–2700	2700–3300	3300–4200	4200–4600	4600–4900
Climate	–	Warm temperate–arid	Temperate–subarid	Cool–subhumid	Cold–subhumid	Frigid–humid
Vegetation	–	Desert, scrub	Steppe, steppe forest	Steppe forest, dwarf scrub	Dwarf scrub, forbs, turf, scree	Scree and rock vegetation
Representative species	–	Capparis spinosa, Haloxylon thomsonii, Isodon rugosus, Juniperus excelsa ssp. polycarpos	Juniperus semiglobosa, Poa sterilis, Koeleria cristata, Artemisia brevifolia, Lonicera microphylla	Juniperus turkestanica, Festuca olgae, Astragalus candolleanus, Betula utilis ssp. jacquemontii	Kobresia capillifolia, Poa attenuata, Leontopodium ochroleucum, Festuca coelestis	Saussurea gnaphalodes, Lagotis globosa, Psychrogeton olgae, Silene gonosperma

(Continued)

Table 10.2 (Continued)

	Colline	Submontane	Montane	Subalpine	Alpine	Subnival
Nanga Parbat						
Altitude	1100–2000	2000–2700	2700–3400	3400–3900	3900–4500	4500–5200
Climate	Hot–arid	Warm temperate–subarid	Temperate–subhumid	Cool–subhumid	Cold–humid	Frigid–humid
Vegetation	Desert, desert steppe, xeromorphic trees	Dwarf scrub, steppe, steppe forest	Steppe, scrub, coniferous forest	Forest, scrub, dwarf scrub, forbs	Grassland (meadow, turf), scrub, dwarf scrub	Rock and scree vegetation
Representative species	Capparis spinosa, Artemisia fragrans, Heliotropium dasycarpum, Stipagrostis plumosa, Pistacia khinjuk	Pinus gerardiana, Juniperus excelsa ssp. polycarpos, Cedrus deodara, Quercus baloot	Pinus wallichiana, Picea smithiana, Abies pindrow, Juniperus semiglobosa, Artemisia brevifolia	Betula utilis ssp. jacquemontii, Juniperus communis ssp. alpina, J. squamata	Kobresia capillifolia, Carex stenocarpa, Salix karelinii, Rhododendron anthopogon	Saussurea gnaphalodes, Carex nivalis, Primula macrophylla

Data basis and sample methods

Minimum–maximum altitude distribution data were determined for vascular plant species of all study sites along the Karakorum transect. Species records lacking altitudinal information were excluded from analyses. Diversity was measured by α (species richness) and β diversity (species turnover, Sørensen coefficient; Whittaker, 1972). Phytogeographical properties ('geo-elements') were defined for all species.

Data quality is heterogeneous with respect to sample size, collecting effort, representative coverage, taxonomic reliability, presence and weighted inclusion of published data. Literature records and historic herbarium data, clearly referable to accepted species, were incorporated. These, however, form a minor fraction as compared to original species lists surveyed for the present analyses. Important new herbarium collections investigated for the present study include Kongur–Oytagh: *G. & S. Miehe, U. Wündisch*, K2-North: *W. B. Dickoré* (Dickoré, 1991), Batura–Hunza: *A. Bosshard, R. Schaffner & F. Klötzli, E. Eberhardt, G. & S. Miehe*, and Nanga Parbat: *M. Nüsser* (Nüsser, 1998; Dickoré and Nüsser, 2000).

In Table 10.1, absolute altitudinal range of records from individual study sites, total number of specimens collected and species encountered are summarised. In addition, estimated fractions of area covered by floristic investigations and fractions of species numbers encountered of a theoretical total are given. Numbers given in Table 10.1 specify spatial completeness and collecting effort. Values are, however, approximate.

Due to difficult accessibility, the highest samples from Kongur–Oytagh are from around 3900 m and thus probably significantly below the upper limit of vascular plant distribution. Persson's (1938) list, referring to the same site, was ignored since it partly confronts taxonomic concepts applied to new and apparently rather comprehensive collections. It has to be noted that the lowest altitudes sampled in K2-North, 3800 m, refer to the topographical baseline of that site. The highest altitudes are possibly under–represented to a certain degree in all study sites.

RESULTS

Total species richness and regional differences

The total number of species of all study sites is 1225. Species richness of individual study sites varies from 134 (K2-North), 416 (Kongur–Oytagh), 420 (Batura–Hunza), to 871 (Nanga Parbat). For the Nanga Parbat Massif, Dickoré and Nüsser (2000) reported 962 accepted species, thus leaving 90.5% to the present study (records lacking altitudinal information excluded).

Species richness is notably higher on the southern declivity (Nanga Parbat) than on the northern declivity (Kongur–Oytagh). At least two-times more species per standard area are to be expected for the western Himalaya as compared to the western Kunlun. Species numbers from the Karakorum (Batura–Hunza, K2-North) cannot directly be compared, due to the much higher baseline elevation.

Altitudinal gradients of species richness

Curves of altitudinal species richness (Figure 10.2) are basically hump-shaped and reveal almost normal distributions. Slope of curve and mode (maximum species number; Nanga Parbat: 350 > Kongur–Oytagh: 250 > Batura–Hunza: 150 > K2-North: 50) are specific to each study site and apparently negatively correlated to aridity or continentality. Altitude of maximum species richness differs only slightly between the outer ranges (Nanga Parbat: 3100 m, Kongur–Oytagh: 3200 m), but is considerably higher towards the Central Karakorum (Batura–Hunza: 3500 m, K2-North: 4400 m).

Species richness decreases monotonically, but not linearly, towards the upper altitudinal limit of plant life in all curves. Sampling effects, however, are observed to variable degrees in all study sites at the highest altitudes.

There is a considerable left skew to the altitudinal gradient (Nanga Parbat, Kongur–Oytagh, less conspicuously so also in Batura–Hunza), representing a near plateau or a very slight increase of species richness at lower

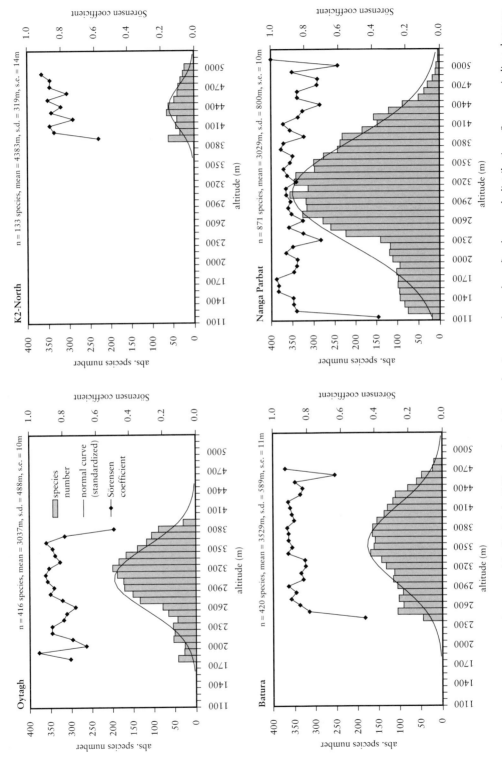

Figure 10.2 Species numbers per altitudinal belt (100 m classes) in study sites: total number, fitted normal distribution, Sörensen indices between altitudinal classes

altitudes, followed by a steep rise. This pattern is spurious in K2-North, probably due to low total and maximum species richness. Further, discontinuities in the K2-North curve are probably related to collecting effects. Widespread desert conditions prevail in this study site. Oases are very local and confined to lower altitudes close to the topographical baseline. Accordingly, the relatively high species number of the 3900 m interval is virtually the effect of a single few hundred square metre spot of turf vegetation surrounding a spring (Dickoré, 1991).

A slight increase of species richness to the left part of the curves, thought of as a baseline plateau, comprises around 100 species on the southern declivity (Nanga Parbat, Batura–Hunza), and only about half that number on the northern declivity (Kongur–Oytagh, K2-North). The upper altitudinal limit of the 'plateau' follows an order of increasing continentality: Nanga Parbat (2300 m) < Kongur–Oytagh (2600 m) < Batura–Hunza (3200 m); interpolated to K2-North (4300 m). The 'plateau'-curve transition corresponds to a major species turnover, as expressed by Sørensen-indices (Figure 10.2), and possibly more pronounced so by geo-element distributions (Figure 10.3).

Taxonomic composition and turnover

All study sites share only 31 out of 1225 species. Regional species turnover is summarised in Table 10.3. Nanga Parbat holds the highest total species number and the highest percentage of specific taxa (i.e. species that are absent in all of the other three study sites). Kongur–Oytagh and Batura–Hunza, with similar species numbers, differ in the much lower percentage of specific taxa in the latter. The outer declivities contain a substantially higher proportion of specific taxa as compared to the inner Karakorum. Almost 20% of Nanga Parbat species also occur in Kongur–Oytagh. Similarity is highest between Nanga Parbat and Batura–Hunza and lowest between Nanga Parbat and K2-North, as revealed by Sørensen coefficients.

Steep gradients of genus and family diversity mirror the generally low species richness

found in the Karakorum Divide. The largest, approximately equal-sized families, *Compositae* and *Gramineae* together account for 25% of the total number of species. Combined to the next two families, *Leguminosae* and *Cruciferae*, and four more to follow in size (*Ranunculaceae*, *Rosaceae*, *Umbelliferae*, *Cyperaceae*) almost 50% of the species are covered. Latitudinal and periphery–central trends are also apparent to the family level (Table 10.4). The generic composition comprises 25 genera with 10 or more species (*Astragalus*: 28, *Carex*: 23, *Taraxacum*: 20), as opposed to 81 genera represented by two, and 269 genera with only one species.

Endemic species

Endemic species, here defined in a wide sense as occupying probably less than 100 000 km^2 in a rectangular section of Western High Asia including western Tian Shan, Hindukush and western Himalaya, are poorly represented in terms of numbers and proportion to the total (Table 10.5). A slight proportional increase towards the inner Karakorum is not significant (chi-square test), and can probably be referred to definition, taxonomic reliability and scale effects. Endemic species are almost evenly distributed over all altitudes, possibly except the highest (Figure 10.3).

Kongur–Oytagh contains 27 endemic species, 25 of which are specific to that study site (e.g. *Oxytropis biflora*, *Aquilegia atrovinosa*, *Saussurea lacostei*, *Eritrichium pamiricum*, *Androsace squarrosula*). A certain geographical isolation of this most humid western part of the Kunlun northern slopes is evident, although species lists to delineate exact distribution areas are still underrepresented. Two endemics only are shared with other study sites: *Androsace flavescens* and *Elymus jacquemontii* with K2-North and Batura–Hunza.

K2-North has 13 endemic species, most of these apparently with a relatively wide but scattered distribution in high altitudes of the Karakorum. Six endemics are specific to that

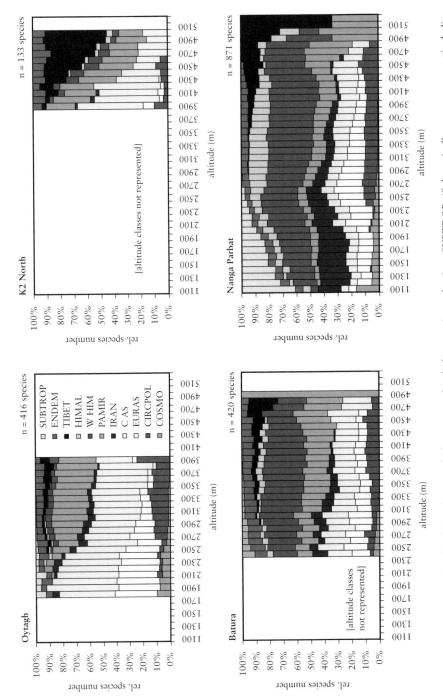

Figure 10.3 Relative altitudinal distribution of phytogeographical elements in study sites. SUBTROP: 'Subtropical' warm-temperate including 'Mediterranean' (N Africa to SW Asia) and occasionally extending through SE Asia; ENDEM: Endemic, microarealophytes and regional endemics; TIBET: Tibetan, Tibetan Plateau alpine, occasionally extending to N Karakorum, E Pamir and Central Tian Shan; HIMAL: (Sino-)Himalayan, humid outer Himalaya (E Afghanistan, Kashmir – SE Tibet, W China); W HIM: West Himalayan, subhumid inner West Himalaya (E Afghanistan, Kashmir – W Nepal); PAMIR: Pamiran, montane winter rain regions, W fringe of High Asia (Tian Shan, Pamir, W Himalaya); IRAN: Irano-Turanian, subarid winter rain region of SW Asia (Turkey/Iran – W Pakistan); C AS: Central Asiatic, arid or montane (Mongolia, Altai, Tian Shan, Pamir, Karakorum, Tibet); EURAS: Eurasiatic, temperate to arctic Europe and Asia (Euro-Siberian), occasionally N Africa; CIRCPOL: Circumpolar, temperate to arctic Eurasia and N America; COSMO: Cosmopolitic, almost world–wide (predominantly temperate)

Table 10.3 Species numbers, absolute number and percentage of specific species, and absolute species numbers shared, and Sørensen coefficients between two study sites (species numbers shared between three study sites are: Kongur–Oytagh, K2-North and Batura–Hunza 20, Kongur–Oytagh, Batura–Hunza and Nanga Parbat 82, Kongur–Oytagh, Nanga Parbat and K2-North 5, and K2-North, Batura–Hunza and Nanga Parbat 14)

Species		Specific to study site		Absolute species numbers shared/ Sørensen coefficients			
Study site	Total	number	%	Kongur–Oytagh	K2-North	Batura–Hunza	Nanga Parbat
Kongur–Oytagh	419	185	44.2	–	76	163	164
K2-North	134	24	17.9	0.27	–	78	57
Batura–Hunza	426	68	16.0	0.39	0.28	–	295
Nanga Parbat	871	518	59.5	0.25	0.11	0.45	–

Table 10.4 Species numbers of the 20 largest families

Family	Kongur–Oytagh	K2-North	Batura–Hunza	Nanga Parbat	Total
Compositae	53	15	70	109	158
Gramineae	59	17	49	99	137
Leguminosae	29	11	25	48	84
Cruciferae	27	8	25	44	68
Ranunculaceae	19	1	12	35	52
Rosaceae	17	11	19	31	48
Umbelliferae	11	2	9	35	45
Cyperaceae	20	13	18	34	44
Labiatae	8	3	15	32	44
Caryophyllaceae	11	4	19	31	39
Chenopodiaceae	20	4	12	16	34
Boraginaceae	9	4	13	25	33
Polygonaceae	10	2	11	22	32
Scrophulariaceae	10	5	8	20	30
Primulaceae	9	3	9	18	27
Gentianaceae	11	5	6	12	23
Salicaceae	5	1	8	11	15
Onagraceae	1	–	4	12	15
Crassulaceae	5	1	7	11	15
Saxifragaceae	1	2	7	12	14

Table 10.5 Spatial distribution of endemic species: total number in study sites and proportion of total number

Endemic species	Number	% of total
Kongur–Oytagh	27	6.4
K2-North	13	9.7
Batura–Hunza	37	8.9
Nanga Parbat	51	5.9
All study sites	109	8.9

study site (e.g. *Carex borii, Eritrichium spathulatum, Sibbaldia olgae*), 7 shared with Batura–Hunza (e.g. *Corydalis adiantifolia, Lagotis globosa*).

In Batura–Hunza 37 endemic species have been reported, only 16 of which are specific (e.g. *Astragalus hoffmeisteri, Eritrichium patens, Epilobium chitralense, Androsace russellii, Scutellaria heydei*). The region shares 10 endemics with Nanga Parbat. The latter mainly belong to a small group of species that are endemic to the (montane) Gilgit–Indus basin (e.g. *Tricholepis tibetica, Haloxylon thomsonii, Rhodiola pachyclados*, the latter also disjunctively in the Safed Koh range).

Nanga Parbat contains 51 endemic species (e.g. *Nepeta adenophyta, Saussurea costus, Colpodium nutans, Puccinellia minuta, Draba cachemirica, Megacarpaea bifida, Primula duthieana*, *Aphragmus obscurus*, *Hedysarum cachemirianum*) which are evenly distributed along the altitudinal gradient. A few endemics confined to relatively lower altitudes of the Indus Gorge (e.g. *Dictyolimon gilesii*, *Haplophyllum gilesii*, *Rhamnella gilgitica*) may, however, rank a higher degree of taxonomic isolation.

Diversity zones and phytogeography

Figure 10.4 maps diversity zones of the Karakorum Divide. Reasonably homogenous diversity zones apparently coincide with the orographical structure. The 1000 species isoline largely follows the Himalayan main ridge, Nanga Parbat appears to be situated already beyond this.

Absolute species numbers of study sites, extrapolated to 10 000 km² standard areas (Evans *et al.*, 1955) range between ca. 200 and around 1000. Species richness is higher, though probably not exceeding 1500 species per 10 000 km² , in the outer ranges of the western Himalaya (Hazara, Kashmir; outside the study area), as suggested by provisional database figures. Our ongoing studies suggest a total number of vascular plant species of less than 1500 for the Karakorum Mountains (180 000 km²).

Phytogeographical boundaries between the Irano–Turanian, the Central Asiatic, and the Sino-Himalayan regions follow diversity depressions to a certain degree, and with characteristic deviations (Figure 10.4). Throughout the Karakorum Mountains, the Central Asiatic–Sino-Himalayan boundary closely adheres to the continental watershed. The Khunjerab Pass (4600 m), though connecting distribution areas of many montane species (Dickoré, 1995), marks a major change in the floristic composition.

Batura–Hunza, geologically situated north of the Karakorum main ridge (Karakorum batholith, Searle, 1991), has many species of the Sino-Himalayan phytogeographical element. The poor and unsaturated flora of this study site corresponds to a depauperate Inner Himalayan flora, sharing almost two third of

the species with Nanga Parbat. There are, however, almost equal proportions of Himalayan and Central Asiatic elements. Also some Irano–Turanian elements are present throughout the Karakorum southern slope. These species would justify a separate recognition of the intermediate region between the main Himalaya ridge and Karakorum, from Chitral to Ladakh, Zanskar, Spiti and beyond.

DISCUSSION

Spatial gradients in the Karakorum Divide

Latitudinal and altitudinal gradients of species richness were found to be group specific (Kessler, 2000b) and depend on a variety of variables that are extensively discussed in the literature. The most obvious and general physical correlations were reported with energy–water balances. Positive correlations were found with high temperatures (Rohde, 1992; Leathwick *et al.*, 1998) and high precipitation or ample supply of liquid water (O'Brien *et al.*, 1998) combined to low climatic seasonality (Givnish, 1999). Rapoport's rule, of monotonic latitudinal productivity/species richness relationship, has been applied to altitudinal gradients by Stevens (1992). More recently, concern was expressed about sampling biases (Colwell and Hurtt, 1994; Rahbek, 1995) and scale problems (Rahbek and Graves, 2000) that affect elevational results. Accordingly, Rahbek (1997) discussed hump-shaped altitudinal patterns with respect to geometrical constraints, while stochastic models were intended to simulate margin effects of latitudinal patterns (Bokma *et al.*, 2001).

Since we made no effort to standardise altitudinal segments, the following discussion of individual species richness/altitude curves and sections thereof in the Karakorum Divide must remain tentative and largely restricted to comparison between individual study sites. Given the extensive altitudinal and climatic gradients of the transect and the large geographical scale of our study sites, some generalisations may be made. The prominent unimodal relationship of

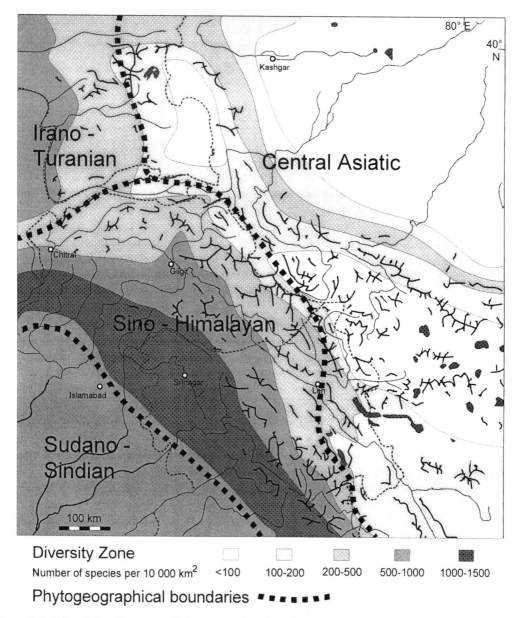

Diversity Zone

Number of species per 10 000 km^2 | <100 | 100-200 | 200-500 | 500-1000 | 1000-1500

Phytogeographical boundaries ▪▪▪▪▪▪▪

Figure 10.4   Map of diversity zones and phytogeographical boundaries in western High Asia

species richness with altitude revealed in the present study seems to represent a typical feature of mountains based in sub-humid to arid zones.

The assumption of hump-shaped curves to describe altitudinal gradients of species numbers correlates well with the sections right of the mode, whereas irregularities are obvious to the left. Taking into consideration species properties ('geo-elements') and turnover rates, both points of inflection seem to define significant ecological changes. In the outer ranges (Kongur–Oytagh, Nanga Parbat), the upper point of inflection coincides with the subalpine belt, i.e. the treeline ecotone of the montane–alpine transition. In the inner ranges (K2-North,

141

Batura–Hunza), it rather corresponds to the transition towards a subnival belt.

The left side discontinuity of the altitudinal gradient of species richness has been related to 'valley effects', per-arid conditions at the bottom of the profile (Schweinfurth, 1956), and to possible relief effects (Dickoré and Nüsser, 2000). This discontinuity can now be traced throughout the Karakorum Divide, and thus may be considered a typical feature of the dry valleys around the Tibetan Plateau. The discontinuity mirrors higher species turnover rates, and probably indicates an effective barrier to the distribution of subtropical or temperate elements into the montane flora. In the inner ranges of the Karakorum, the upper edge of the low species number plateau extends up to the alpine belt. Correlation of curve parameters or sections (mode and points of inflection) with vegetation belts or floristic zones, however, needs to be carefully examined in future.

Correlation of habitat zonation and biogeographical boundaries with altitudinal diversity gradients and turnover patterns, though occasionally obscured by habitat modifications (Davies et al., 1999), was demonstrated in plot-based studies by many authors. According to Kessler (2000a), floristic elevational belts tended to be better defined at strong environmental boundaries in fairly species-poor habitats as compared to species-rich tropical forests at lower elevations. Our impression of the Karakorum Divide is that altitudinal floristic boundaries, though not significant in the landscape aspect, are the sharper the more arid the general environment is.

Regional significance of diversity properties

One of the most obvious characteristics of Southeast Asia is the continuity of the vegetation from tropical rain forests to boreal coniferous forests, with a rich array of evergreen and deciduous temperate forests in between. To the east of the Tibetan Plateau, there are important zones of endemism, especially in east Sichuan/west Hubei and southeast Yunnan/west Guangxi, which have been characterised by equable climates and favourable conditions since the middle Miocene (15 Ma) (Axelrod et al., 1998). Geographical proximity and polarity of favourable and adverse ecological conditions have probably existed since the early Tertiary (Kubitzki and Krutzsch, 1998).

The east Himalayas and southwest China are considered to represent the highest extra-tropical phytodiversity (Barthlott et al., 1996). Highly structured north–south oriented mountain chains which connect tropical Southeast Asia to the temperate Holarctic probably served as contiguous migration pathways of biota in the course of climatic fluctuations. There is a very gradual and largely continuous transition into the dry–cold high plateau of Central Tibet to the northwest. The Tibetan Plateau itself is situated at extremely high altitude, little differentiated by west–east running ridges and valleys.

The peripheral mountain ranges, the Himalayas, and the eastern declivity of the Tibetan Plateau, to a lesser extent also the Kunlun Shan to the north, are narrow continuous belts of more favourable climatic conditions. Throughout the Central Himalayas, the very narrow climate and vegetation zones range from equable (sub)tropical, humid to seasonal cold, arid conditions. These zones probably contribute to a remarkable longitudinal floristic continuity but also to certain east–west divergences (Dickoré, 2001). Longitudinal migration pathways for subtropical and humid temperate floristic elements along the deeply north–south dissected southern slopes of the Himalayas may, however, be fragmented to a higher degree (Miehe, 1990).

The 'chambered system' around the upper Indus, i.e. the highly structured, widely spaced and relatively humid ranges of the inner northwestern Himalayas and southwestern Karakorum, is characterised by moderate species richness and by a small number of endemics. However, several species may occupy only tiny, disjunct spatial increments of their apparent rather wide distribution area, due to the extremely rugged terrain. The

Gilgit–Indus basin may still prove a relatively important refuge area and secondary diversification centre in the Himalayan flora.

The flora of the Karakorum is apparently predominantly young and relatively poor due to the present sub-optimal climate conditions and drastic ground clearance in the past. The Karakorum Mountains are the orogenetically most active and probably fastest growing mountain system on earth (Searle, 1991). Destabilisation of all slopes results in common hazards like rock fall, landslides and lakes dammed by these events or advancing glaciers (Hewitt, 1998). The Karakorum comprises the largest glacier area outside the Arctic and Antarctic regions. Glaciation of the Karakorum during the cold periods of the Quarternary was possibly almost complete. A continuous ice sheet of the Tibetan Plateau was proposed by Kuhle (1988, 1996, 1998) and Gupta and Sharma (1992), with a snowline depression exceeding 1000 m during the last glacial maximum. The Pleistocene glaciation of Tibet is lively debated by several authors who assume a lesser snowline depression of 600–800 m (Hövermann and Lehmkuhl, 1993; Lehmkuhl et al., 1998). Following warm humid periods from around 9000–6500 BP, with steppe vegetation and lakes covering large tracts of the Tibetan Plateau, a trend towards aridity continues until the present (VanCampo and Gasse, 1993; VanCampo et al., 1996). Actual vegetation features have been related to recent desiccation trends by Miehe (1996) and Schickhoff (2000). Dickoré and Nüsser (2000) speculated about the Pleistocene history of the Karakorum flora, generally in favour of a large-scale glaciation. Evidence for climatic trends and history, however, is biased by the present inhospitable conditions of the inner Karakorum Divide. Huge altitudinal shifts of vegetation belts according to climatic variations in space and time seem quite possible.

The generally unfortunate geo-ecological conditions apparently do not preclude a rather speciose, actively evolving and highly adaptable flora of western high Asia, in which vicariance patterns dominate over concentric nestedness. Proximity and polarity of diversity hot and cold spots probably contribute to a phytogeographical key position of the respective high mountain flora. The mountain ranges radiating from the Pamir knot thus may have served as important stepping stones for the evolution and dispersal of the Eurasiatic alpine flora. A close neighbourhood and altitudinal segregation of the contrasting Irano–Turanian, Central Asiatic and Sino-Himalayan elements, and apparent links between the latter and Euro–Siberian elements are the most obvious features.

Endemism, though low in numbers, fits into the image of a geographically and taxonomically central and highly coherent flora of the Karakorum. Apparent weak altitudinal and spatial correlation of endemism, point to the need for an improved understanding of evolution and radiation patterns along complex altitudinal and latitudinal gradients.

The spatial delimitation of diversity zones around the Tibetan Plateau of Barthlott et al. (1996, 1999) seems to be largely inaccurate, especially for the east Himalaya–Yunnan hot spot, but also throughout the present study area (Figure 10.4). Scale constraints and synonymy may blur diversity zones in these high mountains and contribute to at least two times overestimated species numbers in certain regions.

Changes and threats

The present floristic inventory largely defines baseline conditions, due to scattered or almost absent historical data from the literature or herbaria. A first 'red list' of vascular plants of Pakistan (Chaudhri and Qureshi, 1992) includes a considerable number of rare and threatened species of the mountain flora. Taxonomy, however, is often doubtful and requires considerable further international efforts in biosystematic research.

The altitudinal belt of highest species diversity apparently coincides with the most productive (forest belt) and anthropogenically most transformed zone. Threats to plant species are most likely to be expected in this belt. There is, however, also evidence that zones of particular interest and vulnerability do not necessarily mirror diversity hot spots and that the more locally distributed species are easily overlooked in plot-based studies.

Human-caused modifications to the vegetation are obvious throughout western high Asia, though probably still inconspicuous in the most remote regions of the Karakorum Mountains (K2-North). Regional threats concern mainly uncontrolled timber exploitation and degradation through overgrazing and subsequent erosion (Schickhoff, 1995, 1996). Although parts of the Karakorum have been declared National Parks, juniper and other coniferous forests are extremely endangered and may be completely extirpated in the near future. *Juniperus* trees, attaining an age of 2000 years (Esper, 2000), and possibly beyond that, are at present radically cleared.

Large-scale changes in distributions of permafrost, biomes and net primary production are expected on the Tibetan Plateau due to global climate changes. Recent continuous permafrost distribution extends from the Central Karakorum throughout Northern Tibet, southeast to approximately 33° N and 95° E and covers a total area of 1.5 million km^2 in Tibet (Jian Ni, 2000). Modelling results (Zhang *et al.*, 1996; Jian Ni, 2000) are, however, largely contradictory with regard to the predicted increase or decrease of area covered by conifer forests, alpine meadows and deserts.

Watershed management and various ecological troubleshooting measures are already important issues in many parts of the central Asiatic mountains and their forelands. Conserving vegetation cover and its components may soon become an important contribution to monitoring the environment and assessing risks.

ACKNOWLEDGEMENTS

The 'Flora Karakorumensis' project is funded by the German Research Council (DFG), including fieldwork in cooperation with the second author's wife, S Miehe. We owe our gratitude for inspiring discussions, valuable advice, and editorial help to A Berg, H Bruelheide, SR Gradstein, M Kessler (all Göttingen), E Eberhardt (Marburg), M Nüsser (Berlin), L Klimes (Trebon, Czech Republic), T Spribille (Fortine, MT/USA), and to an anonymous reviewer. We especially appreciate the untiring help with data-processing by U Wündisch (Göttingen). The present study would not have been possible without the help of numerous taxonomists and various major herbaria in Europe and overseas.

References

Adhikari BS, Joshi M, Rikhari HC and Rawat YS (1992). Cluster analysis (Dendrogram) of High-altitude (2150–2500 m) forest vegetation around Pindari Glacier in Kumaun Himalaya. *Journal of Environmental Biology*, **13**:101–05

Agakhanjanc O (1980). Die geographischen Ursachen für die Lückenhaftigkeit der Flora in den Gebirgen Mittelasiens. *Peterm. Geogr. Mitt.*, **124**:47–52

Axelrod DI, Al-Shehbaz I and Raven PH (1998). History of the modern flora of China. *Proceedings of the First International Symposium on Floristic Characteristics and Diversity of East Asian Plants*: pp 43–55. Beijing (China Higher Education Press, Springer Verlag)

Barthlott W, Biedinger N, Braun G, Feig F, Kier G and Mutke J (1999). Terminological and methodological aspects of the mapping and analysis of the global biodiversity. *Acta Botanica Fennica*, **162**:103–10

Barthlott W, Lauer W and Placke A (1996). Global distribution of species diversity in vascular plants: towards a world map of phytodiversity. *Erdkunde*, **50**:317–27

Bokma F, Bokma J and Mönkkönen M (2001). Random processes and geographic species richness patterns: why so few species in the north? *Ecography*, **24**:43–9

Braun G (1996). Vegetationsgeographische Untersuchungen im NW-Karakorum (Pakistan).

Kartierung der aktuellen Vegetation und Rekonstruktion der potentiellen Waldverbreitung auf der Basis von Satellitendaten, Gelände- und Einstrahlungsmodellen. *Bonner Geogr. Abh.*, **93**, 156 pp

Chaudhri MN and Qureshi RA (1992). ['1991'] Pakistan's endangered Flora II. A checklist of rare and seriously threatened taxa of Pakistan. *Pakistan Systematics*, 5:1–84

Colwell RK and Hurtt GC (1994). Nonbiological gradients in species richness and a spurious Rapoport effect. *American Naturalist*, **144**: 570–95

Davies ALV, Scholtz CH and Chown SL (1999). Species turnover, community boundaries and biogeographical composition of dung beetle assemblages across an altitudinal gradient in South Africa. *Journal of Biogeography*, **26**:1039–55

Dickoré WB (1991). Zonation of flora and vegetation of the northern declivity of the Karakorum/Kunlun Mountains (SW Xinjiang China). *GeoJournal*, 25:265–84

Dickoré WB (1995). Systematische Revision und chorologische Analyse der Monocotyledoneae des Karakorum (Zentralasien, West-Tibet)/Flora Karakorumensis I. Angiospermae, Monocotyledoneae. *Stapfia*, **39**. Linz, 298 pp

Dickoré WB and Nüsser M (2000). Flora of Nanga Parbat (NW Himalaya, Pakistan) – An annotated inventory of vascular plants with remarks on vegetation dynamics. Englera (Berlin), **19**, 253 pp

Dickoré WB (2001). Observations on some *Saussurea* (*Compositae–Cardueae*) of W Kunlun, Karakorum and W Himalaya. *Edinburgh Journal of Botany*, 58:15–29

Esper J (2000). Paläoklimatische Untersuchungen an Jahrringen im Karakorum und Thien Shan Gebirge (Zentralasien). *Bonner Geogr.*, **103**, 126 pp

Evans FC, Clark PJ and Brandt RH (1955). Estimation of the number of species present in a given area. *Ecology*, **36**:342–3

Givnish TJ (1999). On the causes of gradients in tropical tree diversity. *Journal of Ecology*, 87: 193–210

Gomulkiewicz R, Thompson JN, Holt RD, Nuismer SL and Hochberg ME (2000). Hot spots, cold spots, and the geographic mosaic theory of coevolution. *American Naturalist*, **156**:156–74

Grubov BI (ed.) (1963–). *Plantae Asiae Centralis* 1–. Akademiya Nauk SSSR/Academia Sci. Rossica, Leningrad/St Petersburg

Gupta SK, Sharma P (1992). On the nature of the ice cap on the Tibetan Plateau during the Quarternary. *Global and Planetary Change*, **97**:339–43

Hartmann H (1966). Beiträge zur Kenntnis der Flora des Karakorum. *Botanische Jahrbücher*, **85**:259–328, 329–409

Hartmann H (1968). Über die Vegetation des Karakorum. I. Teil: Gesteinsfluren, subalpine Strauchbestände und Steppengesellschaften im Zentral–Karakorum. *Vegetatio*, **15**:297–387

Hartmann H (1972). Über die Vegetation des Karakorum. II. Teil: Rasen- und Strauchgesellschaften im Bereich der alpinen und der höheren subalpinen Stufe des Zentral-Karakorum. *Vegetatio*, **24**:91–157

Hewitt K (1998). Himalayan Indus streams in the Holocene: Glacier-, Landslide-interrupted fluvial systems. In Stellrecht I (ed.), *Culture Area Karakorum Scientific Studies* 4.1. Köppe, Köln, pp 101–26

Hövermann J and Lehmkuhl F (1993). Bemerkungen zur eiszeitlichen Vergletscherung Tibets. *Mitt. Geogr. Ges. Lübeck*, **58**:137–58

Hsü Jen (1982). The uplift of the Qinghai-Xizang (Tibet) plateau in relation to the vegetational changes in the past. *Acta Phytotax. Sin.*, **20**:385–91

Hsü Jen (1984). Late cretaceous and cenozoic vegetation in China, emphasizing their connections with North America. *Annals of the Missouri Botanical Gardens*, **70**:490–508

Jian Ni (2000). A simulation of biomes on the Tibetan Plateau and their responses to global change. *Mountain Research Development*, **20**(1): 80–9

Kessler M (2000a). Altitudinal zonation of Andean cryptogam communities. *Journal of Biogeography*, 27:275–82

Kessler M (2000b). Elevational gradients in species richness and endemism of selected plant groups in the central Bolivian Andes. *Plant Ecology*, **149**:181–93

Kubitzki K and Krutzsch W (1998). Origins of East and South East Asian plant diversity. *Proceedings of the First International Symposium on Floristic Characteristics and Diversity of East Asian Plants*, pp 56–70. Beijing (China Higher Education Press, Springer Verlag)

Kuhle M (1988). The Pleistocene glaciation of Tibet and the onset of Ice Ages – An autocycle hypothesis. *GeoJournal*, **17**:581–95

Kuhle M (1996). Rekonstruktion der maximalen eiszeitlichen Gletscherbedeckung im Nanga Parbat-Massiv (35° 05′–40′ N/74° 20′–75° E). In Kick W (ed.), Forschung am Nanga Parbat. Geschichte und Ergebnisse. *Beitr. Mat. Reg. Geogr.*, 8:135–55. Berlin

Kuhle M (1998). Reconstruction of the 2.4 million km[2] late Pleistocene ice sheet on the Tibetan Plateau and its impact on the global climate. *Quarternary International*, 45/46:71–108

Leathwick JR, Burns BR and Clarkson BD (1998). Environmental correlates of tree alpha-diversity in New Zealand primary forests. *Ecography*, **21**:235–46

Lehmkuhl F, Owen LA and Derbyshire E (1998). Late quaternary glacial history of Tibet. *Journal of Quarternary Science*, 13:121–42

Miehe G (1990). Langtang Himal. Flora und Vegetation als Klimazeiger und – zeugen im Himalaya. A Prodromus of the Vegetation Ecology of the Himalayas. Mit einer kommentierten Flechtenliste von Josef Poelt. *Diss. Botanica*, **158**, 529 pp

Miehe G (1996). On the connection between vegetation dynamics and climatic changes in High Asia. *Palaeo 3*, **120**:5–24

Miehe G, Winiger M, Böhner J and Zhang Yili (2001). The Climatic Diagram Map of High Asia. Purpose and concepts. *Erdkunde*, **55**(1):94–7

Miehe S, Cramer T, Jacobsen JP and Winiger M (1996). Humidity conditions in the western Karakorum as indicated by climatic data and corresponding distribution patterns of the montane and alpine vegetation. *Erdkunde*, **50**: 190–204

Miehe S and Miehe G (1998). Vegetation patterns as indicators of climatic humidity in the western Karakorum. In Stellrecht I (ed.), *Culture Area Karakorum Scientific Studies* 4,1. Köppe, Köln, pp 101–26

Nasir E and Ali S I (eds) (1970–). *Flora of (West) Pakistan* 1–. University of Karachi, Karachi

Negi HR and Upreti DK (2000). Species diversity and relative abundance of lichens in Rumbak catchment of Hemis National Park in Ladakh. *Current Science*, **78**:1105–12

Nüsser M (1998). Nanga Parbat (NW-Himalaya) – Naturräumliche Ressourcenaustattung und humanökologische Gefügemuster der Landnutzung. *Bonner Geogr.*, Abh. 97, 232 pp

O'Brien EM, Whittaker RJ and Field R (1998). Climate and woody plant diversity in southern Africa: relationships at species, genus and family levels. *Ecography*, 21(5):495–509

Ovczinnikov PN (1957–). *Flora Tadzhikskoy SSR* 1–. Akademiya Nauk SSSR/Academia Sci. Rossica Moskva & Leningrad/St Petersburg

Paffen KH, Pillewizer W and Schneider HJ (1959). Forschungen im Hunza Karakorum. *Erdkunde*, 10:1–33

Pampanini R (1930). La Flora del Caracorùm. In *Spedizione Italiana De Filippi nell'Himalaya, Caracorùm e Turchestàn Cinese* (1913–1914), ser. 2, 10:1–285. Bologna

Pampanini R (1934). Aggiunte alla Flora del Caracorum. In *Spedizione Italiana De Filippi nell'Himalaya, Caracorùm e Turchestàn Cinese* (1913–1914), ser. 2, 11:141–78. Bologna

Persson C (1938). A list of flowering plants from East Turkestan and Kashmir. *Botanical Notes*, 1938:267–317

Rahbek (1995). The elevational gradient of species richness: a uniform pattern? *Ecography*, 18:200–05

Rahbek C (1997). The relationship among area, elevation, and regional species richness in neotropical birds. *American Naturalist*, **149**: 875–902

Rahbek C and Graves GR (2000). Detection of macro–ecological patterns in South American hummingbirds is affected by spatial scale. *Proceedings of the Royal Society of London*, ser. B (Biol. Sci.) **267**:2259–65

Richter M, Pfeifer H and Fickert T (1999). Differences in exposure and altitudinal limits as climatic indicators in a profile from Western Himalaya to Tian Shan. *Erdkunde*, 53:89–107

Rohde K (1992). Latitudinal gradients in species diversity – the search for the primary cause. *Oikos*, 65:514–27

Samant SS, Dhar U and Rawal RS (1998). Biodiversity status of a protected area in West Himalaya: Askot Wildlife Sanctuary. *International Journal of Sustainable Development and World Ecology*, 5:194–203

Schickhoff U (1993). Das Kaghan-Tal im Westhimalaya (Pakistan). Studien zur landschaftsökologischen Differenzierung und zum Landschaftswandel mit vegetationskundlichem Ansatz. *Bonner Geogr Abh.* 87. Bonn, 268 pp

Schickhoff U (1995). Himalayan forest-cover changes in historical perspective: a case study in the Kaghan Valley, northern Pakistan. *Mountain Research and Development*, **15**:3–18

Schickhoff U (1996). Die Wälder der Nanga–Parbat–Region: Standortsbedingungen, Nutzung, Degradation. In Kick W (ed.), Forschung am Nanga Parbat. Geschichte und Ergebnisse. *Beitr. Mat. Reg. Geogr.*, 8:177–89. Berlin

Schickhoff U (2000). The impact of Asian summer monsoon on forest distribution patterns, ecology, and regeneration north of the main Himalayan range (E Hindukush, Karakorum). *Phytocoenologia*, 30:633–54

Schweinfurth U (1956). Über klimatische Trockentäler im Himalaya. *Erdkunde*, **10**:297–302

Searle MP (1991). *Geology and tectonics of the Karakorum Mountains*. Wiley, Chichester

Skrine CP (1925). The Alps of Qungur. *Geographical Journal*, 66:385–409, 480

Spehn, E (2001). Global Mountain Biodiversity Assessment: A report on the first international conference on mountain biodiversity. *Mountain Research and Development*, 21:88

Stevens GC (1992). The elevational gradient in altitudinal range – An extension of Rapoport latitudinal rule to altitude. *American Naturalist*, **140**:893–911

Stewart RR (1972). *Flora of West Pakistan. An annotated catalogue of the vascular plants of West Pakistan and Kashmir*. University of Karachi, Karachi

Tang CQ and Ohsawa M (1997). Zonal transition of evergreen, deciduous, and coniferous forests along the altitudinal gradient on a humid subtropical mountain, Mt Emei, Sichuan, China. *Plant Ecology*, **133**:63–78

Thin NN and Harder DK (1996). Diversity of the flora of Fan si Pan, the highest mountain of Vietnam. *Annals of the Missouri Botanical Gardens*, 83:404–08

Thomson T (1852). *Western Himalaya and Tibet: A narrative of a journey through the mountains of Northern India during the years 1847–48.* Reeve & Co., London. 501 pp

Troll C (1939). Das Pflanzenkleid des Nanga Parbat. Begleitworte zur Vegetationskarte der Nanga Parbat-Gruppe (Nordwest-Himalaia) 1:50 000. *Wiss. Veröff. Dt. Mus. Länderk.* Leipzig, NF 7:149–93

VanCampo E and Gasse F (1993). Pollen- and Diatom-inferred climatic and hydrological changes in Sumxi Lo basin (western Tibet) since 13000 year B.P. *Quaternary Research*, **3** D:300–13

VanCampo E, Cour P and Hang SX (1996). Holocene environmental changes in Bangong Co basin (Western Tibet). 2. The pollen record. *Paleogeography, Paleoclimate and Paleoecology*, **120**:49–63

Vetaas OR (1997). The effect of canopy disturbance on species richness in a central Himalayan oak forest. *Plant Ecology*, **132**:29–38

Walter H and Breckle S-W (1984). *Ökologie der Erde 2: Spezielle Ökologie der Tropischen und Subtropischen Zonen.* UTB, Stuttgart

Weiers S (1995). Zur Klimatologie des NW Karakorum und angrenzender Gebiete. Statistische Analysen unter Einbeziehung von Wettersatellitenbildern und eines Geographischen Informationssystems. *Bonner Geogr.*, Abh. **92**:169 pp

Whittaker RH (1972). Evolution and measurement of species diversity. *Taxon* **21**:213–51

Wu Cheng-Yih (ed.) (1980–1986). *Flora Xizangica 1–5.* Science Press, Beijing

Wu Zheng-yi and Raven PH (eds) (1994). *Flora of China.* Science Press & Missouri Botanical Garden

Zhang XS (1983). The Tibetan Plateau in relation to the vegetation of China. *Annals of the Missouri Botanical Gardens*, **70**:564–70

Zhang XS, Yang DA, Zhou S, Liu CY and Zhang J (1996). Model expectation of impacts of global climate change on biomes of the Tibetan Plateau. In Omasa K, Kai K, Taoda H, Uchijima Z, Yoshino M (eds), *Climate change and plants in East Asia.* Springer, Tokyo. pp 25–38, xxv–xxxiv

Zheng Du (1983). Untersuchungen zur floristisch-pflanzengeographischen Differenzierung des Xizang-Plateaus (Tibet), China. *Erdkunde*, 37:34–47, map

Zheng Du (1988). A study on the altitudinal belt of vegetation in the western Kunlun mountains. *Chinese Journal of Arid Land Research*, **1**:227–37

Zhong XH, Zhang WJ and Luo J (1999). The characteristics of mountain ecosystem and environment in the Gongga Mountain region. *Ambio*, 28:648–654

Zhu H (1997). Ecological and biogeographical studies on the tropical rain forest of south Yunnan, SW China with a special reference to its relation with rain forests of tropical Asia. *Journal of Biogeography*, 24:647–62

Patterns of Plant Species Diversity in the Northeastern Tibetan Plateau, Qinghai, China

11

Qiji Wang, Liu Jianquan and Zhao Xinquan

INTRODUCTION

The Tibetan plateau is the highest plateau in the world and is often called 'the third pole'. This area influences Eurasian atmospheric circulation, the distribution of ecosystems and their structure, function, adaptation, and evolutionary patterns (Chang, 1983). Meadows and shrubs cover 35% of the plateau area but forest, grassland and desert ecosystems are also widely distributed (Zhao and Zhou, 1999). Very few vegetation and biodiversity studies of the Tibetan plateau have been undertaken due to the harsh conditions. Therefore, our knowledge of the distribution of plant communities and biodiversity patterns in this area is still poor. Investigation of spatial patterns of plant distribution is the basis for understanding mechanisms how such patterns were formed (Fisher, 1960; Bazzaz, 1975; Rohde, 1992). Understanding spatial patterns is also the foundation for studies of species diversity which elucidate changes in species number and turnover, and how species diversity correlates with environmental factors (Nicholson and Monk, 1974; Houssard *et al.*, 1980; Rohde, 1992; Rosenzweig, 1995). In general, there are two main approaches to understanding the ecological factors that impact on biodiversity; namely monitoring the effects of experimental manipulation on diversity within communities, and investigating natural gradients in diversity at large scales (Gould and Walker, 1997). On the Tibetan plateau a given vegetation type usually occupies a very large area, which underlines both the homogeneity of plant communities and of environments. However, five distinct vegetation types could be distinguished in the northeastern Tibetan plateau: forest, shrub, meadow, grassland and desert. The ecotone, defined as the transitional area between two distinctive vegetation types, usually provides insight as to the affinity of species to certain communities (Rosenzweig, 1995). Therefore, ecotones were selected as the large-scale survey targets.

The main objectives of the present study were: (1) to describe and compare plant biodiversity across vegetation ecotones of the Tibetan plateau and (2) to detect possible correlations between plant species diversity and ecological factors, with special emphasis on elevational effects. To our knowledge, this paper is the first analysis of plant biodiversity along a transect of the Tibetan plateau.

STUDY AREA AND METHODS

The study area was in the southern Qinghai of the Tibetan plateau, 32° 42′–36° 26′ N and 98° 49′–100° 54′ E (Figure 11.1), at altitudes from 3255 to 4340 m. The transect, from southeast to northwest, covers the forest, shrub, meadow, grassland and desert ecosystems. The non-forest vegetation types are considered to be 'alpine' vegetation. The survey route comprises five geographical units: Banma, Dari, Youyun, Maduo and Zaling Lake. There is a meteorological station within each unit, from which unpublished meteorological data were mainly obtained. The climate is characterised by long, cold winters and short, relatively cool summers. The annual

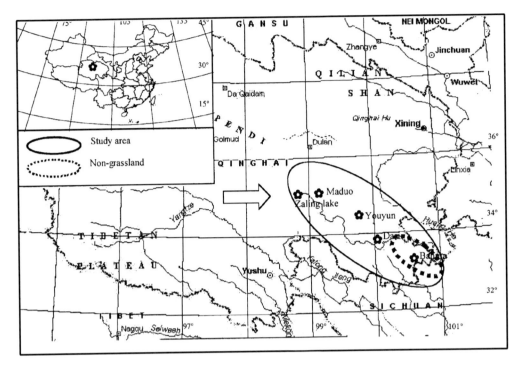

Figure 11.1 Map of the study area on the northeastern Tibetan plateau

average temperature in the five stations decreases from 2.5 °C to –5.0 °C, from the southeast to the northwest of the transect. The annual mean precipitation decreases from 655 mm to 240 mm, from the southeast to the northwest, along the transect.

In the selected transect route, the ecotones between two vegetation types were easily identified. The vegetation types are sometimes mosaics but show a trend from forest, shrub, meadow, grassland to desert type from southeast to northwest. Sixty plots were selected as representative of the ecotones along the study route (Figure 11.1). In each plot, at least two communities were sampled, resulting in a total of 141 samples. The sample sizes per community were: 10×10 m for the forest–shrub ecotone, 5×5 m for the shrub–meadow ecotone, and 1×1 m for the meadow–grassland and the grassland–desert ecotones. These sample sizes were chosen as representing more or less the minimal area required to encompass the species number of each ecotone. Additional to collecting data on species in the sample areas, we surveyed the plot for any further species. In most cases, species in the sample areas contained all the species of each ecotone. Within each sample area, species were identified and the cover and height of each species were measured. Phytodiversity patterns were calculated according to Pielou (1969):

(1) Species richness (s) is the total number of all species in the sample.
(2) Shannon–Wiener Index, $H' = -\Sigma P_I \, \text{Ln} \, P_I$ where $I = 1$, p_i = the relative value of species i calculated according to the cover and height of each species in the sample.

RESULTS

The forest–shrub ecotone contained about 28 species per 100 m^2 and was dominated by

Table 11.1 The diversity index of main ecotones on the northeastern Tibet plateau

Ecotone type	n	Layer*	Richness	Shannon index	Evenness
Forest–shrub		T	1.1 ± 0.11	0.1 ± 0.11	0.08 ± 0.08
		S	6.8 ± 0.36	2.5 ± 0.13	0.96 ± 0.01
		H	19.7 ± 1.12	3.8 ± 0.11	0.87 ± 0.01
Total	9		27.6 ± 1.41	6.4 ± 0.22	–
Shrub–meadow		S	4.6 ± 0.76	1.7 ± 0.26	0.79 ± 0.09
		H	23.7 ± 2.41	4.0 ± 0.19	0.85 ± 0.02
Total	11		28.3 ± 2.79	5.8 ± 0.35	–
Meadow–steppe	22	H	18.3 ± 1.07	3.7 ± 0.09	0.87 ± 0.01
Steppe–desert	19	H	11.0 ± 0.88	2.8 ± 0.13	0.87 ± 0.01

*T: Tree layer, S: Shrub layer, H: Herb layer

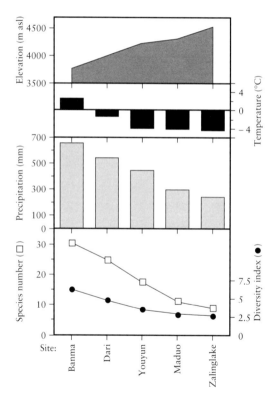

Figure 11.2 The diversity index, species number, annual precipitation and annual average temperature at six sites along the transect

Picea likiangensis on the northern slope and *Juniperus convallium* on the southern slope. The dominant shrubs in this ecotone were *Ribes stenocarpum*, *Ribes glaciale*, *Lonicera rupicola* and *Rubus ptilocapus*, with mainly *Carex atrofusca* and *Adiantum fimbriatum* in the herb layer. In the shrub–meadow ecotone the dominant shrubs were *Potentilla fruticosa*, *Salix oritrepha* and *Daphne* sp., while the herb layer consisted mainly of *Leontopodium nanum*, *Artemisia argyi*, *Carex scabrirostris*, *Bistorta vivipara* and *Festuca ovina*. The shrub–meadow ecotone constituted about 28 species in each sample of 25 m². The meadow–grassland ecotone was dominated by *Kobresia pygmaea*, *Festuca ovina*, *Potentilla nivea*, *Leontopodium nanum*, and a mean of 18 species were found in 1 m². The main species in the grassland–desert ecotone were *Stipa purpurea*, *Carex duriuscula*, *Androsace mariae*, *Arenaria pulvinata* and *Knorringia sibirica*, and there were about 11 species in 1 m².

Plant richness and diversity of each ecotone were compared (Table 11.1). The plant richness of the ecotones decreased from shrub–meadow (28.3 species), to forest–shrub (27.6 species), to meadow–grassland (18.3 species), to grassland–desert (11.0 species). Surprisingly, the forest–shrub ecotone had a higher Shannon–Wiener index (6.4) compared to the shrub–meadow ecotone (5.8) and this was due to the greater evenness of plant species distribution.

The annual average temperature and humidity index decreased from the southeastern Banma to northwestern Zaling Lake (Figure 11.2). All diversity data of each geographical unit were incorporated and

151

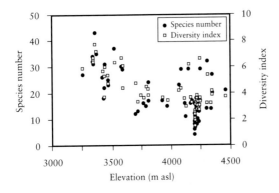

Figure 11.3 The species number and diversity index trend with altitude

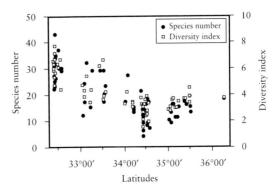

Figure 11.4 The species number and diversity index trend with latitude

compared with the meteorological data of each unit. From this, it was found that changes in plant richness and diversity index correspond to changes in annual average temperature and precipitation (Figure 11.2). This correlation indicates that temperature and water might be the main factors controlling plant species richness on the Tibetan plateau. Both species number and diversity index showed a decreasing trend with increasing elevation and latitude (Figure 11.3, 11.4). However, there are sites at high latitudes which showed relatively high species diversity. This might be due to their comparatively low elevations and high longitudes (i.e. western location?).

DISCUSSION

Our analysis revealed that plant species diversity mirrors change in environmental severity along this high–elevation transect, which is in line with recent reviews on trends in species richness with increasing altitude (Rahbek, 1995; Körner, 1999; Chapter 1). Odland and Birks (1999), working in western Norway, found maximum richness within the upper forest zone at 600–700 m and consistent reductions up to the mid–alpine zone at ca. 1800 m. Mark *et al.*

(2000), who studied Mount Armstrong in New Zealand, found negligible changes in species richness at lower altitudes but a steady reduction above the treeline up to 2200 m. Along our transect on the Tibetan plateau, plant richness and diversity index decreased almost 5-fold with altitude increasing from 3255 m to 4340 m. The reduction in plant species richness with latitude corresponds to the vegetation pattern of the Tibetan plateau and China in general (Hou and Zhang, 1980), but species numbers are very low compared to other mountain regions of the world. This may be the consequence of the combined effects of elevation (low temperature) and low precipitation. Given this small species pool, we would expect to find little functional redundancy within these plant communities, hence making them vulnerable to damage following the loss of only a few species (the insurance hypothesis of biodiversity).

ACKNOWLEDGEMENTS

This research was supported by the CAS Research Projects (KZ951-A1-204, KZ95T-04-03) and the National Key Project for Basic Research (G19980840800) awarded to Wang Q J and Zhao X Q.

References

Bazzaz FA (1975). Plant species diversity in old-field successional ecosystems in southern Illinois. *Ecology*, **56**:485–8

Chang DHS (1983). The Tibetan plateau in relation to the vegetation of China. *Annals of the Missouri Botanic Gardens*, **70**:564–70

Fisher AG (1960). Latitudinal variation in organic diversity. *Evolution*, **14**:64–81

Gould WA and Walker MD (1997). Landscape-scale patterns in plant species richness along an arctic river. *Canadian Journal of Botany*, **75**:1748–65

Hou XY and Zhang XSH (1980). *The Geographic Distribution Pattern of the Vegetation of China*. Science Press, Beijing, pp 731–8

Houssard CJ, Escarr J and Romane F (1980). Development of species diversity in some Mediterranean plant communities. *Vegetatio*, **43**:59–72

Körner Ch (1999). *Alpine Plant Life*. Springer, Berlin

Mark AF, Dickinson KJM and Hofstede RGM (2000). Alpine vegetation, plant distribution, life forms and environments in a perhumid New Zealand region: oceanic and tropical high mountain affinities. *Arctic, Antarctic and Alpine Research*, **32**:240–54

Nicholson JD and Monk CD (1974). Plant species diversity in old-field succession on the Georgia Piedmont. *Ecology*, **55**:1075–85

Odland A and Birks HJB (1999). The altitudinal gradient of vascular plant richness in Aurland, western Norway. *Ecography*, **22**:548–66

Pielou EC (1969). *An Introduction to Mathematical Ecology*. Wiley, New York

Rahbek C (1995). The altitudinal gradient of species richness: A uniform pattern? *Ecography*, **18**:200–05

Rohde K (1992). Latitudinal gradients in species diversity: the search for the primary cause. *Oikos*, **65**:514–27

Rosenzweig ML (1995). *Species Diversity in Space and Time*. Cambridge University Press, Cambridge

Zhao XQ and Zhou XM (1999). Ecological basis of alpine meadow ecosystem management in Tibet: Haibei Alpine Meadow Ecosystem Research Station. *Ambio*, **28**(8):642–7

Factors Influencing the Spatial Restriction of Vascular Plant Species in the Alpine Archipelagos of Australia

12

James B. Kirkpatrick

INTRODUCTION

Throughout glacial periods in the northern hemisphere large ice sheets have covered most of the area occupied by alpine and tundra vegetation during interglacials, making alpine and tundra vegetation more extensive in interglacials than glacials. The opposite tendency has prevailed in the southern land masses (South America may be an exception) where, in interglacial periods, such as the Holocene, alpine areas contract and fragment from relatively widespread distributions at the height of glacial periods. This process of contraction and fragmentation has been particularly marked in Australia, where alpine environments have contracted to more than 100 habitat islands, all of which formed part of a few large habitat islands 18 000–22 000 years ago (Kershaw *et al.*, 1986; Kirkpatrick, 1997; Kirkpatrick and Fowler, 1998; Figure 12.1). This process of contraction and expansion of alpine areas has occurred many times during the Quaternary, providing many opportunities for allopatric speciation and subspeciation, albeit over short time periods of 10 000 years or so. Simpson (1975) and Simpson and Neff (1985) have suggested that contractions and expansions of alpine environments in the Andes might have resulted in high speciation rates, high levels of local endemism and high species richness, suggestions partly supported by Ferrayra *et al.* (1998) for northwestern Patagonia.

The alpine vegetation of Australia is highly variable in its floristic composition and structural attributes with most of the variation being in the island state of Tasmania (Kirkpatrick, 1982, 1983, 1986, 1997). This variation is strongly related to variations in winter temperatures and soil fertility, both of which change along a geographic gradient from far southwestern Tasmania, to northeastern Tasmania, to the alpine areas of New South Wales (Kirkpatrick and Bridle, 1998, 1999).

There are several recognised centres of local endemism within alpine Australia. The Snowy Mountains of New South Wales have many local endemic alpine species (Costin *et al.*, 2000) and, in Tasmania, the Central Plateau and the far southwestern mountains have several alpine species in recognised centres of local endemism (Kirkpatrick and Brown, 1984). Kirkpatrick and Brown (1984) suggested that the centres of local endemism in Tasmania were areas containing environments that were much more widespread in the colder periods of the Quaternary than the present.

The present paper searches for correlates of localised distributions of obligate alpine species within Australia. It also compares the results of analyses based on single mountain species, species occupying less than 10 000 km^2 and the mean range of species in the floras of mountains, and discusses the implications of these analyses for the future of the alpine flora with continued climatic change.

METHODS

Vascular plant species lists were obtained from 75 alpine habitat islands throughout the range of alpine vegetation in Australia. The full species list was divided into species that were confined to the alpine and high subalpine zones (obligate alpine species) and others

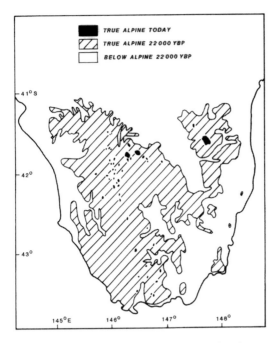

Figure 12.1 Distribution of areas above the climatic treeline for the height of the last glacial (hatched) and the present in Tasmania (black) (Kirkpatrick, 1997). The height of last glacial coastline is shown

(widespread species). The total richness, obligate alpine species richness, widespread species richness and the ratio of obligates to total species were calculated for each mountain. The geographic range of each obligate alpine species was approximated by calculating the area of the ellipse defined by the most northerly, southerly, easterly and westerly occurrences of the species. The mean of the geographic ranges for the obligate alpine flora on each alpine island was calculated. The numbers of obligate alpine species on each alpine island with a geographic range of less than 7857 km^2 (equivalent to the 100 × 100 km range considered to qualify species as rare in Australia (Briggs and Leigh, 1988)) was counted (henceforth called local endemic species), as was the number of obligate alpine species confined to each alpine island.

Alpine islands were divided into three size classes: > 100 km^2; 10–100 km^2, < 10 km^2. They were also divided into two isolation classes: < 10 km from the nearest alpine area;

> 10 km from the nearest alpine area. The mean temperature of the warmest quarter was calculated for the highest point on each habitat island, using data derived from BIOCLIM (McMahon, 1995). The means for 33 variables were calculated for each alpine island for which data were available from quadrats (see Table 12.1 and Kirkpatrick and Bridle, 1998, 1999 for details of methods).

Euclidean distance was used to construct two matrices: environment, using the factor scores for climate and soils; geographic position, using eastings and northings. The Bray–Curtis coefficient was used to construct a distance matrix from the floristic data. Correlations between these matrices were then calculated. The dependent variables (mean geographic range, local endemic species, number of species confined to one habitat island) were correlated with all other variables, except alpine habitat island size and island isolation. The relationships between alpine habitat island size, island isolation and presence or absence of one mountain species, and other variables were determined using one way ANOVA. Some of the latter correlations and ANOVA analyses were repeated for the Tasmanian and mainland Australian subsets. No corrections for multiple comparisons were made, as the chance of a type I or II error in any single comparison does not change if it is part of a larger set of comparisons.

RESULTS

The geographic position and environment distance matrices were both significantly correlated with the floristic composition distance matrix ($r = 0.696$, $p = 0.000$; $r = 0.562$, $p = 0.000$). In the case of the relationship between floristic distance and geographic distance, there was considerable heteroscadicity, with a higher range of floristic distance values in the lower range of the geographic distance values (Figure 12.2). The relationship between floristic distance values and environmental distance values approximated homoscadicity, but had much more dispersion around the line of best fit (Figure 12.3).

156

Table 12.1 Pearson's product moment correlation coefficients (above) and probabilities (below) for the relationships between mean geographic area and independent continuous variables, and number of local endemics and independent continuous variables, for both the data set as a whole and for the mainland and Tasmanian subsets. $* = p < 0.05$, $** = p < 0.01$, $*** = p < 0.001$. See Kirkpatrick and Bridle (1998, 1999) for details of measurement and calculation of the variables

	Mean geographic area			Local endemics		
	All	Mainland	Tasmania	All	Mainland	Tasmania
Soil factor 1	0.132	0.224	0.068	−0.077	0.063	−0.190
Soil factor 2	0.565***	0.446	0.266	0.049	−0.109	0.038
Soil factor 3	0.441**	0.237	0.043	−0.057	−0.340	0.178
Mean soil depth	0.510**	0.331	0.119	−0.057	−0.329	0.131
Maximum soil depth	0.475**	0.220	0.183	−0.060	−0.337	0.173
pH	0.577***	0.483	0.309	0.055	−0.102	0.068
Conductivity	0.214	−0.120	−0.207	0.003	0.073	−0.021
Total P	0.002	0.142	0.522**	−0.165	−0.039	−0.372*
Extractable P	0.554***	0.047	0.446*	0.024	−0.095	−0.289
Extractable K	0.143	0.115	−0.014	0.071	0.142	−0.076
Organic carbon	0.002	0.389	0.168	−0.080	0.010	−0.193
N	0.564***	0.276	0.595**	0.035	0.076	−0.244
Ca	0.246	0.550	0.195	−0.103	−0.228	−0.012
Cu	0.361*	−0.126	−0.124	−0.160	−0.094	−0.308
Fe	0.008	0.308	0.401*	−0.161	−0.211	−0.215
Mn	0.037	0.202	0.148	−0.035	0.029	−0.046
Zn	0.341*	0.371	−0.134	−0.138	−0.364	−0.160
Highest altitude	0.489***	−0.800**	0.228	0.332**	0.808**	−0.171
Climatic factor 1	0.575***	−0.469	−0.669***	0.123	0.812**	0.262
Climatic factor 2	0.670***	0.664*	−0.431*	0.073	−0.583*	0.289
Climatic factor 3	0.125	0.674*	−0.299	−0.278	−0.735**	0.312
Annual temperature	0.299	0.695**	−0.335	−0.157	−0.708**	0.315
Daily max. temp. warm. month	0.702***	0.625*	−0.224	−0.010	−0.684*	0.213
Daily min. temp. cold. month	0.425**	0.596*	−0.479**	−0.266	−0.707**	0.352
Annual temp. range	0.808***	−0.276	0.663***	0.162	0.453	−0.423*
Temp. warmest quarter	0.564***	0.887***	0.034	−0.177	−0.792**	−0.135
Temp. coldest quarter	0.235	0.694**	−0.388*	−0.237	−0.722**	0.343
Annual rainfall	0.728***	−0.620*	−0.685***	0.051	0.501	0.224
Precip. wettest quarter	0.712***	−0.533	−0.658***	0.106	0.464	0.195
Precip. driest quarter	0.688***	−0.247	−0.690***	0.052	0.703**	0.288
Floristic ordination score 1	0.827***	−0.169	−0.709***	−0.054	0.463	0.151
Floristic ordination score 2	0.035	−0.030	−0.445*	0.249	0.381	0.139
Floristic ordination score 3	0.175	0.595*	−0.193	−0.065	−0.438	0.154

The pattern of variation in mean geographic range is strongly geographically patterned, with western and southern Tasmanian mountains having low values and eastern Tasmanian and mainland mountains having high values (Figure 12.4). Numbers of local endemics were high in southern and western Tasmania, and generally relatively low in eastern Tasmania and the mainland mountains (Figure 12.5), a pattern generally similar to that for mean geographic range. They were, however, highest in the Snowy Mountains, the Bogong High Plains, the Central Plateau and the Southern Range, a pattern dissimilar to that found for mean geographic range. Few mountains had species or subspecies confined to them (Figure 12.6). Taxa restricted to one alpine island were most numerous in the Snowy Mountains, followed by the Central Plateau. The only other alpine habitat islands that had endemic taxa were the Bogong High Plains, the Southern Range, Mt Counsel, Mt Field, Mt Wellington and

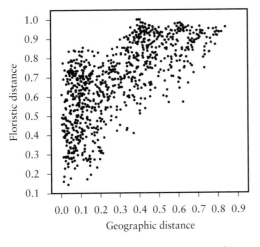

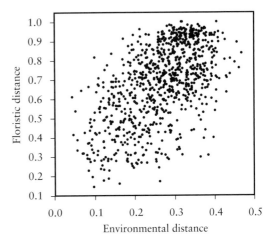

Figure 12.2 Relationship between Bray–Curtis distance between mountain pairs in floristic composition and Euclidean distance between mountain pairs in geographic distance

Figure 12.3 Relationship between Bray–Curtis distance between mountain pairs in floristic composition and Euclidean distance between mountain pairs in environmental distance

Ben Lomond. On the latter three mountains the endemic taxa were relatively weak subspecies.

In the data set as a whole, there were 21 significant correlations of edaphic, vegetation and climatic variables with mean geographic range (Table 12.1). Seven of these significant relationships, all edaphic, were nonsignificant in both the mainland and Tasmanian subsets (Table 12.1), indicating that they were a product of the marked environmental differences between the two regions. The significant relationships in the mainland subset were all with climate or vegetation variables, while in the Tasmanian subset there were significant relationships with all types of variables (Table 12.1). Mountain area was not significantly related to mean geographic area in the data set as a whole ($F = 1.16$, $p = 0.320$) or in the mainland subset ($F = 1.56$, $p = 0.219$), but in the Tasmanian subset there was a significant relationship ($F = 7.80$, $p = 0.009$), with the more extensive alpine islands having larger mean geographic areas than the smaller alpine islands. This relationship is the coincidental product of the concentration of large alpine islands in central and eastern Tasmania. The more isolated mountains had significantly higher mean geographic ranges than the other areas ($F = 5.16$, $p = 0.026$).

There were no significant correlations between the number of local endemics and any of the climatic or edaphic variables for the full data set (Table 12.1). Several climatic variables were significantly correlated with number of local endemics in the mainland subset, while total P and annual temperature range were the only significant (but weak) correlations within the Tasmanian subset (Table 12.1). The only significant relationship with number of local endemics was with habitat island area ($F = 23.15$, $p = 0.000$), with large islands having a mean of 9.7 local endemic species, medium-sized islands having a mean of 1.35 and small islands having a mean of 0.90. The number of local endemics was almost totally independent of isolation ($F = 0.00$, $p = 0.960$). One-mountain endemics were almost totally confined to large habitat islands ($F = 17.24$, $p = 0.000$), and, like local endemics, were not influenced by isolation ($F = 0.41$, $p = 0.525$). The presence or absence of one-mountain endemics was significantly related to only two of the edaphic, climatic and vegetation variables, surface soil Mn ($F = 5.86$, $p = 0.020$), with Mn being higher on habitat islands with one-mountain endemics, and mean temperature of the warmest quarter at the highest elevation ($F = 4.24$, $p = 0.043$), with temperatures being

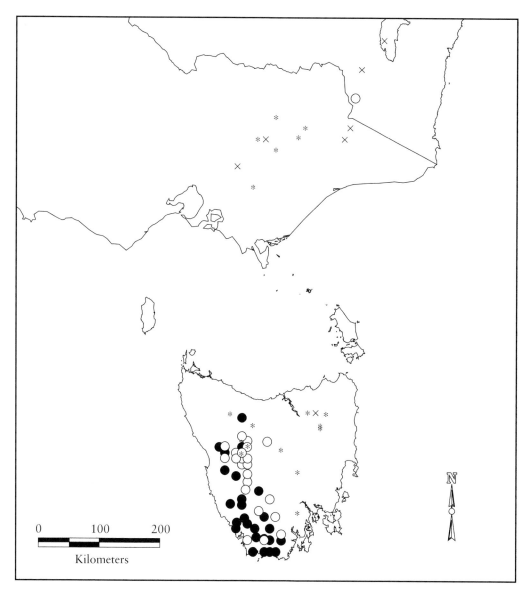

Figure 12.4 Geographic patterns of variation in the mean geographic range of obligate alpine species. Filled circles, < 75 000 km²; open circles, 75 000–100 000 km²; asterisks, 100 000–150 000 km²; crosses, > 150 000 km²

lower on habitat islands with one-mountain endemics. These two variables have a weak, significant negative relationship ($r = -0.338$, $p = 0.029$). Alpine habitat islands with one-mountain endemics had significantly higher obligate alpine species richness ($F = 22.23$, $p = 0.000$) and widespread species richness ($F = 27.01$, $p = 0.000$) than those without one-mountain endemics.

Isolation strongly influenced the ratio of alpine obligate species richness to total species richness ($F = 11.63$, $p = 0.001$), with the less isolated mountains having a mean ratio of 0.51 and the more isolated mountains a mean of 0.42. Area also had a strong effect, with mountains less than 10 km² in area having a mean ratio of 0.43, while the larger mountains had a mean of 0.51 ($F = 8.04$, $p = 0.006$). Total

159

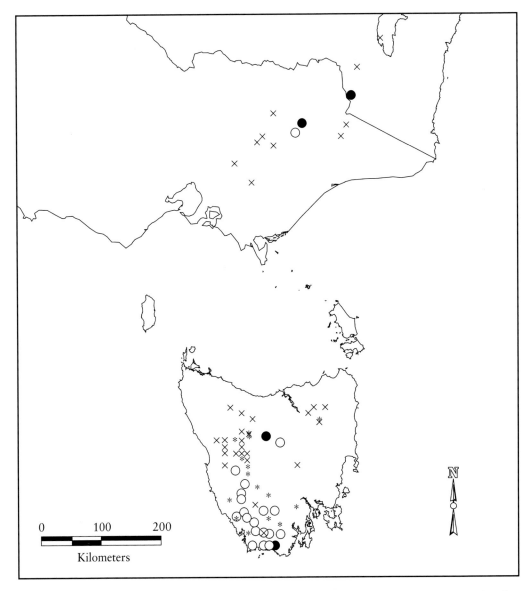

Figure 12.5 Geographic patterns of variation in numbers of local endemics (mean geographic range less than 7857 km²). Filled circles, > 4 species; open circles, 2–4 species; asterisks, 1–3 species; crosses, no species

species richness also showed a strong relationship with area (Figure 12.7, $F = 24.72$, $p = 0.000$), with a mean of 123 species for the larger mountains and 73 for the smaller mountains. A Chi² analysis indicated that the area and isolation classes were statistically independent (Chi² $= 0.856$, d.f. $= 1$, $p = 0.355$). The mean temperatures of the warmest quarter at the summit are strongly differentiated by habitat island area ($F = 33.29$, $p = 0.000$), with the mean for islands less than 10 km² in area being 10.12 °C and the mean for islands larger than 10 km² in area being 8.72 °C.

DISCUSSION

There is no indication that the long-term major disjunction between the Tasmanian and

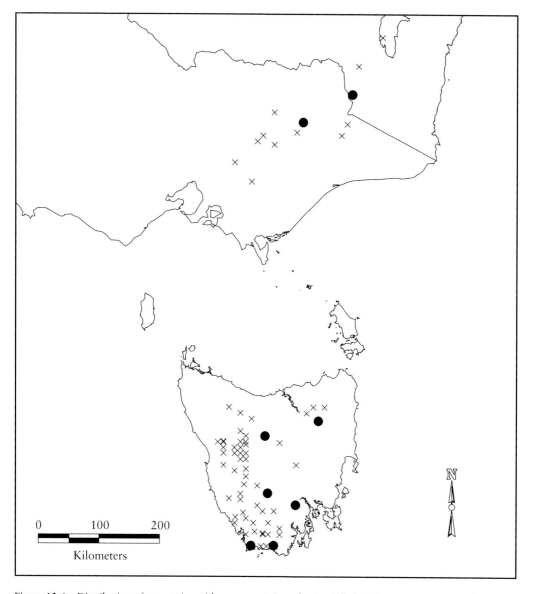

Figure 12.6 Distribution of mountains with one mountain endemics. Filled circles, present; crosses, absent

mainland Australian alpine areas has had a major effect on species composition beyond that occasioned by environmental distance. If there was an effect it would have been expressed by two distinct sets of relationships between floristic distance and environmental distance, one set relating to the distance values between Tasmanian and mainland mountains and the other relating to the distance values within these subsets.

Mean geographic range proved not to be a very useful indicator of concentrations of locally distributed obligate alpine species, because of the influence of the environmental similarities between eastern Tasmania and mainland mountains, which created a dichotomy between the far-western Tasmanian mountains and the rest.

The alpine habitat islands have increasing numbers of species, numbers of obligate alpine species, proportions of obligate alpine species to

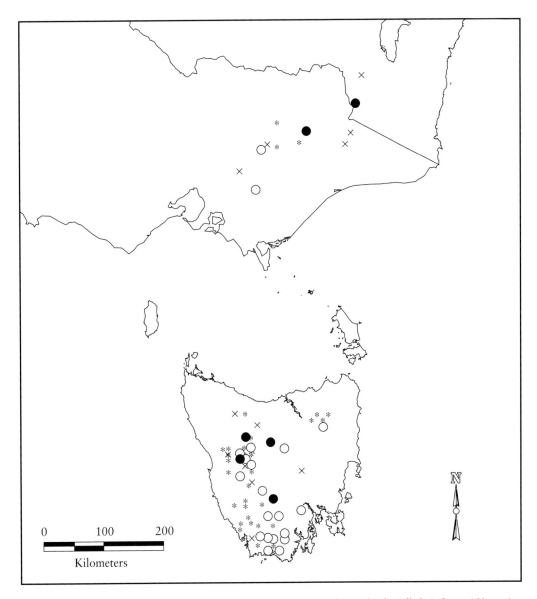

Figure 12.7 Geographic variation in total species richness between alpine islands. Filled circles, > 150 species; open circles, 100–150 spp.; asterisks, 50–99 spp.; crosses < 50 spp

total species, numbers of local endemics and numbers of one-mountain endemics, with increasing area. Increasing isolation reduced the proportions of obligate alpine species to total species, but did not influence the other variables. These results can be interpreted in the context of island biogeographic theory (McArthur and Wilson, 1967) as the result of relaxation (loss of species in response to a decrease in area) of obligate alpine species numbers from the non-insular past, the loss of species through local extinction being more extreme on smaller islands, and with the relaxation of alpine obligates being tempered to some degree by proximity to other alpine islands. However, an interpretation in terms of island biogeographic theory needs to be tempered by the fact that the larger alpine islands tend also to be those that have the most alpine environments, as indicated by summer warmth on their peaks. The relationship between one-mountain endemics and high Mn reflects the relative lack of masking of the weathering products of the substrate in fjaeldmark and snow patch short alpine herbfield, both formations that are largely confined to the areas that are coldest in summer (Kirkpatrick and Bridle, 1999).

The implications of continued increases in summer temperatures, predicted for this century in response to anthropogenic inputs of greenhouse gases, are relatively serious for Australian alpine environments and floras (Good, 1998), as well as those elsewhere (e.g. Guisan *et al.*, 1998; Saetersdal *et al.*, 1998). In Australia, the nival zone is already lacking, many of the islands consist of plateaus that could shift to non-alpine environments with relatively minor summer temperature increases, and many of the obligate alpine species are confined to those parts of alpine islands that have the lowest summer temperatures. All of the ten species most strongly confined to the alpine zone in Tasmania will have no environments left within their present summer temperature range if these temperatures increase by 3.5 °C (Howard, 2000). Island isolation will increase and island areas will decrease with increasing summer temperatures, possibly leading to other more stochastic losses of obligate alpine species from surviving alpine islands.

The strong indication of the importance of relaxation cannot, in itself, prove that no taxonomically expressed divergence has occurred during the Holocene. It may be, for instance, that the single mountain subspecies of *Euphrasia* and *Chionogentias* restricted to Ben Lomond, Mt Field and Mt Wellington have evolved since the last glacial, as these are all relatively isolated alpine islands, as well as being moderately large. However, most local endemic species are found on more than one mountain, and at the environmental extremes of alpine vegetation in Australia, indicating relict distributions.

ACKNOWLEDGEMENTS

Some of the data used in this paper were collected as part of a project funded by the Australian Research Council.

References

Briggs JD and Leigh JH (1988). *Rare or threatened Australian plants.* Special Publication No. 14. Australian National Parks and Wildlife Service, Canberra

Costin AB, Gray M, Totterdell CJ and Wimbush DJ (2000). *Kosciusko alpine flora.* 2nd ed. CSIRO, Melbourne

Ferreyra M, Cingolani A, Ezcurra C and Bran D (1998). High-Andean vegetation and environmental gradients in northwestern Patagonia, Argentina. *Journal of Vegetation Science,* 9:307–16

Good R (1998). The impact of snow regimes on the distribution of alpine vegetation. In Green K (ed.), *Snow, a natural history, an uncertain future.* Australian Alps Liaison Committee, Canberra, pp 98–112

Guisan A, Theurillat J and Keinast F (1998). Predicting the potential distribution of plant species in an alpine environment. *Journal of Vegetation Science,* 9:65–74

Howard RH (2000). *The potential effect of climatic warming on the distributions of Tasmanian obligate alpine vascular plant species.* Grad. Dip. Env. Studies (Hons) thesis, School of Geography and Environmental Studies, University of Tasmania

Kershaw AP, McEwen Mason JR, McKenzie GM, Strickland KM and Wagstaff BE (1986). Aspects of the development of cold-adapted flora and vegetation in the Cenozoic of southeastern mainland Australia. In Barlow BA (ed.), *Flora and fauna of alpine Australasia – ages and origins.* CSIRO, Canberra, pp 147–60

Kirkpatrick JB (1982). Phytogeographical analysis of Tasmanian alpine floras. *Journal of Biogeography,* 9:255–71

Kirkpatrick JB (1983). Treeless plant communities of the Tasmanian high country. *Proceedings of the Ecological Society of Australia,* **12**:61–77

Kirkpatrick JB (1986). Tasmanian alpine biogeography and ecology and interpretation of the past. In Barlow BA (ed.), *Flora and fauna of alpine Australasia: ages and origins.* CSIRO, Melbourne, pp 229–42

Kirkpatrick JB (1997). *Alpine Tasmania: an illustrated guide to the flora and vegetation.* Oxford University Press, Melbourne

Kirkpatrick JB and Bridle KL (1998). Environmental relationships of floristic variation in the alpine vegetation of south-east Australia. *Journal of Vegetation Science*, 9:251–60

Kirkpatrick JB and Bridle KL (1999). Environment and floristics of ten Australian alpine vegetation formations. *Australian Journal of Botany*, 47:1–21

Kirkpatrick JB and Brown MJ (1984). The palaeo-geographic significance of local endemism in Tasmanian higher plants. *Search*, 15:112–13

Kirkpatrick JB and Fowler M (1998). Locating likely glacial refugia in Tasmania using palynological and ecological information to test alternative climatic models. *Biological Conservation*, 85:171–82

MacArthur RH and Wilson EO (1967). *The theory of island biogeography.* Princeton University Press, Princeton

McMahon J (1995). *BIOCLIM – The BIOCLIMatic prediction system, version 3.6.* CRES, ANU, Canberra

Saetersdal M, Birks HJB and Peglar SM (1998). Predicting changes in Fennoscandinavian vascular plant species richness as a result of future climatic change. *Journal Biogeography*, 25:111–22

Simpson BB (1975). Pleistocene changes in the flora of the high tropical Andes. *Paleobiology*, 1:273–94

Simpson BB and Neff JL (1985). Plants, their pollinating bees, and the Great American Interchange. In Stehli FD and Webb SD (eds), *The great American biotic interchange.* Plenum, New York, pp 427–52

Biodiversity of the Subalpine Forest–grassland Ecotone of the Andringitra Massif, Madagascar

13

Urs Bloesch, Andreas Bosshard, Peter Schachenmann,
Hanta Rabetaliana Schachenmann and Frank Klötzli

INTRODUCTION

Purpose and aims

The main purpose of this study, mandated by the WWF, is a biotaxonomical and eco-functional assessment of parts of the protected area of the Andringitra Mountains. In view of future sustainable multiple land use, including conservation, pastoralist and tourist objectives, an integrative, participative interdisciplinary research programme was developed:

(1) Analysis of the forest/grassland ecotone on the subalpine Andohariana plateau complex, its structure, plant biodiversity, dynamic processes, convergence phenomena and human influences (Bosshard and Mermod, 1996; Roger, 1995, 1996).

(2) Evaluation of the biological and scenic values for the conservation and the development of the area (Bosshard and Mermod, 1996; Bosshard and Rakotovao, 1997; Bosshard and Klötzli, 1998; Ralaiarivony, 1999; Schachenmann, 1999).

(3) Installation of long-term experiments to obtain a better understanding of the role of fire and grazing (Bosshard and Mermod, 1996).

(4) Development of sustainable land management tools involving simple methods. Definition and monitoring of ecosystem health and resilience, disturbance capacities and desired limits of acceptable change for major vegetation units (Bloesch et al., 2000).

THE STUDY AREA

Physiographical situation and climate

The Andringitra Mountains range is located in south-central Madagascar (22° 14′ S, 46° 32′ E). Being aligned northwest–southeast and having an elevation range between 500 to 2600 m asl, with numerous granite domes and peaks (Pic Boby 2658 m asl, Pic Bory 2630 m asl). It represents an orographic barrier between the humid and cooler oceanic flank in the east and the dry and hotter continental flank in the west. The Andringitra Mountains are made up of Precambrian granite and gneiss (Paulian et al., 1971). The very long process of physical and chemical weathering has resulted in a great variety of geomorphological features. Our special investigation area is the Andohariana plateau (see Figure 13.1) of about 7000 ha, undulating between 1900 and 2100 m asl. The western boundary is marked by a distinct belt of ancient rockfalls, and behind a steep rocky arena towers up to 600 m above the plateau, while to the east there are low seams dropping abruptly towards cultivated valley bottoms at around 1400 m asl, or into rainforest further east.

Seasonal climatic changes between summer and winter and temperature and humidity variations between day and night, make this area one of the most extreme in the tropics (Koechlin et al., 1974). The dry season, from June to October, coincides largely with the austral winter, where minimum temperatures at night may drop to −16 °C (lowest temperature

Figure 13.1 The central, most open part of the Andohariana plateau showing a complex vegetation mosaic (photograph A. Bosshard)

recorded in 1995) at summital zones above 2500 m asl. During November and December, before the monsoon season starts, temperatures are very high. Lightning frequently starts fires and rainfall remains very variable in space, duration and amount (from floods to droughts). In the monsoon season, between December and April, the prevailing humid climate is from the Indian Ocean, resulting in a steep rainfall gradient from east to west (2000 to 1000 mm). Throughout the year, except for September and October, the treeline on the eastern slope is exposed to daily cloud formation producing, in parts, typical aspects of a subalpine cloud forest (*sylve à lichen*).

Flora and fauna

The Andohariana plateau is complex at the approximate observed treeline (about 2300 masl), which is a transition zone between forest and grassland. This plateau is multi-ecotonal with respect to precipitation, temperature, hydrology and bedrock, giving rise to a vegetation mosaic of punoid[1] to paramoid[2] grass and ericoid bushland, depending on the distance from the rain forest, on relief and elevation.

---

1 Puna: Steppe-like grassland at and above timberline in mountains of the semi-humid to (semi-) arid tropics
2 Paramo: Grassland at and above timberline in mountains of the humid tropics

The 'inselberg' character of the massif, combined with altitudinal zonation, has led to high local endemism. Among the monocotyledons, 40% are native to Madagascar, while 7.7% are endemic to the Andringitra Mountains. Among the dicotyledons, the corresponding proportions are 24% and 3.4%, respectively. Numerous microhabitats are found, with ericoid bush, and subalpine woodland with *Agauria* spp., rich in lichens, dry and humid grasslands, and a high diversity of Ericaceae, Asteraceae, and Poaceae, peaty depressions and rocky outcrops rich in xerophytes. The subalpine prairies are unique for geophytic Orchidaceae, of which over 30 species have been recorded so far. A rapid biodiversity assessment along an altitudinal transect of the eastern slope (Goodman, 1996) also indicated an extraordinary faunal diversity. Among the reptiles, 80% are native to Madagascar, of which 12% are endemic to Andringitra. Among amphibians, 11.4% of the 52% native to Madagascar, are endemic to Andringitra. Fifteen species of primates and insectivores were recorded, all endemic to Madagascar. Uniquely amongst the primates, a race of ringtailed lemurs (*Lemur catta*) has adapted from their typical lowland forest habitats further to the southwest to this high mountain environment, colonising a very special ecological niche of rock 'desert' with succulent vegetation types.

The plateau of Andohariana is distinguished by a number of peculiarities unique for Madagascar and the world. First, the grassland types are the only (altimontane) puna- and paramo- like types in Madagascar (compare also Koechlin, 1972; Koechlin *et al.*, 1974). The total isolation of the southern mountains on the natural and continental level has allowed high adaptation of the flora. This has also brought together into similar niches of all vegetation units a considerable number of converging groups of Asteraceae (*Helichrysum*, *Stoebe*), Ericaceae (*Philippia*), and Poaceae (*Panicum*), with ericoid to cupressoid leaves. The most striking example is *Panicum*, with its cupressoid leaves. At the same time, many of these species are endemic plants, some having their sole stronghold in the plateau area.

The heavy dominance of *Panicum* and the nearly total absence of the genus *Festuca* and *Poa*, which are otherwise often dominant in puna and paramo vegetation (such as e.g. on southeastern African mountains, including the Drakensberg Mountains, but also on other continents), is a great peculiarity of the plateau. This is probably due to its geological history (tectonics), when Madagascar split from Gondwana, carrying grasslands of savanna vegetation on its plains – including the genus *Panicum* – which then underwent a slow lifting to the actual altitude to which *Panicum* had to adapt (Koechlin, 1972). Furthermore, a large number of often locally endemic orchids occurs on prairies (e.g. *Cynosorchis*, *Habenaria*, *Liparis*), and on rocks (*Aerangis* and other geophytes).

History of land use

Until about 150 years ago, the Andringitra Massif was never inhabited permanently but was rather considered important for its mystic and spiritual values and served as refuge from feudal warfare for early settlers of the southern Madagascar highlands. Hence, land use was present but was of a more sporadic nature. According to village legends, the first people to venture into subalpine elevations were cattle herders in search of healthy and productive pastures and more security from early cattle rustling. The subalpine pastures of Andringitra, upwards to about 1900 m asl, have been included as an integral part of an intricate lowland–upland rotational grazing system. Stone walls for cattle paddocks, 'living' rock shelters and burial sites are today vivid testimonies of this traditional form of transhumance that continued until 1927. Fire was used by herders as a tool to control ericoid bush, as well as a method of expanding pastures into laurophyllous mountain forests with *Agauria* spp. This period can therefore be described as the colonisation phase.

Botanical expeditions in the early 1920s by the botanist Humbert, discovered the outstanding biodiversity value of Andringitra Mountains range (see biogeographical convergence zone of south-central Madagascar, e.g. Perrier de la Bathie, 1921; Paulian, 1958; Humbert and Cours Darne, 1965; Koechlin *et al.*, 1974) and approximately 31 000 ha of the massif was set aside as a strict nature reserve in 1927 (Nicoll and Langrand, 1989). After 1927 and up to political independence in 1960, the reserve was patrolled and protected from human impacts by the powerful forest service. Protection, however, focused more on forests than on grasslands, where cattle grazing continued to be tolerated at a moderate level. Wild fires within the reserve boundary were controlled as far as possible. This period may be considered as the consolidation and conservation phase.

Pasturing with cattle probably started when the Betsileo people moved in. They cultivated rice, but also used cattle as working animals and as objects of prestige. Traditional pasturing of lowland sites gave rise to new types of anthropogenic grasslands with *Hyparrhenia rufa* and *Aristida barbicollis* (the recent appearance of these species may also be favoured by global warming, see Bloesch *et al.*, 2000). *Heteropogon contortus* and *Imperata cylindrica* on the more mesic sites and with *Aristida* spp., *Ctenium*, *Trachypogon*, *Elionoris*, *Loudetia* on the dry sites, are also used in rotational pasture (profiles in Koechlin, 1972). Cattle grazing in the highlands normally lasts from December to March, but bulls are also pastured from July to September and calves and cows sometimes remain for the whole year. The carrying capacity of the plateau is low, but was never reached in previous decades.

Wild fires, mainly due to lightning during the early monsoon season, have always been an important natural factor (Bloesch, 1999). With the arrival of man about 300 years ago (not yet permanently settled), the fire regime changed, i.e. the use of fire for forest clearance and renewal of pastures led to considerably higher fire frequency (Paulian, 1958). Fire characteristics and impact on vegetation are described in detail by Bloesch (1999). The traditional fire regime, together with moderate cattle grazing, contributed to the particular vegetation mosaic seen today.

After independence (1960), the forestry service lost influence, means and motivation to patrol and manage the reserve effectively. This resulted in a period of conflict, where the 'owner' (the forest service) had no means of control and the 'user' (the local population) had no rights to continue using the mountain pastures on a traditional basis, leading to haphazard and illicit use of mountain resources (Rabetaliana Schachenmann and Schachenmann, 1999). We may describe this period as the open access phase (from 1960–1993). Finally, in 1993, an integrated conservation and development project was mandated to manage the area and a new National Park was decreed in 1998.

METHODS OF VEGETATION AND LAND USE STUDY

To assess the ecotonal area of the Andohariana plateau (Bosshard and Mermod, 1996), 110 relevés of 5 × 5 m, over the whole ecological spectrum, were taken according to the methodology of the Zurich–Montpellier school (Braun-Blanquet, 1964), supported by additional spatio-structural and edaphic data (Randriamboavonjy, 1996). The multivariate analysis was done using the programme Mulva 5 (Wildi and Orloci, 1996). In addition, transects were established between vegetation units in order to detect ecological transition factors. In the absence of reliable topographic maps, distribution of the 14 vegetation units identified (see later) was described for each of six geographically defined landscape sections by estimating their respective cover. Determination of plant species was based on the floristic inventory of Roger (1995, 1996) and confirmed in the national herbarium (a few taxa may need revision). The traditional adaptive strategy of seasonal pasturing on subalpine grasslands and forests were studied by direct observation, through informal interviews and questionnaires (Rabetaliana, 2002).

The possibility of reversing the ongoing scrub invasion of the Andohariana plateau was explored by: a) controlled burning of one experimental plot in August (rather late intensive fire), and b) clipping off all *Philippia* spp. close to the ground in the other plot. The size of each experimental plot was 10 × 10 m. The survival rate was recorded after one year.

RESULTS

Fourteen vegetation units are clearly distinguishable. The spatial distribution of vegetation units is a complex, small-scale mosaic, varying mainly with the relief and the land use. This distribution pattern is typical for modulated topography on a subalpine level. A total of 250 species was recorded in all vegetation units on the Andohariana plateau. The mean species number per relevé was 22, ranging from 7 to 39 (Table 13.1). The smallest number of species was present in the units growing under intermediately favourable conditions on deep soils with high organic matter content and with annual recurring anaerobic periods (units 4, 6, 7, 8, Table 13.1 and Figure 13.3). Here, the strong dominance of *Panicum* spp. and *Helichrysum* spp. formed dense vegetation layers. Opposed to this, the highest species richness was found in dry or seasonally dry sites, particularly where an open layer of phanerophytes was present (units 2, 11, 13).

Figure 13.2 provides a synoptic picture of the relation between site condition and vegetation type in the southern part of the plateau of Andohariana. Fire and grazing can extend the distribution of herbaceous vegetation types (see Figure 13.3).

Invasion of *Philippia*

Previously, *Philippia* spp. stands, 1–2.5 (3) m high, occurred mainly in parts of vegetation units 9–13 on the Andohariana plateau. However, within a few years, one third of the plateau (about 2000 ha) was invaded by *Philippia* spp. including other vegetation units (mainly 6). This rapid encroachment may have been triggered by the strongly reduced burnt areas and the cessation of almost all pasture activities since the beginning of the project in

Table 13.1 Species diversity and life form spectra of the 14 predominant vegetation units

Vegetation unit	Species number (mean ± S.E.)	Number of relevés	Predominant species/genus*	Prevalent life forms**	Specification
1	21.8 ± 4.6	5	*Xerophyta dasylirioides var andringitrensis* (g, ss)	CH	rock vegetation
2	29.0 ± 3.7	4	*Helichrysum syncephaloides* (ss)	CH	rock vegetation
3.1	1 7.0 ± 0.0	1	*Cyperaceae & Lycopodium inundatum*	HC	on peat
3.2	26.3 ± 1.8	3	*Paspalum commersoni* (gs)	HC	on peat
4	7.3 ± 3.0	3	*Helichrysum* spp. (ss)	CHG	soil depressions
5	24.3 ± 4.6	3	*Arundinella nepalensis* (gs)	HG	riverine vegetation
6	15.8 ± 1.6	8	*Panicum* spp. (gs)	HCG	mountain meadow
7	13.8 ± 1.7	8	*Helichrysum & Panicum andringitrensis* (ss & gs)	CH	mountain meadow
8	17.6 ± 1.6	9	*Helichrysum, Panicum, Aristida* (ss, gs)	HCG	mountain meadow
9	23.5 ± 2.5	16	*Ctenium encinnum* (gs)	HCG	mountain meadow
10	21.7 ± 2.5	9	*Ctenium encinnum & Vaccinium* (gs, ss)	CPH	mountain meadow
11	27.1 ± 2.2	14	*Pteridium & Agrostis* (fern, gs)	PHC	often shruby meadow
12	22.7 ± 1.4	6	*Mundulea barclayi* (t)	PHC	shrub with trees on rock
13	38.8 ± 3.6	6	*Helichrysum melastofolium* (s)	PHG	dense shrub

*Predominant species: t = tree, s = shrub, ss = small shrub, g = graminoid, gs = grasses and sedges.
**Prevalent life forms (> 20% of biomass): P = Phanerophytes, C = Chamaephytes, H = Hemicryptophytes, G = Geophytes, T = Therophytes. Total number of species: ca. 250. Data from Bosshard and Mermod (1996).

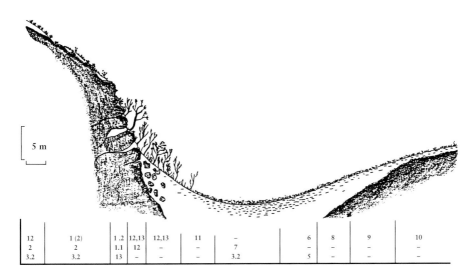

Figure 13.2 Schematic transect through a part of the Andohariana plateau (from Bosshard and Mermod, 1996). Numbers indicate the vegetation units. First row: site with good water run-off; second row: site seasonally waterlogged or with water flow; third row: site permanently waterlogged or with water flow. Stony blocks/rocks are in dotted grey, humiferous soil (Andosol) in dashes

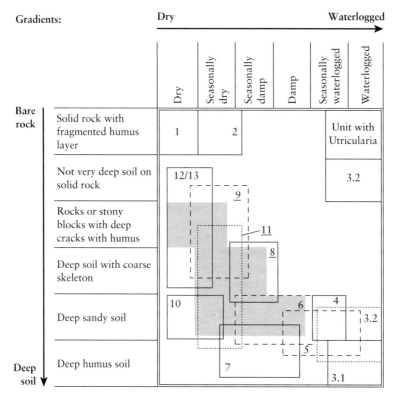

Figure 13.3 Ecogram of the main vegetation units (from Bosshard and Mermod, 1996). Soil parameters: pH (KCl) 3.8–4.8, mean 4.1; C 20–48%; C/N 6–13; P (Olsen) 0.02–0.07 mg/100 g soil. Underlined figures: potential successional stages dependent on recurrent fires. Dotted area: potential woodland sites

1993 (only controlled pasturing has been allowed since the creation of the Park in 1998).

The fire experiment on an encroached site shows that only 10 out of 555 *Philippia floribunda* plants survived. Many additional observations of burnt areas show that burning kills all *Philippia* spp. except *Philippia trichoclada*. We observed vigorous resprouting of *Philippia trichoclada* on large areas burnt by a deliberate (arson) late fire in October 1999.

In our clipping experiment, 103 out of 114 (*Philippia trichoclada*) and 69 out of 86 (*Philippia floribunda*) survived, i.e. the number of plants remained about the same after one year. Furthermore, extensive resprouting from the base dramatically increased the number of stems, i.e. up to ten sprouts per plant. Therefore, the opening of the dense ericoid bush by clipping is only a transitional measure.

DISCUSSION

Comparison with other African mountain ranges (Drakensberg and Kilimanjaro)

Certain typical characteristics of structure and biodiversity in the vegetation on the Andohariana plateau (by comparison with the Drakensberg Mountains and Kilimanjaro, the nearest mountains with an afro-alpine or punoid belt) may be summarised. The Andohariana has an area of only 50–100 km[2] compared to more than 1000 km[2] in the other two mountain areas and differs in latitude, Kilimanjaro is equatorial while the Drakensberg Mountains are further south (about 28° S). Surprisingly, species richness is quite similar, about 250 species in the Andohariana in the subalpine belt compared to 230 species on the

Drakensberg Mountains (increasing to 900 species if the afromontane belt is included) and 125 species on Kilimanjaro (above the treeline, > 3300 m asl, Hedberg 1957, 1964; Klötzli, 1958). The number of species per vegetation unit (plot of 25 m^2) is also very similar in comparable vegetational belts. Taking the main mesic vegetation units on Andohariana in the punoid and paramoid ericaceous belt, including the *Panicum* grasslands, we generally counted between 16 to 24 species per plot, depending on soil and relief. In the Drakensberg Mountains (Killick, 1963), we find 16–20 species (afroalpine *Helichrysum* belt, about 2900–3300 m asl, including *Festuca* grassland), whereas on Kilimanjaro there are about 13 species in the ericaceous belt reaching\from ca. 2800–4000 m asl and 6 species at 4000–4400 m asl in the *Festuca* grassland.

Differences become much more pronounced within certain genera. For example, the genus *Helichrysum* has 44 (mostly endemic) species in the Drakensberg Mountains, 10 species on Kilimanjaro and about 15 species on Andohariana. Furthermore, we obtain very low numbers for Andohariana species common to these other mountains, e.g. only five tree species in Andohariana of the 50 most frequent trees in the afromontane to subalpine belt of the Drakensberg Mountains. Also, Kilimanjaro and the Drakensberg Mountains have few species in common, e.g. only 18 species of the total (but 84 genera are common). Thus, the Andohariana mountains represent very isolated islands with high numbers of endemics.

Shrub invasion as a major threat to biodiversity

Recently, large parts of the Andohariana plateau have been invaded by rapidly spreading *Philippia* spp. The Ericaceae includes other genera (e.g. *Calluna* sp., *Erica* sp., *Rhododendron* sp.) which often build dense scrub stands and which may become very invasive due to changing land use. The outspread of *Philippia* spp. threatens the unique and pristine biodiversity of the plateau, the pasture value and the spectacular scenery (eco-tourism value). Another threat to biodiversity of the

Andohariana plateau is rapid population growth, accelerated by immigration, which has increased land-use pressure on the natural resources and thereby increased the conflict potential between the different actors.

Natural causes for the absence of forest on the plateau

Within the treeline, almost pure open stands of *Agauria salicifolia* occur under the escarpment or the foothills zone of the dome-shaped mountains. *Agauria salicifolia* is an evergreen tree which grows well on stony blocks. Few other tree species, mainly *Dombeya* sp., are associated with it (Humbert and Cours Darne, 1965). Regeneration of *Agauria salicifolia* is quite good and was observed in over 80% of the sites, with 50–160-cm tall saplings up to 50 m away from the next tree. Regeneration of other species is nearly non-existent, with the exception of invading *Pinus khasia* and occasional scleromorphic species. Since 1927 fluctuation in tree groups have been observed (compare also Humbert, 1927, photographic documents; Koechlin, 1972; Koechlin *et al.*, 1974). These changes are partly due to the changing grazing regime and/or fire management, thus providing an indication of fire dynamics on forest stands.

So far, there have been few special investigations on the peculiarity of the treeline in the Andringitra Mountains or the factors leading to this ecotonal pattern. Only Guillaume (in Paulian *et al.*, 1971) stated that it would be highly unrealistic to assume that the whole plateau was covered by forest. Koechlin *et al.* (1974) provided some statements on the natural occurrence of grasslands in the highlands but failed to explain the causes of the forest–grassland mosaic with a variable treeline. However, the fact that some species only occur in open grassland, shows that such meadows existed before man's arrival (a time too short for speciation). Two factors may be considered as the cause of the absence of forest, as evidenced by analysis of afforested and open sites:

- The macro-climatic gradient (i.e. less rain, less mist, longer dry periods, resulting in more critical moments for seedling establishment); and

- The influence of topography (large areas of the plateau are flat and this favours the formation of 'cold air lakes', especially in natural basins).

However, these two factors may not be the only (decisive) limitations. In an analysis of the critical ecological factors of the afro-alpine treeline, Klötzli (1975) showed that different hydric conditions in the soils of grass- and shrubland in the Simien Mountains in Ethiopia were significant for the occurrence of forest:

- The fine-grained, humus-rich Andosols were, in all seasons, unsuitable for regeneration of woody plants. The infiltration rate of water in dry soils is inhibited because of the weak re-wetting speed of this peat-like humus. On the other hand, the same soil in the wet state was normally not sufficiently aerated. Compared to ligneous plants, grass-like species are clearly better adapted to these alternating conditions of humidity because they are able to resprout readily from their roots, from stolons or rhizomes (compare e.g. Walter, 1971).
- The skelet-rich Andosols were suitable for the woody plants *Erica arborea* and *Hypericum revolutum*. Such soils are well drained, their skelet-rich main root horizon allows water to percolate through and the deep plant roots have the chance to make use of subsoil water reserves. Further advantages often include favourable temperatures and nutrient conditions in these rocky sites. Moreover, such sites also offer better regeneration conditions for woody plants. Seedlings trying to establish in a grass sward have little chance to develop or even to germinate, even in grass-rich open stands of ligneous plants (details in Klötzli, 1993).

The more humid prairies (vegetation units 3–5) and most of the *Helichrysum* shrub and wetlands with *Arundinella nepalensis*, *Panicum cupressifolium*, *P. andringitrense* and *P. spergulifolium* (units 7–8) have always been nearly devoid of ligneous plants and are not threatened by ongoing encroachment by *Philippia* spp. Furthermore, basins with 'cold air lakes' (with peat-like humus) are not colonised by trees. In contrast to the statements of Nicoll and Langrand (1989), those open prairies have – under the prevailing conditions – always existed on the Andohariana plateau, but were confined to smaller areas. The opening up of the shrub-rich units is due to grazing and manmade fire and is currently being reversed by encroachment. Meadow and Linder (1993) consider that grassland is a natural form of vegetation of afromontane sites in east and southeastern Africa. Burney (1987a,b) provides evidence that the high plateau of Madagascar was covered with a mosaic of forests, savannas and grasslands prior to the arrival of man (about 1500–2000 years ago) but lacked continuous forests, as suggested by several authors (Perrier de la Bathie, 1921; Humbert, 1927).

Effects of grazing and fire on diversity

Domestic herbivores and man probably led to the extermination of the larger indigenous herbivores. The neglected fact is that grassland existed before man came in with his cattle. The grazers were not only small mammals or an occasional *Aepyornis*, but included a number of greater lemurs – around 10 species – that were about the size of a calf (e.g. *Megaladapis edwardsi* or *Archaeolemur*). These species were able to graze such areas in a sustainable way, considering the size of their jaws (skull length 20–30 cm) and the morphology and size of teeth. A list of such species, probably all exterminated soon after the arrival of man, is given in Mahe (1965). In addition to natural and manmade fire, their grazing must have had a similar effect to cattle or horses. The grazing activity of these lemurs probably triggered the speciation of local endemic grasses.

Fire plays an important role in maintaining the diversity of the plateau vegetation. Due to the low fuel load of these subalpine vegetation types, burning is strongly, patchlike, thereby enhancing the diversity of the mosaic. Furthermore, fire is necessary for keeping the pasture open and promoting palatable species (e.g. West, 1965; Heady and Heady, 1982). Also, pasturing reduces biomass but in a very selective way: contrary to fire or mowing, it gives a better chance for establishment and growth of non-grazed plants and woody plants. Provided any fire is prohibited in the future, as was the case until recently due to the forest policy, most likely the vegetation would develop as follows:

(1) In true puna and paramo-like grasslands competition for water is such that seedlings have no chance to establish. The distribution of grasslands would therefore be more or less stable, but its diversity would drop because of litter accumulation and overshading suppressing the orchid meadows.

(2) The invasion of the Andohariana plateau would further expand. Parts of *Philippia* scrub would be replaced by stands of *Agauria*, excluding basin-like sites where natural *Philippia* scrub would subsist. The small tree islands and rocky groves would gradually grow and replace parts of the scrub, especially on slopes.

(3) Rocky zones would obviously remain in the same form, but there would be a slight tendency to species enrichment. Some woody plants would establish in fissures and crevasses (e.g. *Agauria*), especially along gulleys.

(4) Swampy sites would undergo small changes or transformations. These areas change little under the influence of pasturing animals and fire.

Management recommendations

We propose the following management techniques to tackle the problem of scrub invasion of the Andohariana plateau. Since encroachment of the plateau always starts with the fire-sensitive *Philippia cryptoclada*, we propose controlled burning of the part of the prairie which is about to be encroached (next to dense ericoid bush), thereby arresting further encroachment. In order to re-open the ericoid bush, we propose a combination of clipping and burning. First, all *Philippia* plants should be clipped, then three months later (at the end of the dry season) the dry biomass should be burned. This procedure would lead to increased fire intensity close to the ground, thereby possibly also destroying the rootstocks of *Philippia trichoclada*. This management should be tested in small designated areas before more widespread application.

An efficient and sustainable conservation strategy must consider varying sensitivities of the different vegetation types:

- Units particularly sensitive to fire: open forests, rocky sites;
- Units sensitive to pasturing: forests, scrub and steep slopes, slope wetlands;
- Units sensitive to trampling: wetlands;
- Units particularly sensitive to non-pasturing activities: grasslands of high productivity that would lose their biodiversity, and were primarily adapted to extinct large consumers; and
- Units sensitive to nutrient emissions (waste, faeces, detergent): streams, oligotrophic wetlands.

CONCLUSION

Recurrent fires and assumed high grazing pressure by now extinct large lemurs were critical factors for the evolution of the rich vegetation mosaic of the Andohariana plateau. Traditional land use maintained the high biodiversity of the area. Thus, the subalpine forest/grassland ecotone is not entirely natural, but represents a 'sustainably disturbed' anthropogenic landscape (Pyne, 1992; Schachenmann, 1999). Nowadays, the appropriate use of fire and grazing is a prerequisite to maintain the unique vegetation mosaic of the Andohariana plateau. This land use will contribute to the livelihood of the local population and should support sustainable ecotourism that will also be profitable for the local people.

ACKNOWLEDGEMENT

We would like to thank the WWF of Madagascar for their generous support of this work, and the project director, Joseph Ralaiarivony and his team for their excellent collaboration. We are grateful to E Roger for the determination of most plant species and M Randriambololona for providing the data on the experimental plots.

References

Bloesch U (1999). Fire as a tool in the management of a savanna/dry forest reserve in Madagascar. *Applied Vegetation Science*, 2:117–24

Bloesch U, Edmond R, Rosolonandrasana, BPN, Randriambololona M, Mahatsanga V, Andrianandrasana M and Kulus E (2000). *Proposition d'un plan de gestion des feux sur le plateau d'Andohariana (Parc) et les zones péripheriques nord (Namoly), et ouest (Sahanambo).* Rapport de mission, WWF Madagascar

Bosshard A and Mermod T (1996). *Réserve Nationale Intégrale 5 Andringitra, plateau de l'Andohariana: Description floristique et écologique, mise en valeur, recommandations pour la conservation et l'aménagement.* WWF Madagascar

Bosshard A and Klötzli F (1998). *Plateau d'Andohariana, RNI Andringitra, Madagascar. Analyse écologique et physiologique de la végétation. Stratégies et principes pour un développement soutenable. Recommandations pour une gestion adaptée.* Rapport de mission, WWF Madagascar

Bosshard A and Rakotovao (1997) *Contributions à l'écologie, la valorisation et la protection de la végétation du RNI Andringitra, Madagascar.* Rapport de mission, WWF Madagascar

Braun-Blanquet J (1964). *Pflanzensoziologie, Grundzüge der Vegetationskunde*, 3rd edn. Springer, Wien, New York

Burney DA (1987a). Late quaternary stratigraphic charcoal records from Madagascar. *Quaternary Research*, 28:274–80

Burney DA (1987b). Late Holocene change in Central Madagascar. *Quaternary Research*, 28:130–43

Goodman SV (ed.) (1996). *A floral and faunal inventory of the eastern slopes of the Réserve Naturelle Intégrale d'Andringitra, Madagascar: With reference to elevational variation.* Fieldiana No 85 Field Museum of Natural History, Chicago

Heady HF and Heady EB (1982). Range and wildlife management in the tropics. In Payne WJA (ed.), *Intermediate Tropical Agriculture Series.* Longman, London

Hedberg O (1957). Afroalpine vascular plants. A taxonomic revision. *Symb. Bot. Upsaliensis*, 15:144 pp

Hedberg O (1964). Features of afroalpine plant ecology. *Acta Phytogeogr. Suec.*, 49:144 pp

Humbert H (1927). Destruction d'une flore insulaire par le feu. Principaux aspects de la végétation à Madagascar. *Mémoires de l'Académie Malgache*, 5:1–80

Humbert H (1936). *Flore de Madagascar et des Comores.* Laboratoire de Phanérogamie du Musée National d'Histoire Naturelle, Paris

Humbert H and Cours Darne G (1965). *Corte internationale du topis végétal et des conditions écologiques 1:1 000 000.* Extr. Trav., sect. Scienc. Tech. Inst. Franç. Hors Sér. 6, Pondichéry

Killick DJB (1963). An account of the plant ecology of the Cathedral Park area of the Natal Drakensberg. *Memoires of the Botanical Survey of South Africa*, 34:1–178

Klötzli F (1958). Zur pflanzensoziologie der südhangen der alpinen stufe des Kilimandscharo. *Ben Geobot. Forsch. Inst.*, Zurich, 1958:33–59

Klötzli F (1975). Zur Waldfähigkeit der Gerbirgssteppen Hoch-Semiens (Nordäthiopien). *Beitr naturk Forsch Südwest-Deutschland*, 34:131–47

Klötzli F (1993). *Oekosysteme.* Uni-Taschenbücher 1479. Fischer, Stuttgart

Koechlin J (1972). Flora and vegetation of Madagascar. In Battistini R and Richard-Vindard, G (eds), *Biogeography and ecology of Madagascar.* Junk, The Hague, pp 145–90

Koechlin, J, Guillaumet J-L and Morat P (1974). *Flore et Végétation de Madagascar.* J. Cramer, Vaduz

Mahe J (1965). Les subfossiles malgaches (collection de l'Académie Malgache). *Revue de Madagascar*, 29:51–8

Meadow ME and Linder HP (1993). A palaeoecological perspective of the origin of afromontane grasslands. *Journal of Biogeography*, 20:345–55

Nicoll ME and Langrand O (1989). *Madagascar: Revue de la conservation des aires protégées (AP).* WWF International, Gland

Paulian R (1958), L'Andringitra. *Revue de Madagascar*, nouvelle série, 3:51–4

Paulian R, Betsch J-M, Guillaumet J-L, Blanc C and Griveaud P (1971). Etude des écosystèmes montagnards dans la région malgache. I. Le massif de l'Andringitra. 1970–1971. Géomorphologie, climatologie et groupement végétaux. *Bulletin de la Société d'Ecologie*, II(2–3):198–226

Perrier de la Bathie (1921). *La végétation malgache*. Annales du Musée Colonial de Marseille, 29ème année, 3ème série, 9ème volume

Pyne SJ (1992). Keeper of the flame: A survey of anthropogenic fire. In Crutzen PJ and Goldammer JG (eds), *Fire in the environment*. John Wiley & Sons, Chichester, pp 245–66

Rabetaliana H (2002). *Contribution à la méthodologie d'évaluation du système milieu-sociétés. Analyse et modélisation de la gestion des ressources naturelles dans le nord-Andringitra, Madagascar*. Thèse. Université de Franche-Comté, Besançon

Rabetaliana Schachenmann H and Schachenmann P (1999). Co-ordinating scientific research and practical management to enhance conservation and development objectives in the Andringitra Mountains, Madagascar; lessons learnt! *African Studies Quarterly*, 3:60–7

Ralaiarivony J (1999). Auto-évaluation des opportunités et des contraintes dans le projet de conservation et de développement intégrés d'Andringitra et du Pic d'Ivohibe, Madagascar. In Hurni H and Ramamonjisoa J (eds), *African mountain development in a changing world*. African Mountain Association and Geographica Bernensia, pp 281–8

Randriamboavonjy JC (1996). *Les sols et la végétation de la partie nord de la Réserve Naturelle Intégrale d'Andringitra*. Rapport de mission, WWF Madagascar

Roger E (1995). *Analyses floristiques et écologiques de la "Savane subalpine" d'Andohariana. A Flore*. Rapport de mission, WWF Madagascar

Roger E (1996). *Analyses floristiques et écologiques de la "Savane subalpine" d'Andohariana. A Flore, 2ème récolte*. Rapport de mission, WWF Madagascar

Schachenmann P (1999). Synthèse préliminaire des écosystèmes montagnards du complexe de la Réserve Naturelle Intégrale d'Andringitra et de la Réserve Spéciale du Pic d'Ivohibe, Madagascar. In Hurni H and Ramamonjisoa J (eds), *African mountain development in a changing world*. African Mountain Association and Geographica Bernensia, pp 289–304

Walter H (1971). *Ecology of tropical and subtropical vegetation*. Oliver and Boyd, Edinburgh

West O (1965). *Fire in vegetation and its use in pasture management, with special reference to tropical and sub-tropical Africa*. Mimeo. Publ. N° 1/1965. Commonwealth Bureau of Pastures and Field Crops, Farnham

Wildi O and Orloci L (1996). *Numerical exploration of community patterns. A guide to the use of MULVA-5*. SPB Academic Publishing, Amsterdam

Diversity in Primary Succession: The Chronosequence of a Glacier Foreland

14

Rüdiger Kaufmann and Corinna Raffl

INTRODUCTION

Glacier foreland succession

As a consequence of global warming, rapid deglaciation is currently a world-wide phenomenon in mountain areas. Glacial recession since the end of the so-called Little Ice Age some 150 years ago has formed glacier forelands where barren moraines are colonised and primary succession proceeds. The resulting chronosequences may be regarded as natural experiments and represent model systems for recovery from severe disturbance in the sensitive alpine zone (Matthews and Whittaker, 1987; Matthews, 1992). They show the establishment of communities and the development of diversity in an alpine landscape changing of its own accord in a natural succession process. These model systems may also help to understand how ecosystems recover following anthropogenic degradations of diversity.

The chronosequences resulting from glacial retreat represent ecological gradients with a sequence of different communities. This landscape heterogeneity, termed beta diversity according to Whittaker (1960), may contribute significantly to the diversity patterns of mountain ranges where deglaciation currently occurs. In general, small-scale mosaics and steep gradients are responsible for the great diversity of alpine environments (Körner, 1995, 1999; Haslett, 1997).

In contrast to the extensive knowledge on plant succession, there is a notable scarcity of data about faunal succession in general and primary succession in particular (Miles and Walton, 1993). Similarly, world-wide and comprehensive information exists about plants in glacier forelands (compiled by Matthews, 1992), whereas very few investigations have addressed faunal colonisation (Janetschek, 1949, 1958; Franz, 1969). In most cases, these described early colonising and following species (Janetschek, 1949, 1958) or were restricted to specific taxonomic groups (Gereben, 1995; Paulus and Paulus, 1997), thus precluding any inference about general diversity.

Diversity of invertebrates

The lack of interest in animals may partly be due to the common attitude that animals are passive followers of plants and vegetation structure (Lindroth *et al.*, 1973; MacMahon, 1981; Miles, 1987). The same point of view has been interpreted as 'bottom-up' control by Siemann (1998). In fact, however, faunal elements are essential constituents of the trophic web and an active influence of herbivores and decomposers on succession should be taken into consideration (Schowalter, 1981; Lawton, 1987).

Another reason for the imbalance between studies on vegetation and fauna are the practical difficulties involved in studying faunal diversity. In the case of invertebrates, species determinations are often arduous and costly. A complete species inventory of invertebrates is almost impossible to obtain since most of the relevant groups are difficult to identify, requiring highly skilled specialists. Specific sampling techniques are required for the different strata, i.e. soil, soil surface, herb and dwarf shrub layer, and often the standard techniques have to be modified specifically for alpine habitats. Complicated phenologies (Baars, 1979; McCoy, 1990) and widely fluctuating population dynamics (Meijer, 1989) add to the problems. In summary, diversity studies on invertebrates require careful selection of indicative groups, a pragmatic approach to taxonomic detail, and well planned sample timing (Pearson, 1994; Oliver and Beattie, 1996a; Prendergast, 1997).

Functional groups in diversity assessment

The use of functional groups and guilds instead of species has been extensively discussed (Schulze and Mooney, 1993) and has been advocated with the rationale that, due to species redundancy, functional groups may facilitate a more general view on ecosystem function than individual species could provide. Moreover, in comparisons over a large biogeographic range species will differ even in similarly structured ecosystems. Consequently, functional groups are considered an appropriate tool in advancing from diversity as a mere descriptive quantity towards diversity in the context of ecosystem architecture, function and stability. Obviously, this implies that relations between the diversity of such groups contain relevant information. This line of thought has been followed ever since the classical studies identifying plant structural diversity as a predictor for species numbers of birds (MacArthur and MacArthur, 1961) and lizards (Pianka, 1967). Provided that functional groups indeed represent specific habitat requirements and abilities to cope with environmental constraints, this may be regarded as an organism-based counterpart to Tilman's (1993) resource ratio hypotheses.

Plants are commonly categorised into functional groups by growth form or by physiological traits. For invertebrates, a taxonomic grouping at the level of families or higher taxa was suggested, on the assumption that this might serve as a proxy for functional groups (Simberloff and Dayan, 1991). This assumption, however, has to be tested in each particular case. If it holds, the approach seems practicable and cost effective. Otherwise, a real assessment of functional guilds in invertebrates would require an enormous amount of mostly unavailable species-specific information.

In most investigations of secondary succession relations between the diversity of plant and animal groups reflected the development of trophic structure (Murdoch et al., 1972; Southwood et al., 1979; Siemann, 1998; Siemann et al., 1999). Similar data are unavailable for primary succession on deglaciated terrain, although functional groups of plants have been explored in such habitats (Zollitsch, 1969; Reiners et al., 1971).

Objectives

We studied glacier foreland succession at the species level (Raffl, 1999; Kaufmann, 2001) and analysed the data at the level of overall diversity and at the level of functional groups. Although this is a local study, it can serve as a model for this approach in larger scale studies and even global comparisons. First, we compared the development of plant and animal diversity during succession by considering both total diversity and diversity within plant growth forms and taxonomic groups of invertebrates. Provided these surrogates for functional groups are valid, this should reveal the formation of the communities and their trophic components. We then asked which of the functional or taxonomic groups provide a good indication of overall diversity by analysing similarities and differences among groups.

As local environmental factors exert a strong influence on species composition (Kaufmann, 2001), our aim was to clarify in the present study whether species and functional group diversity is determined by the same environmental factors.

THE ROTMOOS VALLEY CASE STUDY IN COMPARISON TO OTHER GLACIER FORELANDS

Study area and methods

The foreland of the Rotmoos glacier (Figure 14.1) is located on the northern face of the central Alpine main ridge in the Tyrol (46° 52′ N, 11° 02′ E). It is situated above the treeline at 2250–2450 m asl. The snow-free period lasts from July to September. The valley is mostly level, only ascending near the glacier. Thus there

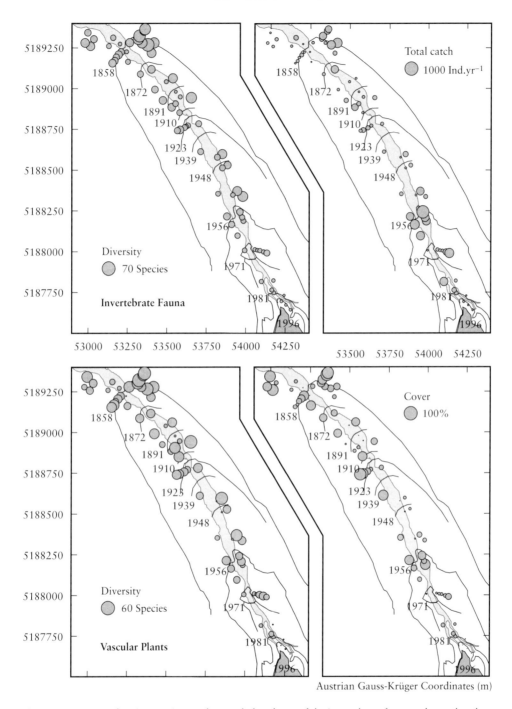

Figure 14.1 Maps showing species numbers and abundance of the invertebrate fauna and vascular plants of 70 sampling sites in the glacier foreland of Rotmoos valley. Glacier extensions with dates known from photographs or reports are indicated. The shaded area marks the central flood plain. Scaling of symbol sizes is indicated. Gauss–Krüger coordinates are scaled in metres

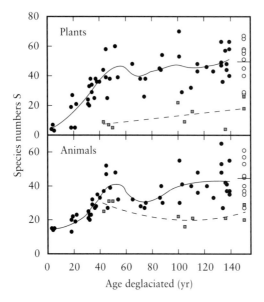

Figure 14.2 Development of plant and animal species numbers with successional age. Solid circles: sites within the chronosequence (trend lines by bi-square smoothing). Open circles: reference sites outside the foreland (means indicated). Squares: Sites in flood plain (broken line: second order fit). Age since deglaciation refers to the year 1997

is no altitudinal gradient confounding the successional chronosequence. Over the last 140 years the glacier has retreated 2 km, currently at a rate of 15–20 m per year. The terminal moraine is dated 1858, the successional chronosequence is well preserved. The glacial stream has formed a central flood plain consisting of gravel or sandy substrate.

Seventy sites were investigated using pitfall traps, recording the vegetation, taking soil samples, and by site characterisation (Figure 14.1). Sites were distributed from the glacier snout to the terminal moraine, including reference sites beyond the glaciated area from the last century. The date of deglaciation of each site was established from reports, photographs, and records of glacial retreat made over the last 100 years (data compiled by Juen, 1998).

Invertebrate fauna was sampled by pitfall trapping (Oliver and Beattie, 1996a). However, this method is biased towards surface-active species, whereas soil fauna and species of the herb layer are underrepresented. The dominant arthropod groups, Coleoptera, Arachnida, Diplopoda, and Chilopoda, were determined to species level. Flying insects were excluded since they are not indicative on a local scale. Even after 5 years of intensive investigations the species list is still incomplete due to the difficulties mentioned above. This also explains why the number of animal species (Figure 14.1) is not much higher than that of plant species, which would be expected in a complete inventory. Vegetation was recorded in the immediate surrounding of the pitfall traps using standard methods (Braun-Blanquet, 1964; Mueller-Dombois and Ellenberg, 1974).

Development of plant and animal diversity

The basic diversity pattern and abundances for surface-active invertebrates and for vascular plants are shown in the maps of Figure 14.1. Succession was characterised by two phases: (1) An initial colonisation with rapid increase in diversity over the first 50 years, and (2) a subsequent development with only little increase in species numbers (Figure 14.2).

Only a few pioneer plant individuals appeared during the early years, whereas the fauna started with about 15 species which were consistently present at the pioneer sites. This initial phase was characterised by a rapid species turnover which also slowed down after 50 years (Kaufmann, 2001). Local variability was high at older sites with consistently lower diversity on the orographic left valley side. Diversity immediately beyond the terminal moraine was similar to that at nearby sites within the foreland, but species composition was different, particularly in plants. This showed that succession is still in progress after 140 years. Plant cover and average faunal yearly catch, as a measure of total abundance, developed in a similar fashion to the respective diversities.

Diversity and abundance were much lower in the central flood plain than at nearby undisturbed sites (Figure 14.1). This disturbance effect was much more pronounced in plants than in animals (Figure 14.2). In essence, these sites frequently revert to the pioneer stage.

The observed two-phase succession pattern seems to be generally valid for vegetation developing towards both alpine grasslands and coniferous forests on glacier forelands (Friedel, 1938; Zollitsch, 1969; Reiners et al., 1971; Birks, 1980; Elven, 1980; Matthews and Whittaker, 1987; Matthews, 1992), and hints of a similar pattern were previously found in faunal assemblages (Janetschek, 1958). An initial phase of 50 years is typical, but it may take much longer in glacier forelands prone to natural disturbances (Zollitsch, 1969).

Animal abundance and, to a lesser degree, also animal species numbers attained a peak after 50 years, which was less conspicuous in plants. Such peaks at a successional age at which pioneers and following colonisers meet have also been reported from other glacier forelands. The interpretation was that a large number of species is able to colonise open space whereas competition limits species numbers in closed vegetation (Matthews, 1992). In faunal communities such peaks are possibly due to an overlap of early colonisers with later appearing herbivores.

Here, as in general in glacier forelands (Matthews, 1992), no decline in diversity was found at mature sites, either in the oldest foreland areas or outside the foreland. Such hump-shaped diversity patterns as predicted by the 'intermediate disturbance hypothesis' may be more typical for secondary successions (Tramer, 1975; Southwood et al., 1979; Brown and Southwood, 1987; Theodose and Bowman, 1997). Moreover, our data support reports that there is little successional trend in evenness (Birks, 1980; Matthews, 1992: p. 188; but see Reiners et al., 1971), thus making species number an appropriate measure of diversity. In other alpine habitats, however, evenness did respond to changes in nutrient availability (Theodose and Bowman, 1997). In summary, this glacier foreland complies largely with previous results from studies using species-level data.

It should be pointed out that a slow-down of successional progress as a certain level of diversity is attained does not imply increased

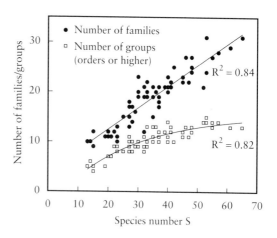

Figure 14.3 Relationship between number of fauna species and higher taxonomic levels. Solid circles: number of families (linear fit). Open squares: number of orders or higher taxa (quadratic fit). Sites of central floodplain excluded

stability. Data on resilience and resistance against specific disturbances would be needed for such an inference.

Diversity of functional and taxonomic groups

The number of animal families or higher order taxa in relation to species numbers are shown in Figure 14.3. Family number was linearly related to species number with a high correlation. Thus, family level is possibly sufficient to describe diversity for this particular situation. Still higher taxa, however, are not. This relationship saturates as later successional species belong to the groups already present.

A more detailed view of specific groups and species numbers is given in Figure 14.4. Harvestmen (Opiliones), beetles (Coleoptera) and spiders (Arachnida) were present from the beginning, species number increasing along the chronosequence. Of the beetles, the carnivorous and mostly mobile ground beetles (Carabidae) were among the first colonisers, followed by rove beetles (Staphylinidae). The herbivorous families (Chrysomelidae and Curculionidae) appeared after 40 years, if at least some vegetation cover was present. But also among spiders (all of them carnivores)

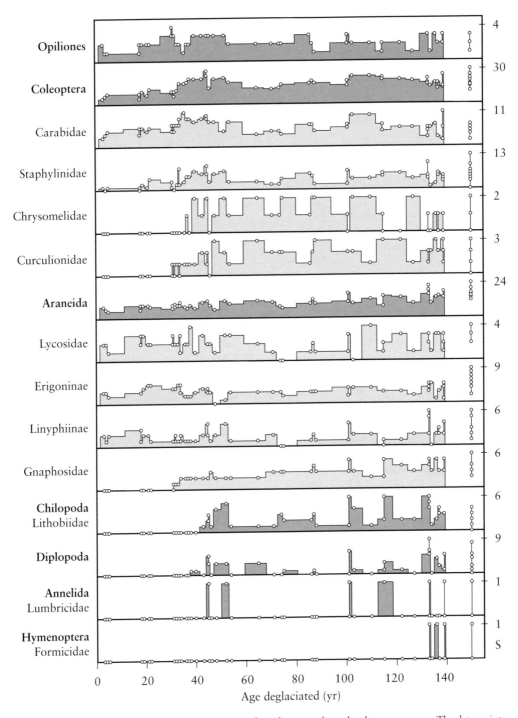

Figure 14.4 Species numbers of the most important faunal groups along the chronosequence. The data points on the extreme right refer to sites immediately beyond the foreland border. Right hand axis indicates scaling of species numbers S

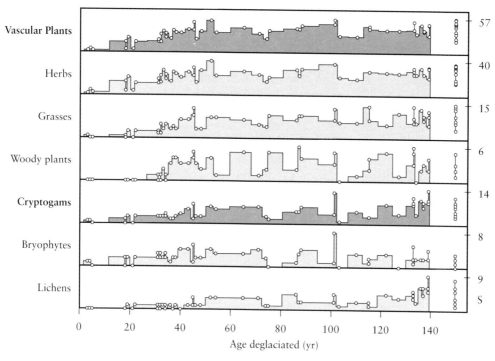

Figure 14.5 Species numbers for plant growth forms along the chronosequence. Same depiction as in Figure 14.4

there was some gradation, ground spiders (Gnaphosidae) appearing later than the other families. Centipedes (Chilopoda) were present after 50 years. At this time, the decomposer guilds also arrived, with millipedes (Diplopoda) and the first earthworms (Lumbricidae). The latter were otherwise restricted to sites older than 100 years, where the formation of true soil with an A horizon had begun (Erschbamer *et al.*, 1999). Ants (Formicidae), finally, have not yet moved into the glacier foreland and were only present near the terminal moraine. Like earthworms, they were represented by single species only. Altogether, this showed that groups arrived and remained, whereby species turnover within groups slowed down with site age. Most groups attained similar species numbers within the glacier foreland as in the mature sites beyond the terminal moraine, where only carabid beetles had a lower, and erigonid spiders a higher, presence.

In the case of animals, a trophic web comprising herbivores and decomposers developed out of a pioneer community totally dominated by carnivores (Figure 14.4). This is in contrast to all experiences from secondary successions, where invertebrate carnivores follow herbivorous invertebrates (Brown and Southwood, 1987). However, similar results are available from primary successions on sand dunes (Lowrie, 1948) and volcanic deposits (Ashmole and Ashmole, 1987, 1997; Sugg and Edwards, 1998) which, in many respects, are comparable to glacier forelands. High carnivore/herbivore ratios may even be a more general characteristic of harsh or strongly limited environments such as the subnival zone (Meyer and Thaler, 1995) and arctic tundra (Chernov, 1995).

In contrast to the species numbers of higher plants, which remained almost constant after the first 50 years, cryptogams, specifically lichens, increased steadily along the entire chronosequence, with a maximum immediately behind the terminal moraine (Figure 14.5). Such a continued increase in lichen numbers was also observed on another Central Alpine

183

glacier foreland (Zollitsch, 1969). Growth forms of vascular plants occurred in the order herbs, grasses (comprising graminoids and sedges), and woody species. After 30 years, all groups were present, with high local variability in the occurrence of woody dwarf shrubs.

A gradual build-up of the canopy architecture, taking about 50 years until vegetation begins to close and shrubs are present (Figure 14.5), is typical for successions in the alpine zone (e.g. Zollitsch, 1969). In glacier forelands below the treeline, shrubs can colonise very quickly and trees may invade after 50 years (Birks, 1980; Matthews, 1992). A shift from annual plants to perennials, as reported from both primary and secondary successions (Brown and Southwood, 1987; Tsuyuzaki, 1996), could not be detected here due to the scarcity of annuals in alpine habitats (Körner, 1995).

INDICATIVE VALUE OF FUNCTIONAL GROUPS

Small-scale patterns of groups

The chronologies of the Rotmoos glacier foreland (Figures 14.4 and 14.5) provided a good overview of the trends on the landscape scale (≈ 1 km), and showed that during the colonisation phase the time elapsing since deglaciation was the determinant of diversity. The question remained whether the pronounced local fluctuations (< 50 m) at later stages, where average diversity was almost constant, were random or whether there were correlations among groups indicating small-scale environmental determinants of community structure. An analysis of this correlation structure is also required to identify which groups or which combinations of groups form a minimum set approximating total diversity.

These questions may be assessed by multivariate statistical techniques frequently used in ecology for such problems. Interdependencies between plant and animal groups obtained by such an analysis (principal components analysis, PCA) are shown in Figure 14.6 in relation to total species richness. A three-dimensional

depiction was required, with the vertical component (axis 1) representing total diversity (S_{tot}). Groups oriented close to the vertical reference line correlate strongly with overall diversity, whereas deviations from the vertical direction indicate group-specific patterns. Thus, species counts of groups appearing near the horizontal plane are almost unrelated to total diversity (e.g. carabid beetles and erigonid spiders). Among the groups similar directions indicate similar site preferences, increasing angles meaning increasing differences, and obtuse angles showing contrasting preferences. Long lines indicate strong relationships; short lines (e.g. bryophytes) indicate low predictability and large random components.

Plant and animal diversities show some divergence but are still highly correlated (Figure 14.6a). The different growth forms of plants (Figure 14.6b) have specific diversity patterns. In particular, cryptogams, lichens and mosses are only weakly related to any other group. The surface-active arthropods, beetles, spiders, and the two myriapod groups (the carnivorous centipedes and the detritivorous millipedes) show still more conspicuous differences (Figure 14.6c). Only centipedes and millipedes prefer similar sites, otherwise correlations between the groups are rather low. Families within these groups again exhibit quite specific habitat requirements, which is apparent from the wide fan out of the most species-rich families of beetles and spiders (Figure 14.6d). In summary, almost all the groups are to some extent related to overall diversity, but none alone sufficiently strongly to be useful as an indicator of total diversity. A combination of beetles and spiders, however, would serve this purpose for the invertebrates. For an interpretation of local patterns within the successional chronosequence, family level is required.

In general, the interdependencies between functional groups will be different across landscapes or biogeographic regions. Therefore, a survey of the correlation structure as outlined above should be helpful in planning large-scale comparisons or screening studies. A lack of correlation between the diversity of different

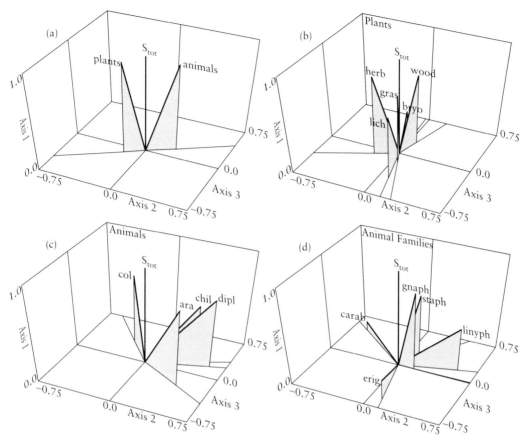

Figure 14.6 Correlation structure of species numbers of the major plant and animal groups in relation to total species richness S_{tot}. Similar directions of the groups indicate similar site preferences, groups near the vertical centre line are strongly correlated with total species richness (for further explanations see text). (a) Overview for plants and animals. (b) Plant growth forms herbs (herb), grasses (gras), woody plants (wood), lichens (lich), and bryophytes (bryo). (c) Animal groups beetles (col), spiders (ara), millipedes (dipl), and centipedes (chil). (d) Beetle families ground beetles (carab) and rove beetles (staph), spider families erigonids (erig), linyphiids (linyph), and ground spiders (gnaph). First three axes from principal components analysis (PCA) rotated for alignment with S_{tot} on axis 1, one analysis split into four graph panels

groups of organisms, often invertebrates, has been noted as disturbing in quite a few cases (e.g. Baur *et al.*, 1996; Oliver and Beattie, 1996a; Prendergast, 1997). Presumably, spatial scale is the most relevant factor in this respect, which is important for experimental designs. Low coherence is more likely to be found within local patterns (Murdoch *et al.*, 1972).

Environmental factors modifying succession

Glacier forelands represent a spatially ordered chronosequence and a mosaic structured by local environmental conditions and disturbances (Matthews and Whittaker, 1987; Gereben, 1995; Helm and Allen, 1995). In extreme cases, the mosaic may even obliterate the successional pattern (Matthews, 1992). The Rotmoos glacier foreland, with a well preserved chronosequence, facilitated analysis of both aspects, whereby at younger sites age was the determinant and at older sites local conditions became important.

It was previously shown that, apart from site age and concomitant soil formation, the local moisture regime and sun exposure were the most important factors for faunal species

Table 14.1 Spearman correlations R_s of fauna and plant species numbers with environmental factors along the entire glacier foreland chronosequence (S_{fauna}, S_{plants}) and for local variations of species numbers (*local* S_{fauna}, *local* S_{plants}). *local* S_{fauna} and *local* S_{plants} correlations were obtained using residual species numbers after accounting for the large-scale trend modelled by exponential relationships with site age: ($S_{fauna} = 43.5-33.7 \, e^{-0.023 \, age}$, $S_{plants} = 48.9-53.7 \, e^{-0.032 \, age}$). Significant correlations ($p < 0.05$) shown in boldface. Sampling sites in the flood plain were excluded (N = 61)

	S_{fauna}	*local* S_{fauna}	S_{plants}	*local* S_{plants}
Site age	**0.72**	0.03	**0.70**	0.07
	$p = 0.000$	$p = 0.808$	$p = 0.000$	$p = 0.594$
Sun exposure	**0.37**	0.02	**0.49**	0.24
	$p = 0.003$	$p = 0.890$	$p = 0.000$	$p = 0.060$
Soil organic content	**0.59**	0.06	**0.65**	0.16
	$p = 0.000$	$p = 0.663$	$p = 0.000$	$p = 0.230$
Moisture	0.23	**0.29**	0.20	0.25
	$p = 0.076$	$p = 0.025$	$p = 0.125$	$p = 0.052$
Snow cover duration	**−0.41**	−0.22	**−0.40**	−0.19
	$p = 0.001$	$p = 0.089$	$p = 0.001$	$p = 0.143$
Valley side	**−0.31**	**−0.47**	−0.25	**−0.32**
	$p = 0.014$	$p = 0.000$	$p = 0.051$	$p = 0.011$
Vascular plant cover	**0.51**	0.17		
	$p = 0.000$	$p = 0.198$		

composition, and that valley sides differed in plant succession (Kaufmann, 2001). This was in agreement with earlier indications that moisture is a factor leading to multiple pathways in glacier foreland succession (Friedel, 1938; Janetschek, 1949; Zollitsch, 1969; Jochimsen, 1970).

The clear relationship between diversity and site age is documented in Figure 14.2. Many of the environmental conditions were also related to site age. Obviously, soil organic content increased as soil formation and succession proceeded, but sun exposure, indicative of the temperature regime, and snow cover duration showed a trend along the foreland too. In consequence, these factors also correlated with species richness (Table 14.1) which is a trivial result. Since we were particularly interested in local effects, we determined the deviation of local species number from the general trend and correlated the deviation with environmental factors. Apart from the difference between valley sides already mentioned, only weak environmental correlations remained (Table 14.1). Moisture had the only significant or nearly significant effect on species numbers. Plant diversity is just about enhanced on sunny sites, and animal diversity reduced by extended snow cover.

Resorting to organismal groups, including families of invertebrates, clarified these relationships. Again, using multivariate techniques, we were able to show that sun exposure together with snow cover constituted a determining factor, and that the difference between the valley sides was predominantly a moisture effect. These effects were statistically significant, but a large fraction of total variability remained unexplained and appeared as random fluctuations, which is not uncommon in ecological studies (e.g. del Moral et al., 1995). Here again, no single group of organisms could elucidate the environmental relationships, but an integrated analysis of all groups did.

Thus, analyses on the levels of species, functional groups, and total diversity basically yielded the same results. For the detailed analysis of local patterns, however, total diversity yielded cursory indications only. Functional groups provided reasonable results when using families for the fauna, albeit not with the precision of species-level analysis. Taxonomic groups as proxy to functional groups of fauna provided a link with environmental factors and thus may be encouraged, at least for studies in alpine areas above the treeline.

CONCLUSIONS AND RECOMMENDATIONS

The case study of the 140-year chronosequence of the Rotmoos glacier foreland showed that species richness increased rapidly over the first 50 years, with high species turnover in both plants and animals. While only a few pioneer plants appeared within the first years, an astonishingly diverse animal community dominated by predators was rapidly established. In older stages there was little further increase in diversity, which was already similar to adjacent non-glaciated sites beyond the terminal moraine. On the other hand, there was no decrease in diversity in these mature sites, as suggested by the intermediate disturbance hypothesis. Changes in species composition across the terminal moraine, however, showed that development of a climax community will take much longer than the time span typically covered by chronosequences of alpine glacier forelands. Regarding this situation as a model for severe disturbance, the important corollary is that rapid assembly of initial communities does not imply rapid full recovery in alpine ecosystems. This fact may not be apparent from species counts alone.

Concerning plant canopy structure, herbs and bryophytes established first, followed by grasses, lichens, and woody plants. Among the animals, beetles and spiders were present from the beginning, while centipedes and millipedes arrived later. After 50 years all major groups were present, with the exception of earthworms and ants. Different taxonomic groups showed markedly different diversity patterns, particularly on a small scale. Even if overall diversity was governed by site age, it also correlated with local environmental conditions. These environmental influences appeared more clearly when groups and families were analysed.

The extensive data set of this case study allows some general conclusions about the inferences possible in alpine habitats on the basis of total diversity, diversity within functional groups, and species level analyses. Overall diversity, which is relatively easy to obtain for vascular plants, is a good indicator of basic ecosystem processes such as initial colonisation. Similarly, it should be a useful indicator of severe ecosystem degradations by natural or human disturbances. However, causal inferences are unlikely to be possible on this level.

The use of functional groups allowed us to gain an overview of community architecture and trophic structure. It also assisted in defining small-scale patterns within this heterogeneous and highly structured alpine landscape. The employment of faunal taxonomic groups as a surrogate for functional guilds seems to have been successful. Family level may be required in order to gain information from different groups, such as beetles or spiders in the present case. Identification of the most important environmental factors seems feasible with this approach. Nonetheless, species level analysis will often be required when the ecological mechanisms responsible for diversity patterns are a concern. Rates of species turnover, species interactions, and specific ecological demands may serve as examples for such more detailed questions.

It was clearly shown that none of the faunal groups alone could serve as an indicator of overall species richness. Thus we recommend concentrating on the dominating faunal groups, preferably covering the entire trophic cascade, but to restrict taxonomic resolution to the absolutely necessary. Using families instead of species meant no serious loss of information with regard to diversity in the glacier foreland, which is in agreement with results from other biomes (Balmford *et al.*, 1996a, 1996b). Morphospecies or other taxonomic surrogates (Oliver and Beattie, 1996b; Roy and Foote, 1997) should also be taken into consideration whenever more information can be gained by increasing sample numbers than by obtaining species data.

Such case studies demonstrating the usefulness of diversity studies on the level of functional and taxonomic groups are not only helpful when aiming at basic questions of ecosystem functionality, but also in connection with applied issues of land-use change, restoration or conservation. The most important practical aspect is that in large-scale monitoring and screening programmes a substantial cost reduction may be achieved by informed employment of higher taxa indicators. Ultimately, this aims at early diagnosis of environmental changes at a stage where they are not yet severe. Ratios and relationships between species richness of organismal groups can be expected to reflect habitat suitability according to group specific requirements as well as trophic interactions. Thus, they indicate the state of an ecosystem on a more integrated level than analysis of individual species, but are more specific than total diversity.

ACKNOWLEDGEMENTS

This investigation was supported by the Austrian Academy of Sciences, Commission for Interdisciplinary Ecological Studies. The following colleagues contributed species determinations of invertebrates, without them it would have been impossible to compile the data base for this synthetic analysis: Anita Juen (Coleoptera), Irene Schatz (Coleoptera: Staphylinidae), Karl-Heinz Steinberger (Araneida), Erwin Meyer (Diplopoda), Karma Moser (Chilopoda), Erhard Christian (Diptera: Chionea spp.), Florian Glaser (Hymenoptera: Formicidae).

References

Ashmole NP and Ashmole MJ (1987). Arthropod communities supported by biological fallout on recent lava flows in the Canary Islands. *Entomologia Scandinavica*, Suppl 32:67–88

Ashmole NP and Ashmole MJ (1997). The land fauna of Ascension Island: new data from caves and lava flows, and a reconstruction of the prehistoric ecosystem. *Journal of Biogeography*, 24:549–89

Baars MA (1979). Catches in pitfall traps in relation to mean densities of carabid beetles. *Oecologia*, 41:25–46

Balmford A, Green MJB and Murray MG (1996a). Using higher-taxon richness as a surrogate for species richness: I. Regional tests. *Proceedings of the Royal Society of London B*, 263:1267–74

Balmford A, Jayasuriya AHM and Green MJB (1996b). Using higher-taxon richness as a surrogate for species richness II. Local applications. *Proceedings of the Royal Society of London B*, 263:1571–5

Baur B, Joshi J, Schmid B, Hänggi A, Borcard D, Star´y J, Pedroli-Christen A, Thommen GH, Luka H, Rusterholz HP, Oggier P, Ledergerber S and Erhardt A (1996). Variation in species richness of plants and diverse groups of invertebrates in three calcareous grasslands of the Swiss Jura mountains. *Revue Suisse de Zoologie*, 103:801–33

Birks HJB (1980). The present flora and vegetation of the moraines of the Klutlan glacier, Yukon territory, Canada: A study in plant succession. *Quaternary Research*, 14:60–86

Braun-Blanquet J (1964). *Pflanzensoziologie*. Springer, Wien

Brown VK and Southwood TRE (1987). Secondary succession: patterns and strategies. In Gray AJ, Crawley MJ and Edwards PJ (eds), *Colonization, succession and stability*. Blackwell Scientific, Oxford, pp 295–314

Chernov YI (1995). Diversity of the arctic terrestrial fauna. In Körner C and Chapin III FS (eds), *Arctic and Alpine Biodiversity: Patterns, Causes and Ecosystem Consequences*. Springer, Berlin-Heidelberg, pp 81–95

del Moral R, Titus JH and Cook AM (1995). Early primary succession on Mount St Helens, Washington, USA. *Journal of Vegetation Science*, 6:107–20

Elven R (1980). The Omnsbreen glacier nunataks – a case study of plant immigration. *Norwegian Journal of Botany*, 27:1–16

Erschbamer B, Bitterlich W and Raffl C (1999). Die Vegetation als Indikator für die Bodenbildung im Gletschervorfeld des Rotmoosferners (Obergurgl, Ötztal, Nordtirol). *Ber nat-med Verein Innsbruck*, 86:107–22

Franz H (1969). Besiedlung der jüngst vom Eise freigegebenen Gletschervorfelder und ihrer Böden durch wirbellose Tiere. Neue Forschungen im Umkreis der Glocknergruppe. *Wissenschaftliche Alpenvereinshefte*, 21:291–8

Friedel H (1938). Die Pflanzenbesiedlung im Vorfeld des Hintereisferners. *Zeitschrift für Gletscherkunde*, 26:215–39

Gereben BA (1995). Co-occurrence and microhabitat distribution of six *Nebria* species (Coleoptera: Carabidae) in an alpine glacier retreat zone in the Alps, Austria. *Arctic and Alpine Research*, 27:371–9

Haslett JR (1997). Insect communities and the spatial complexity of mountain habitats. *Global Ecological and Biogeography Letters*, 6:49–56

Helm DJ and Allen EB (1995). Vegetation chrono-sequence near Exit Glacier, Kenai Fjords National Park, Alaska, USA *Arctic and Alpine Research*, **27**:246–57

Janetschek H (1949). Tierische Successionen auf hochalpinem Neuland. *Ber nat-med Verein Innsbruck*, **48/49**:1–215

Janetschek H (1958). Über die tierische Wieder-besiedlung im Hornkees-Vorfeld (Zillertaler Alpen). *Schlern-Schriften*, **188**:209–46

Jochimsen M (1970). *Die Vegetationsentwicklung auf Moränenböden in Abhängigkeit von einigen Umweltfaktoren.* Veröffentlichungen der Univer-sität Innsbruck, Innsbruck

Juen A (1998). *Artenzusammensetzung und Verteilung von Käfern im Gletschervorfeld des Rotmoostales (Ötztaler Alpen, Tirol).* Diploma Thesis, University of Innsbruck, Innsbruck

Kaufmann R (2001). Invertebrate succession on an Alpine glacier foreland. *Ecology*, **82**:2261–78

Körner C (1995). Alpine plant diversity: a global survey and functional interpretations. In Körner C and Chapin III FS (eds), *Arctic and Alpine Biodiversity: Patterns, Causes and Ecosystem Consequences.* Springer, Berlin, Heidelberg, pp 45–62

Körner C (1999). *Alpine Plant Life: functional plant ecology of high mountain ecosystems.* Springer, Berlin, Heidelberg, New York

Lawton JH (1987), Are there assembly rules for suc-cessional communities? In Gray AJ, Crawley MJ and Edwards PJ (eds), *Colonization, succession and stability.* Blackwell Scientific, Oxford, pp 225–44

Lindroth CH, Andersson H, Bödvarsson H and Richter SH (1973). *Surtsey, Iceland. The Develop-ment of a New Fauna, 1963–1970.* Terrestrial Invertebrates. Munksgaard, Copenhagen

Lowrie D (1948). The ecological succession of spiders of the Chicago dune areas. *Ecology*, **29**:334–51

MacArthur RH and MacArthur JW (1961). On bird species diversity. *Ecology*, **42**:594–8

MacMahon JA (1981). Successional processes: com-parison among biomes with special reference to probable roles of and influences on animals. In West DC, Shugart HH and Botkin DB (eds), *Forest succession: concepts and applications.* Springer, Berlin, Heidelberg, New York, pp 277–304

Matthews JA (1992). *The ecology of recently-deglaciated terrain. A geoecological approach to glacier forelands and primary succession.* Cambridge University Press, Cambridge

Matthews JA and Whittaker RJ (1987). Vegetation succession on the Storbreen glacier foreland, Jotunheimen, Norway: a review. *Arctic and Alpine Research*, **19**:385–95

McCoy ED (1990). The distribution of insects along elevational gradients. *Oikos*, **58**:313–22

Meijer J (1989). Sixteen years of fauna invasion and succession in the Lauwerszeepolder. In Majer JD (ed.), *Animals in primary succession. The role of fauna in reclaimed lands.* Cambridge University Press, Cambridge, pp 339–69

Meyer E and Thaler K (1995). Animal diversity at high altitudes in the Austrian Central Alps. In Körner C and Chapin III FS (eds), *Arctic and Alpine Biodiversity: Patterns, Causes and Ecosystem Consequences.* Springer, Berlin, Heidelberg, pp 97–108

Miles J (1987). Vegetation succession: Past and present perceptions. In Gray AJ, Crawley MJ and Edwards PJ (eds), *Colonization, succession and stability.* Blackwell Scientific, Oxford, pp 1–29

Miles J and Walton DWH (1993). Primary succes-sion revisited. In Miles J and Walton DWH (eds), *Primary succession on land.* Blackwell Scientific, Oxford, pp 295–302

Mueller-Dombois D and Ellenberg H (1974). *Aims and methods of vegetation ecology.* Wiley & Sons, New York

Murdoch WW, Evans FC and Peterson CH (1972). Diversity and pattern in plants and insects. *Ecology*, **53**:819–29

Oliver I and Beattie AJ (1996a). Designing a cost-effective invertebrate survey: A test of methods for rapid assessment of biodiversity. *Ecological Applications*, **6**:594–607

Oliver I and Beattie AJ (1996b). Invertebrate morphospecies as surrogates for species: a case study. *Conservation Biology*, **10**:99–109

Paulus U and Paulus HF (1997). Die Zönologie von Spinnen auf dem Gletschervorfeld des Hornkees in den Zillertaler Alpen in Tirol (Österreich) (Arachnida, Araneae). *Ber nat-med Verein Innsbruck*, **80**:227–67

Pearson DL (1994). Selecting indicator taxa for the quantitative assessment of biodiversity. *Proceed-ings of the Royal Society of London* B, **345**:75–9

Pianka ER (1967). Lizard species diversity. *Ecology*, **48**:333–51

Prendergast JR (1997). Species richness covariance in higher taxa: Empirical tests of the biodiversity indicator concept. *Ecography*, **20**:210–16

Raffl C (1999). *Vegetationsgradienten und Sukzes-sionsmuster in einem zentralalpinen Gletschervor-feld (Ötztaler Alpen, Tirol).* Diploma Thesis, University of Innsbruck, Innsbruck

Reiners W, Worley I and Lawrence D (1971), Plant diversity in a chronosequence at Glacier Bay, Alaska. *Ecology*, **52**:55–69

Roy K and Foote M (1997). Morphological approaches to measuring biodiversity. *Trends in Ecology and Evolution*, **12**:277–81

Schowalter TD (1981). Insect herbivore relationship to the state of the host plant: biotic regulation of ecosystem nutrient cycling through ecological succession. *Oikos*, 37:126–30

Schulze ED and Mooney HA (1993). *Biodiversity and Ecosystem Function*. Springer, Berlin-Heidelberg, New York

Siemann E (1998). Experimental tests of effects of plant productivity and diversity on grassland arthropod diversity. *Ecology*, 79:2057–70

Siemann E, Haarstad J and Tilman D (1999). Dynamics of plant and arthropod diversity during old field succession. *Ecography*, 22:406–14

Simberloff D and Dayan T (1991). The guild concept and the structure of ecological communities. *Annual Review of Ecology and Systematics*, 22:115–43

Southwood TRE, Brown VK and Reader PM (1979). The relationship of plant and insect diversities in succession. *Biological Journal of the Linnean Society*, 12:327–48

Sugg PM and Edwards JS (1998). Pioneer aeolian community development on pyroclastic flows after the eruption of Mount St. Helens, Washington, USA. *Arctic and Alpine Research*, 30:400–07

Theodose TA and Bowman WD (1997). Nutrient availability, plant abundance, and species diversity in two alpine tundra communities. *Ecology*, 78:1861–72

Tilman D (1993). Community diversity and succession: the roles of competition, dispersal, and habitat modification. In Schulze ED and Mooney HA (eds), *Biodiversity and Ecosystem Function*. Springer, Berlin, Heidelberg, New York, pp 327–44

Tramer EJ (1975). The regulation of plant species diversity on an early successional old-field. *Ecology*, 56:905–14

Tsuyuzaki S (1996). Species diversity analyzed by density and cover in an early volcanic succession. *Vegetatio*, 122:151–6

Whittaker RH (1960). Vegetation of the Siskiyou Mountains, Oregon and California. *Ecological Monographs*, 30:279–338

Zollitsch B (1969). Die Vegetationsentwicklung im Pasterzenvorfeld. *Wissenschaftliche Alpenvereinshefte*, 21:267–90

Status and Trends in Diversity of Alpine Vertebrates in the Northwestern United States

15

Martin G. Raphael, Michael J. Wisdom and Barbara C. Wales

CURRENT PATTERNS OF VERTEBRATE DIVERSITY

As part of a broad-scale ecological assessment, we examined the distribution of vertebrate species along an elevational gradient from the Columbia River to the surrounding Cascade and northern Rocky Mountains. The area covers a 58-million ha area in the northwestern United States and supports highly diverse terrestrial communities and associated plant and animal species (Figure 15.1). Marcot *et al.* (1997) identified 547 terrestrial vertebrate species known to occur within the study area. Many of these are only occasional visitors; if those species are excluded the revised total is 487 year-round or migratory species that complete all or a major portion of their life cycle in the area. We based subsequent analyses on this revised total, which includes 28 species of amphibians, 27 reptiles, 301 birds, and 131 mammals.

Habitat associations (occurrence within plant community types) of each of these species have been documented in a series of databases (Marcot *et al.*, 1997; Wisdom *et al.*, 2000). Species richness of these vertebrates varies by plant community, with the greatest diversity associated with riparian habitats (Figure 15.2). Richness is similar among plant communities along an elevational gradient from herbaceous through subalpine communities but is much lower in disturbed habitats (communities dominated by agricultural lands or exotic species) and rock-dominated or barren habitats; richness is lowest in the alpine community (Figure 15.2).

Species richness varied with geographic extent of each community (Figure 15.3). We computed a species–area curve $S = CA^z$, where S = species richness, A = land area, and C and z were regression parameters (Connor and McCoy, 1979; Higgs and Usher, 1980; Usher, 1985). In this case, C = 18.69 and z = 0.23 ($R^2 = 0.54$, $p = 0.016$). Alpine habitats supported the fewest species (n = 46) but also extended over the smallest area among the various plant communities (951 km^2). The 46 species occurring in alpine habitat is lower than predicted from the species–area relationship (S = 89).

Despite the relatively low species richness associated with alpine habitats, the set of species occurring in these habitats is unique and thus makes a strong contribution to overall biodiversity in the area. As noted by Higgs and Usher (1980), the number of species represented by the combination of two areas relative to the number of species in each area is a function of the proportion of species in common between the two areas. Accordingly, we used Jaccard's index as a measure of similarity of vertebrate species composition among all plant communities (Pielou, 1984) and performed a hierarchical cluster analysis to display the results. The alpine community had low overlap in vertebrate species composition with all other communities (Table 15.1) and stands out as supporting a distinct vertebrate community (Figure 15.4).

How many species are endemic to alpine habitats in the area? To investigate this question, we tallied the number of plant communities used by each species and contrasted the

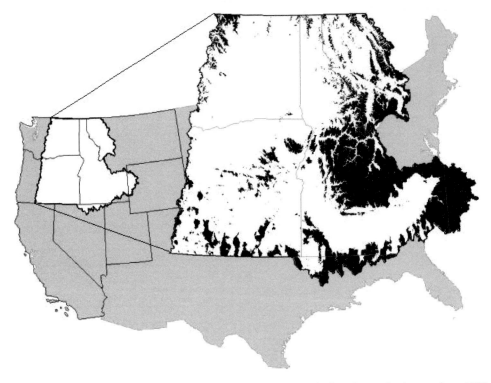

Figure 15.1 Location of the study area in the northwestern USA. Shading denotes land areas above 1700 m asl

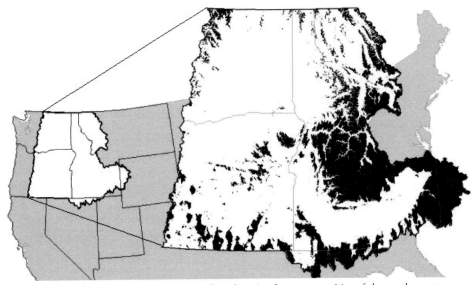

Figure 15.2 Number of species along an elevational gradient in plant communities of the northwestern USA

Erratum: please note the correct image, which should accompany Figure 15.2 is shown below.

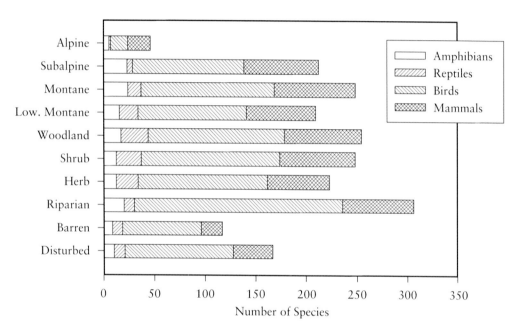

Figure 15.2 Number of species along an elevational gradient in plant communities of the northwestern USA

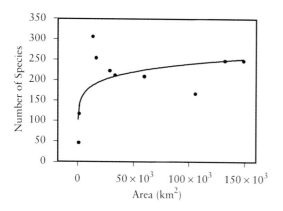

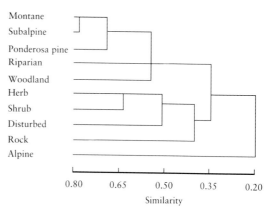

Figure 15.3 Vertebrate species richness in relation to areal extent of plant communities, northwestern USA

Figure 15.4 Dendrogram of similarity in vertebrate species composition among plant communities, north-western USA. Similarity was computed using Jaccard's index (Pielou, 1984, Table 15.1)

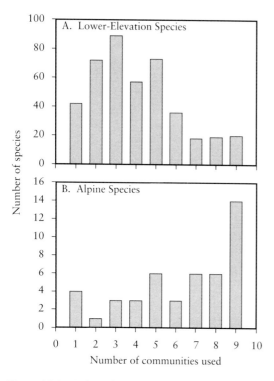

Figure 15.5 Habitat breadth of vertebrates associated with (A) lower-elevation plant communities (n = 441), and (B) alpine-associated vertebrates (n = 46 species). Breadth is portrayed as the number of communities in which a species occurs

resulting frequency distributions between alpine-associated species and the remaining species (Figure 15.5). Alpine species occurred in a greater variety of other plant communities than did lower-elevation species. The median number of communities used by alpine species was seven, whereas the median for lower-elevation species was four; the frequency distributions are clearly more skewed toward greater numbers of plant communities for the alpine species (Figure 15.5). These results indicate that many alpine species use a variety of other plant communities and occur over a wide range of elevation zones. Among the alpine species, we identified 11 that occurred in four or fewer plant communities and four that occurred only in alpine habitats (Table 15.2). The low similarity of species composition in alpine communities in relation to other communities (including adjacent subalpine forest) is due to the occurrence of these four endemics (the first four species listed in Table 15.2) that occur in alpine habitat and nowhere else in this study area.

TEMPORAL TRENDS IN DIVERSITY

Conservation of biological diversity and an evaluation of the role of alpine habitats in conservation, requires consideration of the current status of species as well as their habitat and population trends, both from the past to present and from present to future. Vegetation trends in the area were estimated by Hann et al. (1977) and Hemstrom et al. (2001), who

Table 15.1 Similarity (Jaccard Index) of vertebrate species among plant communities in the northwestern USA

Community	Alpine	Rock	Upland herb	Upland shrub	Upland woodlands	Low montane	Montane	Subalpine	Riparian
Alpine									
Rock	0.132								
Upland herb	0.140	0.435							
Upland shrub	0.114	0.285	0.597						
Upland wood.	0.115	0.208	0.411	0.554					
Low montane	0.128	0.190	0.267	0.325	0.580				
Montane	0.135	0.162	0.233	0.272	0.476	0.725			
Subalpine	0.157	0.192	0.236	0.247	0.421	0.571	0.769		
Riparian	0.107	0.282	0.418	0.417	0.481	0.497	0.543	0.493	
Disturbed[a]	0.092	0.303	0.506	0.446	0.332	0.237	0.193	0.203	0.359

[a]Includes agricultural land and lands dominated by exotic plants

Table 15.2 Number of plant communities used by terrestrial vertebrates associated with alpine habitats in the northwestern USA. Species are listed if they used three or fewer plant community types in addition to alpine plants

Common name	Scientific name	No. of communities used
American pipit	Anthus rubescens	1.00
Black rosy finch	Leucosticte arctoa	1.00
Grey-crowned rosy finch	L. tephrocotis	1.00
White-tailed ptarmigan	Lagopus leucurus	1.00
American pika	Ochotona princeps	2.00
Hoary marmot	Marmota caligata	3.00
Northwestern salamander	Ambystoma macrodactylum	3.00
Wolverine	Gulo gulo	3.00
Marsh wren	Cistothorus palustris	4.00
Northern bog lemming	Synaptomys borealis	4.00
Snowy owl	Nyctea scandiaca	4.00

Table 15.3 Projected amounts of primary habitat within distributional ranges of selected alpine vertebrates in historical, current, and future periods in the northwestern USA. Primary habitat estimates taken from Wisdom et al. (2000) and from databases developed by Raphael et al. (2001). Scientific names in Table 15.2.

Species	Primary habitat (km² × 1000)		
	Historical	Current	Future
Wolverine	88.9	105.6	105.5
Black rosy finch	1.5	1.5	1.5
Grey-crowned rosy finch	2.3	2.3	2.3

described likely historical vegetation conditions and projected the most probable conditions 100 years in the future under proposed land management scenarios. Their results indicate that relatively little broad-scale change has occurred or will occur in alpine habitats, owing to the absence of many intensive management practices such as timber harvesting and road construction. Specifically, cover of alpine communities has remained nearly constant from historical (estimated cover was 953 km² prior to settlement by

Europeans) to current (951 km²) times and is projected to remain so in the future (951 km²) in the absence of any effects of global warming. Global warming could change this pattern and would reduce the extent of alpine habitats (see below). In contrast, cover of seral stages of many other vegetation types has changed dramatically. For example, expansion of agriculture and introduction of exotic plants has caused a 75% reduction in native grasslands from historical to present times. Shrublands have declined by 25% due to the combined effects of agriculture, grazing, and fire exclusion. In forestlands, a dramatic reduction in amount of old and young forest has occurred in relation to intensive timber harvest, fire exclusion, and other major changes in historical disturbance regimes. In the subalpine zone, mature forest currently represents about one-third of its historical extent; in the lower-elevation montane community, mature single-story ponderosa pine (*Pinus ponderosea*) has declined by 60% (USDA, 1996).

To examine temporal trends in specific habitats for alpine species, we examined those species with narrower habitat requirements (from Table 15.2) and selected the subset that was analysed in detail by Wisdom *et al.* (2000). In that analysis, the authors identified all key cover types and seral stages that served as primary habitats for a set of 91 species. These species were selected using a number of criteria, but the primary factor was evidence that the species or its habitat had either declined from historical times or was projected to decline in the future. Three of the species in Table 15.2 were so analysed by Wisdom *et al.* (2000): the two rosy finches, both of which occur only in alpine habitats, and the wolverine, which occurs in both alpine and subalpine habitats (see Table 15.2 for scientific names). For these species, habitat trends appear stable from historical to current to future times (Table 15.3), again assuming no effects of global warming.

Although the broad-scale abundance (geographic extent or estimated amount) of alpine habitats in the area has not changed over the past century, nor is it predicted to change in the

future due to proposed management actions (Hemstrom *et al.*, 2001), the quality of these areas likely has changed, and probably will continue to change, in relation to fine-scale human activities. Alpine habitats are ecologically sensitive and show significant impacts after minor human uses (HaySmith and Hunt, 1995). Negative effects of mining (Brown and Johnston, 1978) and grazing by domestic sheep (Braun, 1980; Paulsen, 1960) have declined substantially with the reduction in these uses in alpine areas, but their long-term effects remain (Hann *et al.*, 1997). Herbicide treatments to increase forage for domestic sheep also have had lasting effects on alpine habitats by changing the composition of vegetation (Smith and Ally, 1966; Thilenius *et al.*, 1974). Non-native fish have been introduced into alpine lakes worldwide, with negative effects on populations of native fish and amphibians (Bahls, 1992; Bradford, 1989; Bradford *et al.*, 1998; Brönmark and Edenham, 1994; Knapp and Matthews, 2000; Tyler *et al.*, 1998). Recreational activities that dominate alpine environments, such as hiking, camping, hunting, horse-back riding, snowmobiling, and skiing, have degraded the land, water, and wildlife resources by simplifying plant communities, increasing animal mortality, displacing and disturbing wildlife, and introducing garbage (Boyle and Samson, 1985; Knight and Gutzwiller, 1995).

Ozone depletion (Blaustein *et al.*, 1994, 1995) and global warming (Bolin *et al.*, 1986) also have the potential to dramatically alter alpine conditions, especially for amphibians. Some species of amphibians, especially pond-breeding frogs, have been shown to suffer reproductive failure due to increased UV exposure, a consequence of ozone depletion (Blaustein *et al.*, 1994, 1995). Global warming may cause upward elevational shifts in distribution of habitats. In the Greater Yellowstone ecosystem of Wyoming, Montana, and Idaho, treeline is predicted to rise an additional 300 m in association with global warming (Romme and Turner, 1991). Such a change would not only decrease habitat abundance but also increase fragmentation, causing further isolation of habitats that are distributed naturally as discontinuous islands along mountain ranges (Fitzgerald *et al.*, 1994).

CONCLUSIONS

We compared and contrasted patterns of species richness, species endemism, and habitat trends of terrestrial vertebrates in alpine versus lower-elevation habitats in the northwestern United States. Species richness was lower in alpine habitats but contained more unique species than most other habitats at lower elevations. Projected trends in alpine habitats indicated little past or future change in the broad-scale extent of these habitats. By contrast, substantial reductions in amount of young and old forest have occurred at lower elevations, and continued change in extent of lower-elevation habitats is projected for the future.

Alpine environments make a unique contribution to overall biodiversity in this alpine area. Although vegetation and habitat trends have changed little over time, increased human activities in alpine environments, and potential effects of global warming, pose a current and future management challenge. Management of recreational activities, combined with potential threats posed by global warming, ozone depletion, mining, and invasions of exotic species, are important factors that will determine the future condition of the alpine environment in the western states.

ACKNOWLEDGEMENTS

We thank Bruce Marcot, Terrell Rich, Mary Rowland, and Richard Holthausen for contributions to the original analyses reported here. Alan Ager, Beth Galleher, and David Hatfield assisted with data analyses. We thank Diane Evans and two anonymous reviewers for suggestions on an earlier draft. Funding from the Pacific Northwest Research Station and the Interior Columbia Basin Ecosystem Management Project supported this work.

References

Bahls PF (1992). The status of fish populations and management of high mountain lakes in the western United States. *Northwest Science*, 66:183–93

Blaustein AR, Edmond B, Keisecker JM, Beatty JJ and Hokit DG (1995). Ambient ultraviolet radiation causes mortality in salamander eggs. *Ecological Applications*, 5:740–3

Blaustein AR, Hoffman PD, Hokit DG, Keisecker JM, Walls SC and Hays JB (1994). UV repair and resistance to solar UV-B in amphibian eggs: a link to population declines? *Proceedings of the National Academy of Sciences* (USA), 91:1791–5

Bolin B, Doos BR, Jager J and Warrick RA (eds) (1986). *The greenhouse effect, climatic change, and ecosystems*. SCOPE Volume 29. John Wiley & Sons, New York

Boyle SA and Samson FB (1985). Effects of nonconsumptive recreation on wildlife: a review. *Wildlife Society Bulletin*, 13:110–16

Bradford DF (1989). Allotropic distribution of native frogs and introduced fishes in high Sierra Nevada lakes of California: implication of the negative effect of fish introductions. *Copeia*, 1989:775–8

Bradford DF, Cooper SD, Jenkins TM, Kratz K, Sarnelle O and Brown AD (1998). Influences of natural acidity and introduced fish on faunal assemblages in California alpine lakes. *Canadian Journal of Fisheries and Aquatic Sciences*, 55:2478–91

Braun CE (1980). Alpine bird communities of western North America: implications for management and research. In de Graff RM and Tilgman NG (compilers), *Workshop proceedings of the management of western forests and grasslands for nongame birds*. US Forest Service General Technical Report, INT No. 86. pp 280–91

Brönmark C and Edenhamn P (1994). Does the presence of fish affect the distribution of tree frogs (*Hyla arborea*)? *Conservation Biology*, 8:841–5

Brown RW and Johnston RW (1978). Rehabilitation of a high elevation mine disturbance. In Kenny ST (ed.), *High altitude revegetation workshop* No. 3. Colorado Water Resource Research Institute, Colorado State University, Fort Collins, Infor. Series 28. pp 116–30

Connor EF and McCoy ED (1979). The statistics and biology of the species–area relationship. *American Naturalist*, 113:791–833

Fitzgerald JP, Meaney CA and Armstrong CA (1994). *Mammals of Colorado*. University Press of Colorado, Niwot, CO. 467 pp

Hann WJ, Jones JL, Karl MG, Hessburg PF, Keane RE, Long DG, Menakis JP, McNicoll CH, Leonard SG, Gravenmier RA and Smith BG (1997). Landscape dynamics of the basin. In Quigley TM and Arbelbide SJ (eds), *An assessment of ecosystem components in the interior Columbia Basin and portions of the Klamath and Great Basins*. Chapter 3. General Technical Report PNW-GTR-405. US Department of Agriculture, Forest Service, Pacific Northwest Research Station, Portland, OR

HaySmith L and Hunt JD (1995). Nature tourism: impacts and management. In Knight RL and Gutzwiller KJ (eds), *Wildlife and Recreationists*. Island Press, Washington DC. pp 203–19

Hemstrom MA, Korol JJ and Hann WJ (2001). Trends in terrestrial plant communities and landscape health indicate the effects of alternative management strategies in the interior Columbia River basin. *Forest Ecology and Management*, **153**:105–26

Higgs AJ and Usher MB (1980). Should nature reserves be large or small? *Nature*, **285**:568–9

Hornocker MG and Hash HS (1981). Ecology of the wolverine in northwestern Montana. *Canadian Journal of Zoology*, **59**:1286–301

Knapp RA and Matthews KR (2000). Non-native fish Introductions and the decline of the mountain yellow-legged Frog from within protected areas. *Conservation Biology*, **4**:428–38

Knight RL and Gutzwiller KJ (eds) (1995). *Wildlife and recreationists, coexistence through management and research*. Island Press, Washington, DC, USA

Lehmkuhl JF, Raphael MG, Holthausen RS, Hickenbottom JR, Naney RH and Shelly JS (1997). Historical and current status of terrestrial species and the effects of the proposed alternatives. In Quigley TM, Lee KM and Arbelbide SJ (eds), *Evaluation of EIS alternatives by the Science Integration Team*. General Technical Report PNW-GTR-406. US Department of Agriculture, Forest Service, Pacific Northwest Research Station, Portland, OR. pp 537–730

Marcot BG, Castellano MA, Christy JA, Croft LK, Lehmkuhl JF, Naney RH, Rosentreter RE, Sandquist RE and Zieroth E (1997). Terrestrial ecology assessment. In Quigley TM and Arbelbide SJ (eds), *An assessment of ecosystem components in the interior Columbia Basin and portions of the Klamath and Great Basins*, Volume III. General Technical Report PNW-GTR-405. US Department of Agriculture, Forest Service, Pacific Northwest Research Station, Portland, OR. pp 1497–713

Paulsen HH Jr (1960). Plant cover and foliage use on alpine sheep ranges in the central Rocky Mountains. *Iowa State College Journal of Science*, **34**:731–48

Pielou EC (1984). *The interpretation of ecological data: a primer on classification and ordination*. John Wiley and Sons, New York

Raphael MG, Marcot BG, Holthausen RS and Wisdom MJ (1998). Terrestrial species and habitats. *Journal of Forestry*, **96**:22–7

Raphael MG, Wisdom MJ, Holthausen RS, Wales BC, Marcot BG and Rich TD (2001). Status and Trends of Habitats of Terrestrial Vertebrates in Relation to Land Management in the Interior Columbia River Basin. *Forest Ecology and Management*, **153**:63–68

Romme WH and Turner MG (1991). Implications of global climatic change for biogeographic patterned in the Greater Yellowstone Ecosystem. *Conservation Biology*, **5**:373–86

Smith DR and Ally HP (1966). Chemical control of alpine avens. *Journal of Range Management*, **19**:376–8

Thilenius JF, Smith DR and Brown GR (1974). Effects of 2,4-D on composition and production of an alpine plant community in Wyoming. *Journal of Range Management*, **27**:140–2

Tyler T, Liss WJ, Ganio LM, Larson GL, Hoffman R, Deimling E and Lomnicky G (1998). Interaction between introduced trout and larval salamanders (*Ambystoma macrodactylum*) in high-elevation lakes. *Conservation Biology*, **12**:94–105

Usher MB (1985). Implications of species–area relationships for wildlife conservation. *Journal of Environmental Management*, **21**:181–91

US Department of Agriculture (USDA) Forest Service (1996). *Status of the Interior Columbia Basin*. General Technical Report PNW-GTR-385. Pacific Northwest Research Station, Portland, OR

US Department of Agriculture, Forest Service, US Department of the Interior, Bureau of Land Management (2000). *Interior Columbia Basin supplemental draft environmental impact statement*. BLM/OR/WA/Pt-00/019+1792. US Department of the Interior, Bureau of Land Management, Portland, OR. [irregular pagination]. Available on-line at http:\\www.icbemp.gov

Wisdom MJ, Holthausen RS, Wales BC, Hargis CD, Saab VA, Lee DC, Hann WJ, Rich TD, Rowland MM, Murphy WJ and Eames MR (2000). *Source habitats for terrestrial vertebrates of focus in the Interior Columbia Basin: broad-scale trends and management implications*. General Technical Report, PNW-GTR-485. US Department of Agriculture, Forest Service, Pacific Northwest Research Station, Portland, OR

Biodiversity of Human Populations in Mountain Environments

16

Cynthia M. Beall

INTRODUCTION

The human dimensions of global mountain biodiversity include well-known aspects such as human impacts on biodiversity (dealt with elsewhere in this volume) and surprisingly poorly known aspects such as the diversity of our own species. Perhaps because human activities are central to biodiversity losses, there is a tendency to consider humans primarily as actors upon their ecosystems rather than as integral components subject to environmental influence. However, humans evolve and adapt to their environments and therefore another human dimension of global mountain biodiversity is the diversity of the human species itself. This Chapter focuses on biodiversity among humans in mountain ecosystems. Because chronic, life-long hypoxia is a stressor unique to mountain ecosystems, the Chapter focuses on human adaptation to high-altitude hypoxia, rather than other features of mountain ecosystems including cold and ultraviolet radiation. High-altitude hypoxia can significantly affect the success of human populations by influencing health, physical work capacity, and reproductive success. Numerous physiological homeostatic responses are involved in offsetting hypoxia, however, humans at high altitude exhibit qualitative and quantitative variation in these responses. This Chapter provides information on variation in some of those physiological responses, their consequences, causes and significance.

Human population size and density decreases with increasing altitude. Some 389.4 million people live above 1500 m altitude, at a population density of about 31 people per 1 km^2 of ice-free land. Just 94.3 million people live above 2500 m at a population density of 18.5 people

per km^2 and only 22.4 million live above 3900 m at a population density of about 8.2 people per km^2 of ice-free land (Cohen and Small, 1998; pers. comm., May 1999). Barometric pressure decreases with increasing altitude and consequently fewer oxygen molecules are inspired into the lungs per volume of air and fewer are available for diffusion into the bloodstream for transport to cells. Animals at high-altitude must adapt to this unavoidable, life-long stress of limited oxygen availability relative to sea level, called high-altitude or hypobaric hypoxia, and still sustain aerobic metabolism. The stress is substantial. At 1500 m the partial pressure of oxygen (i.e. the constant 20.93% concentration of oxygen multiplied by the barometric pressure) is 84% of sea level, at 2500 m the partial pressure of oxygen falls to 75% and by 3900 m is down to 63% of the sea level value.[1]

The 94 + million people residing above 2500 m (an altitude often used as an arbitrary boundary between low and high altitude) manifest an unexpectedly wide range of variation in biological traits thought to offset hypoxia and ensure adequate delivery of oxygen. The people and their variation can be described in terms of three dimensions: time, vertical, and horizontal (Figure 16.1, based on Körner, pers. comm., September 2000). The time dimension

---

1 The partial pressure of oxygen at a given altitude is slightly higher at the equator. The difference is large enough at altitudes above 6000 m that it can influence the performance of mountaineers such as those seeking to summit Mt Everest at 8828 m (Ward *et al.*, 2000). In the latitude range where people live above 2500 m, the latitudinal difference in barometric pressure is similar in magnitude to seasonal variation in one locale.

2 Portions of this manuscript are based on Beall CM (2000), *Annual Review of Anthropology*, 16:423–55.

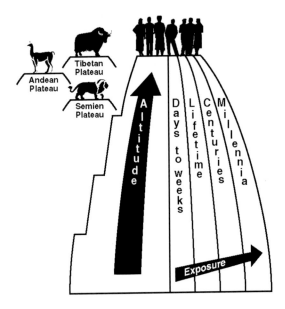

Figure 16.1 Human dimensions of global mountain biodiversity: time, vertical altitude transect, and horizontal continental contrast

compares human populations with different lengths of exposure to high-altitude hypoxia. The time scale ranges from days to weeks of exposure (sojourners such as workers and tourists travelling to high altitudes), to a lifetime of residence (upward migrants such as the Chinese Han on the Tibetan Plateau), a few centuries of residence (such as European settlers on the Andean Plateau or in the Rocky Mountains), and to millennia of residence under hypoxic stress (such as indigenous populations of Andean and Tibetan highlanders). The vertical dimension compares populations living along altitude transects in one mountain system, for example from sea level on the Pacific coast of South America to 4500 m in the Andes. The scale of this dimension extends as high 5000–5500 m for a few indigenous populations and as high as 8828 m for a few, brief, sojourners at the summit of Mt Everest. The horizontal dimension compares indigenous populations living in different mountain systems on different continents such as the Andean, Tibetan and Semien (East African) Plateaus.

There is an evolutionary logic to these dimensions. The time dimension may reveal that populations with brief exposure adapt qualitatively

or quantitatively differently, perhaps less effectively or less efficiently, from populations with millennia of exposure and opportunity for natural selection to improve adaptation. The vertical dimension may detect a dose–response relationship between ambient hypoxia and physiological responses that can identify useful traits for study. The horizontal dimension allows testing of the hypothesis that natural selection has acted on the same traits and with the same outcomes in geographically distant populations adapting to the same environmental stress.

OXYGEN SATURATION OF ARTERIAL HAEMOGLOBIN AT HIGH ALTITUDE

A measure of the physiological stress of high-altitude hypoxia is oxygen saturation of haemoglobin, the percent of arterial haemoglobin that carries oxygen. Oxygen saturation falls abruptly upon exposure to high altitude and then increases somewhat over a three-week sojourn, although it does not return to low-altitude baseline. For example, the oxygen saturation of a group of young men transported from sea level to 4300 m on Pikes' Peak, CO in the US fell significantly to 81%, from 97% at sea level, and then gradually increased significantly to 87% after three weeks of residence but remained significantly below sea level (Wolfel *et al.*, 1991). These data illustrate some of the initial homeostatic responses of a sea-level native population whose prior exposure to hypoxia consists of brief, transient episodes owing to reversible factors, including pulmonary sepsis, sleep, and anaemia. Data describing North Americans and Europeans resident at high altitude illustrates the response of humans with longer exposure to high altitude. Length of residence at the higher altitudes could range from a portion of an individual lifetime to a few generations because those populations have recent, known low-altitude ancestors. Figure 16.2 presents the mean oxygen saturations of samples of healthy adult North American and European residents measured at many altitudes. Combining the results of many studies yields a description of oxygen saturation

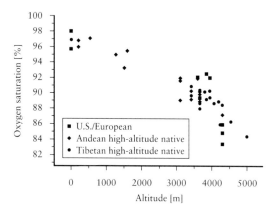

Figure 16.2 Published mean values of percent of oxygen saturation of arterial haemoglobin at various altitudes. The figure is based on the means of 44 samples of ten or more people between 15 and 49 years of age, measured with pulse oximetry and reported in the following references: Antezana *et al.* (1994); Banchero *et al.* (1966); Beall and Goldstein (1990); Beall *et al.* (1992, 1997, 1999); Beall (2000); Brutsaert (2000); Chen *et al.* (1997); Chronos *et al.* (1988); Curran *et al.* (1995); Decker *et al.* (1989); Ge *et al.* (1995); Huang *et al.* (1984, 1992); Hultgren *et al.* (1965); Jansen *et al.* (1999); Kryger et al. (1978); Lawler *et al.* (1988); Leon-Velarde *et al.* (1991, 1994); Marcus *et al.* (1994); Martin *et al.* (1989); Miyachi, 1992; Moore *et al.* (1982a,b); Moore (1990); Reeves (1993); Schlaepher *et al.* (1992); Sun *et al.* (1996); Tucker *et al.* (1984); Weil *et al.* (1968); Zhuang *et al.* (1993, 1996)

across a vertical transect and illustrates a gradual decrease in oxygen saturation that corresponds to the decrease in partial pressure of inspired oxygen at higher altitudes. Figure 16.2 provides information on a still longer time perspective by including the mean oxygen saturations of samples of indigenous populations living along an altitude transect in the Andes, where resident populations have been exposed to the opportunity for natural selection for the past 11 000 years or so (Dillehay, 1999). Figure 16.2 further extends the comparison in the horizontal dimension by presenting the mean oxygen saturations of samples of indigenous populations from the Tibetan Plateau, where resident populations have been exposed to the opportunity for natural selection for the past 7000 years or so (Chang, 1992). A comparative study found that Andean highlanders had significantly higher oxygen saturation than their Tibetan counterparts at ~ 4000 m (Beall *et al.*, 1997, 1999). More explicitly,

comparative studies are necessary to determine whether there is a consistent geographic population difference in oxygen saturation.

Figure 2 illustrates that oxygen saturation decreases with increasing altitude and that there is considerable variation in oxygen saturation at any one altitude, both within a single population such as Andean highlanders and between populations such as Andean and Tibetan highlanders. Oxygen saturation is used to quantify the degree of physiological stress at high altitude and these data reflect variation in that physiological stress despite uniform ambient stress. Determining the causes of this variation is an important area for future investigation.

Haemoglobin concentration at high altitude

Arterial oxygen content is a function of both oxygen saturation and haemoglobin concentration. One hypothetical way to offset the arterial hypoxia described above is to increase haemoglobin concentration. The young men described above did not exhibit any changes in haemoglobin concentration upon arrival at Pike's Peak, but after three weeks of residence their haemoglobin concentrations had increased significantly to 15.4 gm dl^{-1} from 13.6 gm dl^{-1} at sea level and their sea level oxygen content had been restored (Wolfel *et al.*, 1991). Extending the time dimension to lifetimes reveals that mean haemoglobin concentration of residents increases with altitude in the US (Figure 16.3). The longer time scale provided by indigenous populations also extends to higher altitudes the general trend toward higher haemoglobin concentration with altitude. Considering the horizontal dimension, the long-term US/European residents and the Andean highlanders have significantly higher mean haemoglobin concentrations than Tibetans. This was revealed by analyses of samples above 3000 m, where all three populations have been studied. An analysis of covariance on the means of the 53 samples above 3000 m, controlled for the effect of altitude and tested for the effect of population on haemoglobin concentration. At a mean altitude of 3859 m the estimated mean haemoglobin concentration

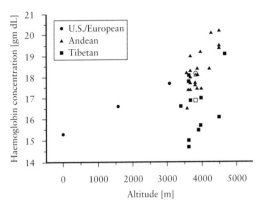

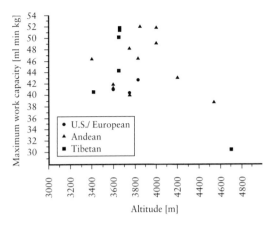

Figure 16.3 Published mean values of haemoglobin concentration at various altitudes. The figure is based on the means of 56 samples of ten or more men between 15 and 49 years of age reported in the following references: Adams and Strang (1975); Arnaud *et al.* (1981); Banchero *et al.* (1966); Beall *et al.* (1987, 1990, 1998); Beall and Goldstein (1990); Bharadwaj (1973); Brutsaert (2000); Cudkowicz *et al.* (1972); Frisancho (1975); Garruto (1976, 1983); Ge *et al.* (1994, 1995); Guleria *et al.* (1971); Huang *et al.* (1992); Hultgren (1965); Leon-Velarde *et al.* (1991); Mazess (1969); Moret *et al.* (1972); Okin *et al.* (1966); Penaloza *et al.* (1963); Ruiz (1973); Samaja (1979); Santolaya (1981); Schoene *et al.* (1990); Spielvogel (1997); Sun *et al.* (1990); Tarazona-Santos (2000); Tufts *et al.* (1985); Vincent *et al.* (1978); Winslow *et al.* (1981, 1988, 1990); Zhuang *et al.* (1993, 1996)

Figure 16.4 Published mean values of maximal physical work capacity VO$_{2max}$ at various altitudes. The figure is based on the means of 20 samples of ten or more men between 15 and 49 years of age reported in the following references: Baker (1969); Brutsaert (1999); Brutsaert (2000); Frisancho *et al.* (1973, 1995); Ge *et al.* (1994, 1995); Hochachka *et al.* (1991); Huang *et al.* (1992); Hurtado (1964); Kollias (1968); Maresh *et al.* (1983); Mazess (1969); Spielvogel *et al.* (1996); Sun *et al.* (1990); Weitz (1984)

of the US/European samples was 18.2 gm dl^{-1} and of the Andean samples was 18.1 gm dl^{-1} as compared with 16.9 gm dl^{-1} for Tibetan samples (F$_{population}$ = 9.0, $p < 0.05$). At the same time, there is a very wide range of variation within a population at a given altitude that needs to be addressed in future studies. For example, urban samples tend to have higher haemoglobin concentrations than rural samples (Beall *et al.*, 1990), although it is not clear why.

Physical work capacity at high altitude

The functional consequences of variation in traits such as oxygen saturation and haemoglobin concentration have been addressed using physical work capacity. Physical work capacity, measured as maximal aerobic capacity (VO$_{2max}$), has been used as an integrated measure of the ability to deliver oxygen to working muscle. It was likely an important factor in the success of early colonising populations that probably arrived on foot and subsisted by dint of their own muscular efforts. Physical work capacity decreased significantly in a group of young men exposed to Pikes' Peak. It fell significantly to 69% of their sea level baselines on the day of arrival and recovered significantly, but only to 76% of sea level by the fifteenth day of residence (Horstman *et al.*, 1980).

There is relatively little information about physical work capacity of residents at various altitudes in the US. More information is available comparing European or Chinese Han migrants, with years of residence at high altitude, to Andean and Tibetan highlanders. However, the findings are not consistent. Some find that upward migrants can attain the same work capacity as natives (Brutsaert *et al.*, 1999), while others do not (Frisancho *et al.*, 1995; Sun *et al.*, 1990). Physical work capacity is substantially influenced by level of habitual physical activity and physical conditioning. Those confounding factors may contribute to the difficulty in determining the influence of the time dimension on physical work capacity.

On the horizontal dimension of population comparison, Figure 16.4 illustrates that, despite the possible contrast in oxygen saturation and significant contrast in haemoglobin concentration, there is no consistent Andean–Tibetan contrast in physical work capacity measured in men across a range of altitudes. While there are too few samples to provide acceptable statistical power to conduct a formal statistical analysis, comparing the three samples, the data in Figure 16.4 suggest that the two patterns are equally effective at oxygen delivery to working muscle. As with other measures, substantial variation within and between populations remains to be explained.

REPRODUCTION AT HIGH ALTITUDE

High altitude populations sustain themselves by reproduction, there is, however, evidence of stress. Birth weight is an index of the effectiveness of maternal oxygen delivery to the foetus. A decrease in average birth weight with increasing altitude has been extensively documented in the US (Figure 16.5). Birth weight is a classic example of normalising selection in human populations because both low and high birth weights are associated with higher infant mortality. The lower birth weights at higher altitudes in the US were associated historically with elevated infant mortality rates, however, modern hospital practices have fully compensated for the lower birth weights (Unger *et al.*, 1988). Figure 16.5 also illustrates that US populations have relatively low birth weights at high altitudes compared with their Andean and Tibetan counterparts with longer residence. Indigenous women appear to deliver more oxygen to their foetuses. Figure 16.5 further illustrates no consistent difference between Andean and Tibetan birth weights, indicating that, despite the population differences in saturation and haemoglobin concentration, the two populations are equally effective at delivering oxygen to the foetus. An analysis of covariance of the 28 samples above

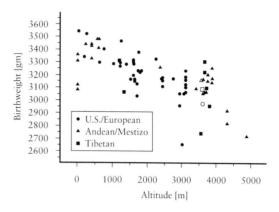

Figure 16.5 Published mean values of birth weight at various altitudes. The figure is based on the means of 86 samples of ten or more newborns reported in the following references: Beall (1981); Cotton *et al.* (1980); Grahn and Kratchman (1963); Guerra-Garcia (1965, 1969); Haas *et al.* (1977, 1989); Haas (1980); Jensen and Moore (1997); Kashiwazaki (1988); Khalid (1997); Leibson *et al.* (1989); Lichty *et al.* (1957); Lubchenco *et al.* (1963); Macedo-dianderas (1966); McClung (1969); McCullough *et al.* (1977); Moore *et al.* (1982, 1984, 1986); Moore (1982, 1990); Mortola (2000); Niermeyer *et al.* (1993, 1995); Palacios-V (1964); Saco-Pollitt (1981); Smith (1997); Thilo *et al.* (1991); Unger *et al.* (1988); Usher and McLean (1969); Weinstein and Haas (1977); Wiley (1994); Yancey *et al.* (1992); Yip (1987); Zamudio *et al.* (1993)

3000 m, where all three populations have been studied, controlled for altitude and tested for the effects of population of origin ($F_{population} = 4.2$, $p < 0.05$). At a mean altitude of 3896 m, the US/European samples' estimated mean birth weight of 2975 gm was significantly lower than the 3162 gm estimated for the Andean samples. The Andean estimated mean did not differ from the Tibetan samples' estimated mean of 3090 gm. The substantial variation within any one population at a given altitude suggests that other factors are operating to influence birth weights.

Another indicator of reproductive stress is the finding that pregnant women at 3100 m in the US have a five-fold higher incidence of pre-eclampsia than women at 1260 m (Palmer *et al.*, 1999). Pre-eclampsia is a condition of late pregnancy characterised by high blood pressure, fluid accumulation and protein in the urine. It is a risk factor for maternal mortality.

CAUSES OF HUMAN BIODIVERSITY AT HIGH ALTITUDE

A variety of possible proximate causes for the population contrasts have been investigated, particularly health and nutritional status. Individual studies controlling for these by study design that excludes ill people or that excludes people with iron deficiency still find the same general trends. Similarly, environmental differences other than hypoxia seem unlikely to explain the differences across the time dimension or the horizontal dimension (see Beall, 2000; Moore *et al.*, 1998 for reviews).

An evolutionary perspective requires considering the hypothesis that some of the variation is heritable. The hypothesis of genetic differences has been difficult to investigate because these are complex traits governed by unknown loci and alleles as well as under the influence of age and sex. One comparative study applied quantitative genetic techniques to begin to analyse genetic factors underlying oxygen saturation and haemoglobin concentration. That comparative study of Andean and Tibetan highlanders at 4000 m found no significant genetic variance in oxygen saturation in the Andean sample, while there was significant genetic variance in the Tibetan sample (Beall *et al.*, 1997, 1999). Evidence for a major gene (an inferred allele with a large quantitative effect at a segregating autosomal locus) was detected in the Tibetan sample. Those inferred major gene is an autosomal dominant allele for 5–6% higher oxygen saturation. The findings provide indirect evidence that the Tibetan sample may have oxygen saturation alleles not present in the Andean sample. Furthermore, because genetic variance is necessary for natural selection to act, one could hypothesise that natural selection may be operating on oxygen saturation in the Tibetan population, but not the Andean. Both Andean and Tibetan samples exhibit high genetic variance of haemoglobin concentration (Beall *et al.*, 1999) indicating there is potential for natural selection to operate on that trait in both populations.

Two different models have been proposed to explain the horizontal contrast. One hypothesises that Tibetans are better adapted to high-altitude hypoxia than Andean highlanders because Tibetans have been living at altitude for longer (Moore *et al.*, 1992). Another hypothesises that both are well adapted using different combinations of traits that resulted from chance differences in the gene pools of the founding populations (Beall, 2000). There is insufficient evidence on the length of habitation and the past or present gene pools of the two plateaus to evaluate this further. There is evidence of habitation of the Andean plateau by hunters and gatherers after 10 000 BP (Dillehay, 1999). However, the earliest chronometrically dated site on the Tibetan Plateau is a village site at an estimated date of 7000 BP (Chang, 1992). It is not known whether there were earlier hunters and gatherer or nomads residing on the Tibetan plateau. Neither is it known whether present-day residents are descended from those early residents.

Human biodiversity and mountain sickness

The preceding discussion dealt with healthy individuals, however, a couple of illnesses are distinctive to high-altitude locales and illustrate another facet of human biodiversity. The illnesses are acute mountain sickness (AMS) and chronic mountain sickness (CMS). They differ in the time dimension as well as other features. With respect to AMS, some sojourners at high-altitudes acclimatise successfully, some have transient symptoms of AMS and some develop life-threatening adverse responses. AMS is usually diagnosed by the presence and severity of symptoms of headaches, anorexia or nausea, dizziness, or shortness of breath that occur in healthy people during the first few days at altitude. The symptoms usually resolve after five days at altitude (Ward *et al.*, 2000). The vertical dimension is involved because sojourners' risk of acute mountain sickness increases with altitude. A study of climbers found that AMS increased with altitude from 9% at 2850 m to 13% at 3050 m, to 34% at 3650 m and 53% at 4559 m (Maggiorini et al., 1990).

With respect to CMS, individuals with long exposure who were apparently adjusted successfully to their high-altitude hypoxic environment may become maladjusted. CMS is characterised by pathologically low oxygen saturation and very high haemoglobin concentration (Ward et al., 2000). Considering the horizontal dimension reveals that it is known among high-altitude residents of the US (Kryger et al., 1989) and the Andes, but is very rare among Tibetans (Ward et al., 2000). The symptoms can be relieved by moving to low altitude. Identifying the presently unknown factors influencing who does or does not develop these two illnesses is an important area for further study.

Significance of human biodiversity at high altitudes

The time and horizontal dimension loom important for understanding human biodiversity at high altitudes. Indigenous populations seem to have better functional capacity in terms of traits that might influence long-term survival in high mountain ecosystems. This suggests that it would take extensive biomedical intervention to replace any high-altitude adapted population without suffering a loss of fitness and function. Yet, the wide range of variation within and between the indigenous Andean and Tibetan populations raises questions about the range of possible variation and combinations of traits enabling successful life at high altitude.

IMPLICATIONS OF HUMAN BIODIVERSITY FOR THE STUDY OF GLOBAL MOUNTAIN BIODIVERSITY

Mountain ecosystems are thought to be especially informative for studies of plant biodiversity because of the rapid change in environment over short distances (Barthlott et al., 1996). Mountain ecosystems could prove to be informative for studies of animal biodiversity too, because the animals depend upon the plants and because the animals must adapt to the environmental stresses of cold and ultraviolet radiation as well as the unavoidable hypoxic stress. Therefore, it is vital to conserve high-altitude species and variants of species for studies integrating molecular, physiological, and phylogenetic aspects of mountain biodiversity. Plans to conserve the high-altitude flora and fauna will need to take into consideration the population dynamics and environmental exploitation practices of the high- altitude human populations.

The data on human biodiversity presented above raise the possibility of similar diversity within other species that have colonised more than one mountain ecosystem. For example, findings that different strains of laboratory rodents exhibit quantitatively different haematological and respiratory traits under hypoxic stress (Ou et al., 1992) parallel the findings in humans and suggest that investigation of wild rodent species at high altitude could be informative.

CONCLUSION

In summary, investigating the origins of biodiversity in a single, well-known, globally distributed species – humans – has the potential to offer insight into the workings of evolution and adaptation. It may be unexpected that studies of humans, a species with a long lifespan, great mobility and no likelihood of reproductive isolation and speciation, can provide insight into biodiversity. However, understanding diversity within a species is an informative and essential level of investigation for studies of biodiversity because, ultimately, phenotypic and genetic variation within species is the necessary raw material for the evolution of species and ecosystems.

References

Adams WH and Strang LJ (1975). Hemoglobin in persons of Tibetan ancestry living at high altitude (38952). *Proceedings of the Society for Experimental Biology*, 149:1036–9

Antezana A-M, Kacimi R, Le Trong J-L, Marchal M, Abousahl I, Dubray C and Richalet J-P (1994). Adrenergic status of humans during prolonged exposure to the altitude of 6,542 m. *Journal of Applied Physiology*, 76:1055–9

Arnaud J, Quilici JC and Riviere G (1981). High-altitude haemotology: Quechua-Aymara comparisons. *Annals of Human Biology*, 8:573–8

Baker PT (1969) Human adaptation to high altitude. *Science*, 163:1149–56

Banchero N, Sime F, Penaloza D, Cruz J, Gamboa R and Marticorena E (1966). Pulmonary pressure, cardiac output, and arterial oxygen saturation during exercise at high altitude and at sea level. *Circulation*, 33:249–62

Barthlott W, Lauer W and Placke A (1996). Global distribution of species diversity in vascular plants: towards a world map of phytodiversity. *Erdkunde*, 50:317–27

Beall CM (1981). Optimal birthweights in Peruvian populations at high and low altitudes. *American Journal of Physical Anthropology*, 56:209–16

Beall CM (2000). Oxygen saturation increases during childhood and decreases during adulthood among high altitude Tibetans residing at 3800–4200 m. *High Altitude Medicine and Biology*, 1:25–32

Beall CM, Almasy LA, Blangero J, Williams-Blangero S, Brittenham GM, Strohl KP, Decker M, Vargas E, Villena M, Soria R, Alarcon A and Gonzales C (1999). Percent of oxygen saturation of arterial hemoglobin of Bolivian Aymara at 3900–4000 m. *American Journal of Physical Anthropology*, 108:41–51

Beall CM, Brittenham GM, Macuaga F and Barragan M (1990). Variation in hemoglobin concentration among samples of high-altitude natives in the Andes and the Himalayas. *American Journal of Human Biology*, 2:639–51

Beall CM, Brittenham GM, Strohl KP, Blangero J, Williams-Blangero S, Goldstein MC, Decker MJ, Vargas E, Villena M, Soria R, Alarcon AM and Gonzales C (1998). Hemoglobin concentration of high-altitude Tibetans and Bolivian Aymara. *American Journal of Physical Anthropology*, 106:385–400

Beall CM and Goldstein MC (1990). Hemoglobin concentration, percent oxygen saturation and arterial oxygen content of Tibetan nomads at 4850 to 5450 m. In Sutton JR, Coates G and Remmers JE (eds), *Hypoxia: The Adaptations*. BC Decker, Inc, Toronto, pp 59–65

Beall CM and Goldstein MC, The Tibetan Academy of Social Sciences (1987). Hemoglobin concentration of pastoral nomads permanently resident at 4850-5450 metres in Tibet. *American Journal of Physical Anthropology*, 73:433–8

Beall CM, Strohl KP, Gothe B, Brittenham GM, Barragan M and Vargas E (1992). Respiratory and hematological adaptations of young and older Aymara men native to 3600 m. *American Journal of Human Biology*, 4:17–26

Beall CM, Strohl KP, Blangero J, Williams-Blangero J, Brittenham GM and Goldstein MC (1997). Quantitative genetic analysis of arterial oxygen saturation in Tibetan highlanders. *Human Biology*, 69:597–604

Bharadwaj H, Singh AP and Malhotra MS (1973). Body composition of the high-altitude natives of Ladakh: A comparison with sea-level residents. *Human Biology*, 45:423–34

Brutsaert TD, Spielvogel H, Soria R, Caceres E, Buzenet G and Haas JD (1999). Effect of developmental and ancestral high-altitude exposure to VO_2 peak of Andean and European/North American natives. *American Journal of Physical Anthropology*, 110:435–55

Brutsaert TD, Araoz M, Soria R, Spielvogel H and Haas JE (2000). Higher arterial oxygen saturation during submaximal exercise in Bolivian Aymara compared to European Sojourners and Europeans born and raised at high altitude. *American Journal of Physical Anthropology*, 113:169–82

Chang K-C (1992). China. In Ehrich RW (ed.), *Chronologies in Old World Archaeology*. The University of Chicago Press, Chicago and London pp 409–15 (vol. 1), 385–404 (vol. 2)

Chen Q-H, Ge R-L, Wang X-Z, Chen H-X, Wu T-Y, Kobayashi T and Yoshimura K (1997). Exercise performance of Tibetan and Han adolescents at altitudes of 3417 and 4300 m. *Journal of Applied Physiology*, 83:661–7

Chronos N, Adams L and Guz A (1988). Effect of hyperoxia and hypoxia on exercise-induced breathlessness in normal subjects. *Clinical Science*, 74:531–7

Cohen JE and Small C (1998). Hypsographic demography: The distribution of human population by altitude. *Proceedings of the National Academy of Sciences*, 95:14009–014

Cotton FK, Hiestand M, Philbin GE and Simmons M (1980). Re-evaluation of birth weights at high altitude. *American Journal of Obstetrics and Gynecology*, 138:220–2

Cudkowicz L, Spielvogel H and Zubieta G (1972). Respiratory studies in women at high altitude (3600 m or 12 200 ft and 5200 m or 17 200 ft). *Respiration*, 29:393–426

Curran LS, Zhuang J, Droma T, Land L and Moore LG (1995). Hypoxic ventilatory responses in Tibetan residents of 4400 m compared with 3658 m. *Respiratory Physiology*, 100:223–30

Decker MJ, Hoekje PL and Strohl KP (1989). Ambulatory monitoring of arterial oxygen saturation. *Chest*, 95:717–22

Dillehay TD (1999). The Late Pleiostocene Cultures of South America. *Evolutionary Anthropology*, 7:206–16

Frisancho AR (1975). Functional adaptation to high altitude hypoxia. *Science*, 187:313–19

Frisancho AR, Martinez C, Velasquez T, Sanchez J and Montoye H (1973). Influence of developmental adaptation on aerobic capacity at high altitude. *Journal of Applied Physiology*, 34:176–80

Frisancho AR, Frisancho HG, Milotich M, Brutsaert T, Albalak R, Spielvogel H, Villena M, Vargal E and Soria R (1995). Developmental, genetic, and environmental components of aerobic capacity at high altitude. *American Journal of Physical Anthropology*, 96:431–42

Garruto RM (1976). Hematology. In Baker PT, Little MA (eds), *Man in the Andes. A Multidisciplinary Study of High-Altitude Quechua.* Dowden, Hutchinson, & Ross, Stroudsburg, PA, pp 261–82

Garruto RM and Dutt JS (983). Lack of prominent compensatory polycythemia in traditional native Andeans living at 4200 metres. *American Journal of Physical Anthropology*, 61:355–66

Ge R-L, Chen Q-H, Wang L-H, Gen D, Ping Yang, Keishi K, Fujimoto K, Matsuzawa Y, Yoshimura K, Takeoka M and Kobayashi T (1994). Higher exercise performance and lower VO$_{2max}$ in Tibetan than Han residents at 4700 m altitude. *American Journal of Applied Physiology*, 77:684–91

Ge R, He Lun G, Chen Q, Li HL, Gen D, Kubo K, Matsuzawa Y, Fujimoto K, Yoshimura K, Takeoka M and Kobayashi T (1995). Comparisons of oxygen transport between Tibetan and Han residents at moderate altitude. *Wilderness and Environmental Medicine*, 6:391–400

Grahn D and Kratchman J (1963). Variation in neonatal death rate and birth weight in the United States and possible relations to environmental radiation, geology and altitude. *Am Journal of Human Genetics*, 15:329–52

Guerra-García R, Velásquez A and Whittembury J (1965). Urinary testosterone in high altitude natives. *Steroids*, 6:351–5

Guerra-García R, Velásquez A and Coyotupa J (1969). A test of endocrine gonadal function in men: urinary testosterone after the injection of HCG. II. A different response of the high altitude native. *Journal of Clinical Endocrinology*, 29:179–82

Guleria JS, Pande JN, Sethi PK and Roy SB (1971). Pulmonary diffusing capacity at high altitude. *Journal of Applied Physiology*, 31:536–43

Haas JD (1980). Maternal adaptation and fetal growth at high altitude in Bolivia. In Greene LS and Johnston FE (eds), *Social and Biological Predictors of Nutritional Status, Physical Growth, and Neurological Development.* Academic Press, New York, London, pp 257–90

Haas JD, Baker PT and Hunt EE (1977). The effects of high altitude on body size and composition of the newborn infant in southern Peru. *Human Biology*, 49:611–28

Haas JD, Conlisk EA and Frongillo J (1989). *Fetal growth and neonatal mortality at high and low altitudes in Bolivia.* 58th Annual Meeting of the American Association of Physical Anthropologists, San Diego, CA

Hochachka PW, Stanley C, Matheson GO, McKenzie DC, Allen PS and Parkhouse WS (1991). Metabolic and work efficiencies during exercise in Andean natives. *Journal of Applied Physiology*, 70:1720–30

Horstman D, Weiskopf R and Jackson RE (1980). Work capacity during 3-wk sojourn at 4300 m: effects of relative polycythemia. *Journal of Applied Physiology*, 49:311–18

Huang SY, Ning XH, Zhou ZN, Gu ZZ and Hut ST (1984). Ventilatory function in adaptation to high altitude: studies in Tibet. In West JB and Lahiri S (eds), *High altitude and man.* American Physiological Society, Bethesda, MD, pp 173–8

Huang SY, Sun S, Droma T, Zhuang J, Tao JX, McCullough RG, McCullough RE, Micco AJ, Reeves JT and Moore LG (1992). Internal carotid arterial flow velocity during exercise in Tibetan and Han residents of Lhasa (3658 m). *Journal of Applied Physiology*, 73:2638–42

Hultgren HN, Kelly J and Miller H (1965). Pulmonary Circulation in Acclimatized Man at High Altitude. *Journal of Applied Physiology*, 20:233–8

Hurtado A (1964). Animals in high altitudes: resident man. In Dill DB (ed.), *Handbook of Physiology.* Section 4: Adaptation to the Environment. American Physiological Society, Washington, DC, pp 843–59

Jansen GF, Krins A and Basnyat B (1999). Cerebral vasomotor reactivity at high altitude in humans. *Journal of Applied Physiology*, 86:681–6

Jensen GM and Moore LG (1997). The effect of high altitude and other risk factors on birthweight: independent or interactive effects. *American Journal of Public Health*, 87:1003–07

Kashiwazaki H, Suzuki T and Takemoto T-I (1988). Altitude and reproduction of the Japanese in Bolivia. *Human Biology*, 60:831–45

Khalid MEM, Ali ME and Ali KZM (1997). Full-term birth weight and placental morphology at high and low altitude. *International Journal of Gynaecology and Obstetrics*, 57:259–65

Kollias J, Buskirk ER, Akers RF, Prokop EK, Baker PT and Picon-Reategui E (1968). Work capacity of long-time residents and newcomers to altitude. *Journal of Applied Physiology*, 24:792–9

Kryger M, McCullough R, Doekel R, Collins D, Weil JV and Grover RF (1978). Excessive polycythemia of high altitude: role of ventilatory drive and lung disease. *American Review of Respiratory Disease*, 118:659–67

Lawler J, Powers SK and Thompson D (1988). Linear relationship between VO_2 max and VO_2 max decrement during exposure to acute hypoxia. *Journal of Applied Physiology*, 64:1486–92

Leibson C, Brown M, Thibodeau S, Stevenson D, Vreman H, Cohen R, Clemons G, Callen W and Moore LG (1989). Neonatal hyperbilirubinemia at high altitude. *American Journal of Diseases of Childhood*, 143:983–7

Leon-Velarde F, Monge CC, Vidal A, Carcagno M, Criscuolo M and Bozzini CE (1991). Serum immunoreactive erythropoietin in high altitude natives with and without excessive erythrocytosis. *Experimental Hematology*, 19:257–60

Leon-Velarde F, Arregui A, Vargas M, Huicho L and Acosta R (1994). Chronic mountain sickness and chronic lower respiratory tract disorders. *Chest*, 106:151–5

Lichty JA, Ting RY, Bruns PD and Dyar E (1957). Studies of babies born at high altitude. *AMAJ Diseases of Childhood*, 93:666–77

Lubchenco LO, Hansman C, Dressler M and Boyd E (1963). Intrauterine growth as estimated from liveborn birth-weight data at 24 to 42 weeks of gestation. *Pediatrics*, 32:793–800

Macedo-Dianderas J (1966). Peso, Talla, Pulso y Presion Arterial del Recion Nacido en las Grandes Alturas. *Archivos del Instituto de biologia Andina*, 1:234–7

Maggiorini M, Bühler B, Walter M and Oelz O (1990). Prevalence of acute mountain sickness in the Swiss Alps. *British Medical Journal*, 301:853–5

Marcus CL, Glomb WB, Basinski DJ, Ward SLD and Keens TG (1994). Developmental pattern of hypercapnic and hypoxic ventilatory responses from childhood to adulthood. *Journal of Applied Physiology*, 76:314–20

Maresh CM, Noble BJ, Robertson KL and Sime WE (1983). Maximal exercise during hypobaric hypoxia (557 torr) in moderate-altitude natives. *Medicine, Science, and Sports Exercise*, 15:360–355

Martin TW, Weisman IM, Zeballos RJ and Stephenson SR (1989). Exercise and hypoxia increase sickling in venous blood from an exercising limb in individuals with sickle cell trait. *American Journal of Medicine*, 87:48–56

Mazess RB (1969). Exercise performance of Indian and White high altitude residents. *Human Biology*, 41:494–518

McClung J (1969). *Effects of High Altitude on Human Birth. Observations on Mothers, Placentas, and the Newborn in Two Peruvian Populations*. Harvard University Press, Cambridge

McCullough RE, Reeves JT and Liljegren RL (1977). Fetal growth retardation and increased infant mortality at high altitude. *Archives of Environmental Health*, 32:36–9

Miyachi M and Shibayama H (1992). Ventilatory capacity and exercise-induced arterial desaturation of highly trained endurance athletes. *Annals of Physiological Anthropology*, 11:263–7

Moore LG (1982). The incidence of pregnancy induced hypertension is increasing among Colorado residents. *American Journal of Obstetrics and Gynecology*, 144:123–9

Moore LG (1990). Maternal O_2 transport and fetal growth in Colorado, Peru and Tibet high-altitude residents. *American Journal of Human Biology*, 2:627–37

Moore LG, Jahnigen D, Rounds SS, Reeves JT and Grover RF (1982a). Maternal hyperventilation helps preserve arterial oxygenation during high-altitude pregnancy. *Journal of Applied Physiology*, 52:690–4

Moore LG, Rounds SS, Jahnigen D, Grover RF and Reeves JT (1982b). Infant birth weight is related to maternal arterial oxygenation at high altitude. *Journal of Applied Physiology*, 52:695–9

Moore LG, Newberry MA, Freeby GM and Crnie LS (1984). Increased incidence of neonatal hyperbilirubinemia at 3100 m in Colorado. *American Journal of Diseases of Childhood*, 138:157–61

Moore LG, Brodeur P, Chumbe O, D'Brot J, Hofmeister S and Monge C (1986a). Maternal hypoxic ventilatory response, ventilation, and infant birth weight at 4300 m. *Journal of Applied Physiology*, 60:1401–06

Moore LG, Niermeyer S and Zamudio S (1998). Human adaptation to high altitude: regional and life-cycle perspectives. *Yearbook of Physical Anthropology*, 41:25–64

Moret P, Covarrubias E, Coudert J and Duchosal F (1972). Cardiocirculatory adaptation to chronic hypoxia. Comparative study of coronary flow, myocardial oxygen consumption and efficiency between sea level and high altitude residents. *Extrait des Acta Cardiologica*, 27:283–305

Niermeyer S, Shaffer EM, Thilo E, Corbin C and Moore LG (1993). Arterial oxygenation and pulmonary arterial pressure in healthy neonates and infants at high altitude. *Journal of Pediatrics*, **123**:767–72

Niermeyer S, Yang P, Shanmina, Drolkar, Zhuang J and Moore L (1995). Arterial oxygen saturation in Tibetan and Han infants born in Lhasa, Tibet. *New England Journal of Medicine*, **333**:1248–52

Okin JT, Treger A, Overy HR, Weil JV and Grover RF (1966). Hematologic response to medium altitude. *Rocky Mountains Medicinal Journal*, **63**:44–7

Ou LC, Chen J, Fiore E, Leiter JC, Brinck-Johnsen T, Birchard GF, Clemons G and Smith RP (1992). Ventilatory and hematopoietic responses to chronic hypoxia in two rat strains. *Journal of Applied Physiology*, **72**:2354–63

Palacios-VL (1964). *Hierro Serico en el recien nacido de la altura*. Tesis de Br #6073, Universidad Nacional Mayor de San Marcos, Lima

Palmer SK, Moore LG, Young D, Cregger B, Berman JC and Zamudio S (1999). Altered blood pressure course during normal pregnancy and increased preeclampsia at high altitude (3100 metres) in Colorado. *American Journal of Obstetrics and Gynecology*, **180**:1161–8

Penaloza D, Sime F, Banchero N, Gamboa R, Cruz J and Marticorena E (1963). Pulmonary hypertension in healthy men born and living at high altitudes. *American Journal of Cardiology*, **11**:150–7

Reeves JT, McCullough RE, Moore LG, Cymerman A and Weil JV (1993). Sea-level P_{CO2} relates to ventilatory acclimatization at 4300 m. *Journal of Applied Physiology*, **75**:1117–22

Regensteiner JG and Moore LG (1985). Migration of the elderly from high altitudes in Colorado. *JAMA*, **253**:3124–8

Ruiz L (1973). *Epidemiologia de la hipertension arterial y de la cardiopatia isquemica en las grandes alturas*. Universidad Peruana Cayetano Heredia, Lima

Saco-Pollitt C (1981). Birth in the Peruvian Andes: physical and behavioral consequences in the neonate. *Child Development*, **52**:839–46

Samaja M, Veicsteinas A and Cerretelli P (1979). Oxygen affinity of blood in altitude Sherpas. *Journal of Applied Physiology*, **47**:337–41

Santolaya R, Araya-CJ, Vecchiola-CA, Prieto-PR, Ramirez RM and Alcayaga-AR (1981). Hematocrito, hemoglobina y presion de oxigeno arterial en 270 hombres y 266 mujeres sanas residentes de altura (2800 mts). *Revista Hospital Roy H. Glover*, **1**:17–24

Schlaepher T, Bartsch P and Fisch H (1992). Paradoxical effects of mild hypoxia and moderate altitude on human visual perception. *Clinical Science*, **83**:633–6

Schoene RB, Roach RC, Lahiri S, Peters RM, Hackett PH and Santolaya R (1990). Increased diffusion capacity maintains arterial saturation during exercise in the Quechua Indians of Chilean altiplano. *American Journal of Human Biology*, **2**:663–8

Smith C (1997). The effect of maternal nutritional variables on birthweight outcomes of infants born to Sherpa women at low and high altitudes in Nepal. *American Journal of Human Biology*, **9**:751–63

Spielvogel H, Caceres E, Koubi H, Sempore B, Sauvain M and Favier R (1996). Effects of coca chew on metabolic and hormonal changes during graded incremental exercise to maximum. *Journal of Applied Physiology*, **80**:643–9

Spielvogel H (1997). Body fluid homeostasis and cardiovascular adjustments during submaximal exercise: influence of chewing coca leaves. *European Journal of Applied Physiology*, **75**:400–06

Sun S, Oliver-Pickett C, Ping Y, Micco AJ, Droma T, Zamudio S, Zhuang J, Huang SY, McCullough RG, Cymerman A and Moore LG (1996). Breathing and brain blood flow during sleep in patients with chronic mountain sickness. *Journal of Applied Physiology*, **81**:611–18

Sun SF, Droma TS, Zhang JG, Tao JX, Huang SY, McCullough RG, Reeves CS, Reeves JT and Moore LG (1990). Greater maximal O_2 uptakes and vital capacities in Tibetan than Han residents of Lhasa. *Respiratory Physiology*, **79**:151–62

Tarazona-Santos E, Lavine M, Pastor S, Giori G and Pettener D (2000). Hematological and pulmonary responses to high altitude in Quechuas: A multivariate approach. *American Journal of Physical Anthropology*, **111**:165–76

Thilo EH, Berman ER and Carson BS (1991). Oxygen saturation by pulse oximetry in healthy infants at an altitude of 1610 m (5280 ft). *American Journal of Diseases of Childhood*, **145**:1137–40

Tucker A, Stager JM and Cordain L (1984). Arterial O_2 saturation and maximum O_2 consumption in moderate-altitude runners exposed to sea level and 3050 m. *JAMA*, **252**:2867–71

Tufts DA, Haas JD, Beard JL and Spielvogel H (1985). Distribution of hemoglobin and functional consequences of anemia in adult males at high altitude. *American Journal of Clinical Nutrition*, **42**:1–11

Unger C, Weiser JK, McCullough RE, Keefer S and Moore LG (1988). Altitude, low birth weight, and infant mortality in Colorado. *JAMA*, **259**:3427–32

Usher R and McLean F (1969). Intrauterine growth of liveborn Caucasian infants at sea level: standards obtained from measurements in 7 dimensions of infants born between 25 and 44 weeks of gestation. *Journal of Pediatrics*, 74:901–10

Vincent J, Hellot MF, Vargas E, Gautier H, Pasquis P and LeFrancois R (1978). Pulmonary gas exchange, diffusing capacity in natives and newcomers at high altitude. *Respiratory Physiology*, 34:219–31

Vogel JHK, Weaver WF, Rose RL, Blount Jr SG and Grover RF (1962). Pulmonary hypertension on exertion in normal man living at 10 150 feet (Leadville, Colorado). *Medical Thoracics*, 19: 269–85

Ward MP, Milledge JS and West JB (2000). *High Altitude Medicine and Physiology*. Oxford University Press, London

Weil JV, Jamieson G, Brown DW, Grover RF, Balchum OJ and Murray JF (1968). The red cell mass-arterial oxygen relationship in normal man. *Journal of Clinical Investigation*, 47:1627–39

Weil JV, Kryger MH and Scoggin CH (1978). Sleep and breathing at high altitude. In Guilleminault C and Dement WC (eds), *Sleep Apnea Syndromes*. Alan R. Liss, Inc, New York, pp 119–36

Weinstein RS and Haas JD (1977). Early stress and later reproductive performance under conditions of malnutrition and high altitude hypoxia. *Medical Anthropology*, 1:25–54

Weitz CA (1984). Biocultural adaptations of the high altitude Sherpas of Nepal. In Lukacs JR (ed.), *The People of South Asia. The Biological Anthropology of India, Pakistan, and Nepal*. Plenum Press, New York and London, pp 387–420

Wiley AS (1994). Neonatal and maternal anthropometric characteristics in a high altitude population of the Western Himalaya. *American Journal of Human Biology*, 6:499–510

Winslow RM, Chapman KW, Gibson CG, Samaja M, Blume FD and Goldwasser E (1988). Hematologic response to hypoxia in Sherpas and Quechua Indians. *FASEB Journal*, 2:A1721

Winslow RM, Chapman KW and Monge CM (1990). Ventilation and the control of erythropoiesis in high-altitude natives of Chile and Nepal. *American Journal of Human Biology*, 2:653–62

Winslow RM, Monge CC, Statham NJ, Gibson CG, Charache S, Whittembury J, Moran O and Berger RL (1981). Variability of oxygen affinity of blood: human subjects native to high altitude. *Journal of Applied Physiology*, 51:1411–16

Wolfel EE, Groves BM, Brooks GA, Butterfield GE, Mazzeo RS, Moore LG, Sutton JR, Bender PR, Dahms TE, McCullough RE, McCullough RG, Huang SY, Sun SF, Grover RF, Hultgen HN and Reeves JT (1991). Oxygen transport during steady-state submaximal exercise in chronic hypoxic. *Journal of Applied Physiology*, 70: 1129–36

Yancey MK, Moore J, Brady K, Milligan D and Strampel W (1992). The effect of altitude on umbilical cord blood gases. *Obstetrics and Gynecology*, 79:571–4

Yip R (1987). Altitude and birth weight. *Journal of Pediatrics*, 111:869–76

Zamudio S, Droma T, Norkyel KY, Acharya G, Zamudio JA, Niermeyer SN and Moore LG (1993). Protection from intrauterine growth retardation in Tibetans at high altitude. *American Journal of Human Biology*, 91:215–24

Zhuang J, Droma T, Sun S, Janes C, McCullough RE, McCullough RG, Cymerman A, Huang SY, Reeves JT and Moore LG (1993). Hypoxic ventilatory responsiveness in Tibetan compared with Han residents of 3658 m. *Journal of Applied Physiology*, 74:303–11

Zhuang J, Droma T, Sutton JR, Groves BM, McCullough RE, McCullough RG, Sun S and Moore LG (1996). Smaller alveolar-arterial O_2 gradients in Tibetan than Han residents in Lhasa (3658 m). *Respiratory Physiology*, 103:75–82

Part III

Climatic Changes and Mountain Biodiversity

Potential Effects of Climate Change on Alpine and Nival Plants in the Alps

17

Michael Gottfried, Harald Pauli, Karl Reiter and Georg Grabherr

INTRODUCTION

In the international literature the term *alpine* is commonly used to describe the uppermost vegetation zone of high mountain systems, from the treeline upwards to the limits of plant life (see Körner, 1999). In contrast, European authors dealing with the Alps have divided this unit into an *alpine* and a *nival* zone (e.g. Braun-Blanquet, 1954; Ozenda, 1985; Ellenberg, 1996; Grabherr, 1997). According to these authors, the alpine zone is confined to the dwarf shrub heaths of the lower alpine zone and the closed grasslands of the upper alpine zone. The nival zone, which is situated above the alpine zone, is dominated by rocks and screes. In this latter zone, only scree specialists and chasmophytes find wider distribution, and grassland fragments can colonise only favourable habitats. The so-called permanent snowline was usually taken as the upper limit of the alpine zone. It was defined as the imaginary line of climate conditions where the snow falling in winter does not melt completely during summer (Schröter, 1926; Reisigl and Pitschmann, 1958; Landolt, 1983; Ozenda, 1985). This hypothetical line, however, is hardly discernible in the landscape. The alpine/nival ecotone is often rather broad and patchy with respect to relief and vegetation patterns (e.g. Gottfried *et al.*, 1998). Moreover, this ecotone is not as clearly marked by life-form turnover as the alpine treeline. Aside from dwarf shrubs, all plant life forms of the alpine zone can also be found in the nival zone. The problems outlined above might be the main reasons why the term *nival* was not generally accepted. To find a more applicable definition,

Reisigl and Pitschmann (1958) described the alpine/nival transition as the subnival zone, where grasslands become patchy and are inter-mingled with nival scree communities. They pointed out, in great detail, the unique features of the subnival and nival zone according to their floristic pool and vegetation patterning. For clarity, we will use the term nival for species which have their distribution centre in the subnival zone (i.e. the alpine/nival ecotone) or the nival zone. In contrast, alpine species are considered to be more or less restricted to the closed alpine grassland, finding their upper limits at the alpine/nival ecotone and extending only rarely into the nival zone. By compiling standard floras of the Alps based on this criterion, we can classify about 30 vascular plant species as nival, with seven of them being endemics of the Alps.

Especially when dealing with issues of bio-diversity research in the face of global change, it is crucial to differentiate between alpine and nival plants. That the mountain flora is moving upwards, with competitors reaching the habitats of less competitive species, has already been proved (Gottfried *et al.*, 1994; Grabherr *et al.*, 1994; Pauli *et al.*, 1996; Moiseev, 2001). Although not observed until now (Grabherr *et al.*, 2001), genetic species losses are the most likely result (Grabherr *et al.*, 1995; Gottfried *et al.*, 1999). If the life zone above the treeline is treated as a single unit, and its organisms are considered as a single group of 'alpine' plants, then such changes can hardly be quantified. In this Chapter, we attempt to show that nival and alpine plant distributions can be separated, not only according to community patterns, but can also be correlated to basic habitat features, e.g. temperature and snowcover duration. From the resulting species–habitat relationships we can generalise the shapes of their contrasting microclimatic niches. On this basis, scenarios of climate change impacts on these two species pools are described. Finally, the question whether these scenarios are valid solely for the European Alps, or whether they might also be relevant for other high mountain systems is discussed.

CAN WE DISTINGUISH ALPINE FROM NIVAL SPECIES POOLS IN TERMS OF TEMPERATURE AND SNOWCOVER DURATION?

Temperature is the outstanding driving factor which triggers the altitudinal turnover of species and vegetation types in mountain systems (Walter, 1985). The various aspects of the temperature regime, however, influence mountain plant distribution patterns differently. While adaptation to low temperature extremes is a key factor, temperature sums (or averages) play an important role as well (Körner and Larcher, 1988). Growth processes of high mountain plants have been reported to be especially sensitive to nighttime temperatures (Körner and Pelaez Menendez-Riedl, 1989).

A second very discriminating factor is the length of the snow period (e.g. Billings and Bliss, 1959; Friedel, 1961; Tranquillini, 1979; Galen and Stanton, 1995; see Körner, 1999 for more references). Most of these authors, however, focused either on the alpine treeline, on dwarf shrubs, or on alpine grassland species. Although some ecophysiological knowledge is available on nival plants themselves (e.g. Brzoska, 1971; Moser et al., 1977; Arnone and Körner, 1997; Larcher et al., 1997), the main differences between alpine and nival species have not yet been sufficiently taken into account. In this section, we investigate whether there is a continuum in the microclimatic requirements of plants living above the treeline, and whether we can clearly distinguish alpine and nival plants according to microclimate.

METHODS

Study area

The study area is on the southern slope system of Schrankogel (47° 02′ N, 11° 06′ E; Tyrol, Austria, Central Eastern Alps), from the upper alpine zone at around 2900 m, to the alpine/nival ecotone at around 3100 m, into the nival zone up to the summit (3497 m). Although the border between closed grasslands and open nival vegetation patterns is locally very sharp, the altitudinal coverage of the ecotone is about 200-m high due to extremely varying relief conditions. About 250 vascular plant species inhabit the alpine zone of Schrankogel (Dullinger, 1998). Some 60 of these species reach the alpine/nival ecotone. Thirty of these can be found in the nival zone. The summit, where seven vascular plant species still exist in the uppermost 20 m, does not extend well above the altitudinal limit of vascular plant life, as also found for other summits of the Central Eastern Alps (Gottfried et al., 1994; Pauli et al., 1996). The highest vascular plant of the Alps was reported at 4450 m for Dom mountain (4545 m, Western Alps, Switzerland; Anchisi, 1986). In 1994 a monitoring network was established at Schrankogel, consisting of nearly 1000 permanent plots of 1 m² each, to focus on plant migration effects in the course of climate change (Gottfried et al., 1998; Pauli et al., 1999).

Microclimatic measurements

Since 1997, miniature dataloggers have measured air temperature every 1.5 h at 32 plots out of the total set of permanent plots (see above). The positions were selected in order to best represent the different habitat types of the study area. They cover wide topographic and microclimatic ranges, from crests with early snowmelt to hollows where snow persists deep into summer; their altitudinal distribution covers about 600 m (Figure 17.1). The loggers are mounted 1–3 cm above the surface (i.e. at average vegetation canopy height) and are shaded against direct solar radiation. A variety of temperature indices can be derived from the data

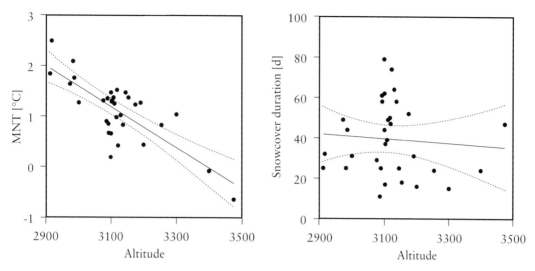

Figure 17.1 Climate indices as measured by 32 sensors (dots). *Left*: mean nighttime temperature (MNT) between June 1, 1998 and July 31, 1998. *Right*: number of days between May 1, 1998 and July 31, 1998, at which the respective sensor was covered by snow (Snow cover duration). Solid lines: altitudinal gradients. Dashed lines: 95% confidence intervals

series, including minima, maxima, daytime and nighttime temperatures. Moreover, the duration of snow cover at these sites can be calculated because the diurnal temperature amplitude immediately drops when snow covers the sensor of the datalogger. Within these 32 permanent plots, where both vegetation data as well as temperature measurements are available, we selected 15 vascular plant species that were present in at least five of the plots (Table 17.1). Pauli et al. (1999) analysed the role of these species in the vegetation of Schrankogel, using classical phytosociological methods (Braun-Blanquet, 1964; Hill, 1979), clearly demonstrating their alpine or nival distribution according to their associations to community patterns. From this, we defined an alpine species group and a nival species group (Table 17.1) and used this grouping for subsequent analyses (Figures 17.2, 17.3).

For the analysis of species–temperature relationships we used the mean nighttime temperature (MNT) for the period June–July 1998. This period provided the best results, i.e. optimal dispersion of species responses along the gradient. It covers the time of year when most biomass is produced in this environment (Grabherr et al., 1980). The MNT values (ranging from –0.64 to 2.49 °C at the selected sites) were condensed into seven classes in steps of 0.5 °C. For each class, the probability of presence of each species (as number of sites where the species is present divided by the number of sites whose MNT fall into the respective class) was calculated. The number of snowfree days in the period May–July 1997 was used to analyse the species relationships to snow cover duration. This is the main snowmelt period in the study area (Gottfried et al., 1999). Climatic and species data were condensed in a manner analogous to the temperature analysis. Regression models were fitted to the species distribution patterns along the two climatic gradients. Linear, sigmoidal and beta-functions were used for the three distribution patterns (see ter Braak and Looman, 1995, for regression techniques and Austin et al., 1994, for beta functions). Previously, the altitudinal gradients of temperature and snow cover duration were examined using linear regression. The nomenclature of species follows Ehrendorfer (1973).

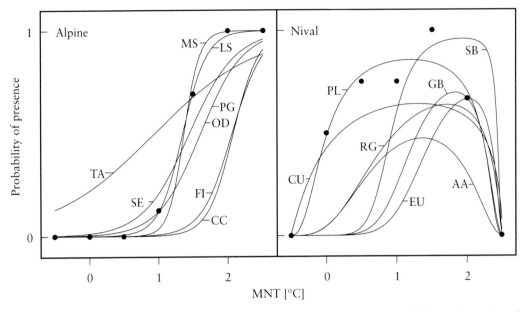

Figure 17.2 Response patterns of species with respect to mean nighttime temperature (MNT). *Left*: members of the alpine species group. *Right*: members of the nival species group. Dots: Underlying data of LS and PL (omitted for the other species, for reasons of clarity). For abbreviations of species names, see Table 17.1

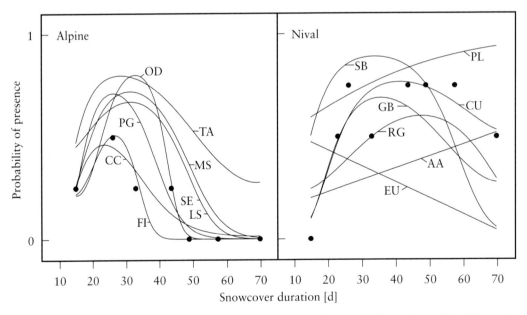

Figure 17.3 Response patterns of species with respect to snow cover duration. *Left*: members of the alpine species group. *Right*: members of the nival species group. Dots: Underlying data of CC and CU (omitted for the other species, for reasons of clarity). For abbreviations of species names, see Table 17.1

Table 17.1 Frequent plant species of the alpine–nival ecotone at Schrankogel. Classification into alpine (A), intermediate (I), and nival (N) species according to Pauli *et al.* (1999), and with respect to responses to mean nighttime temperature (MNT) and snow cover duration (Snow), based on the shapes of response curves. R^2: correlations of the fitted response curves (Figures 17.2, 17.3) with the underlying data

Species	Abbrev.	Pauli et al. 1999	MNT	Snow	R^2_{MNT}	R^2_{Snow}
Carex curvula	CC	A	A	A	0.94	0.85
Festuca intercedens	FI	A	A	A	0.92	0.76
Luzula spicata	LS	A	A	A	1.00	0.88
Minuartia sedoides	MS	A	A	A	0.93	0.73
Oreochloa disticha	OD	A	A	A	0.95	0.80
Primula glutinosa	PG	A	A	A	0.95	0.83
Silene exscapa	SE	A	A	A	0.94	0.57
Tanacetum alpinum	TA	I	I	I	0.73	0.58
Erigeron uniflorus	EU	N	N	I	1.00	0.50
Androsace alpina	AA	N	N	N	0.94	0.22
Cerastium uniflorum	CU	N	N	N	0.90	0.76
Gentiana bavarica	GB	N	N	N	0.70	0.69
Poa laxa	PL	N	N	N	0.96	0.55
Ranunculus glacialis	RG	N	N	N	0.91	0.39
Saxifraga bryoides	SB	N	N	N	0.99	0.93

RESULTS AND DISCUSSION

Altitudinal gradients

A clearly defined altitudinal temperature gradient existed over the measurement period ($R^2 = 0.6$; Figure 17.1, left). However, at $-0.4\,°C\,100\,m^{-1}$ it is much lower than the average altitudinal lapse rate of $-0.7\,°C\,100m^{-1}$ reported in the literature. This is not too surprising because temperature differences along an altitudinal gradient are generally lower at night than during the day in mountain systems (Barry, 1981; for a comparison of temperature gradients on Schrankogel during all seasons, see Gottfried *et al.*, 1999).

In contrast, the altitudinal gradient of snow cover duration showed no significant trend (Figure 17.1, right). This is due to the wide topographical variation at equal altitudes, where exposed ridges with early snowmelt alternate with deep hollows where snow can persist even into the summer. This result supports the view that the alpine/nival transition is not clearly visible in the landscape in terms of a permanent snowline; with sharp snow melting patterns.

Temperature

Two different response patterns of the species under consideration with respect to MNT can be clearly distinguished (Figure 17.2). The members of the alpine group show sigmoidal responses, demonstrating that they have their optimal habitats at higher MNT than those actually measured at the studied sites (Figure 17.2, left). The nival group (Figure 17.2, right) show gaussian response patterns, where the group members have their optima in the study area and tolerate colder nights than those of the alpine group. All modelled response patterns of the involved species fit the data very well (see R^2_{MNT} in Table 17.1). Thus, the grouping of species is based on the different shapes of their response curves rather than on certain threshold values.

Snow cover duration

The alpine species also show very uniform response patterns to snow cover duration (Figure 17.3, left), finding optimal sites where snow cover lasted only 30 days (of the three-month measurement period). The species tend

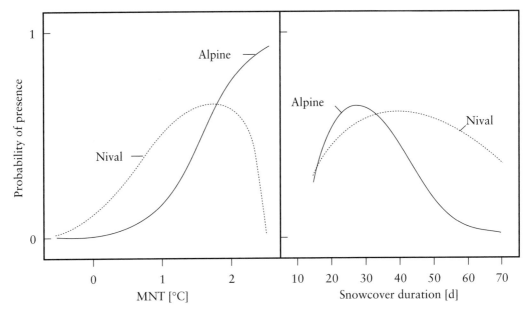

Figure 17.4 Generalised niche schemes of alpine species (solid) and nival species (dashed) with respect to mean nighttime temperature (MNT; *left*) and snowcover duration (*right*)

to avoid those places where snow cover lasted for 50 to 70 days beyond May 1, i.e. well into the mid-summer. All R^2 values of the fitted curves indicate a good representation of the underlying data (see R^2_{Snow} in Table 17.1).

The nival species group (Figure 17.3, right) reacts in a somewhat more complicated manner to snow cover. The underlying data were, in some cases, noisier when compared to the species reactions to MNT, which led to lower R^2 of the fitted curves. Nevertheless, all of the nival species tend to have a higher tolerance to long-lasting snow than alpine species.

The alpine and the nival species group

The results presented demonstrate that the plant species living around the alpine/nival ecotone can clearly be divided in two groups according to their different association to plant communities, as well as according to their microclimatic habitats. A number of ecological details concerning the studied species have been reported (Gottfried *et al.*, 1998; Pauli *et al.*, 1999). Here, we stress the fact that the classification into alpine and nival species holds true when community structures, as well as

ecophysiological requirements of the species, are analysed (Table 17.1, columns 3–5). Out of the 15 species studied, only two did not fit this scheme exactly. These latter two species showed somewhat intermediate responses and were therefore classified as intermediate in the microclimatic analyses. One of them was also presented as intermediate by Pauli *et al.* (1999).

GENERALISED NICHE SCHEMES

Generalising the response patterns of the species (see above, Figures 17.2 and 17.3), the niches can be characterised with respect to temperature and snow cover duration: (1) The alpine species pool is restricted to habitats of higher temperature in which snow melts relatively early, thus providing a comparatively long vegetation period which starts in late spring or early summer. (2) The nival species pool prefers sites with a lower temperature regime and long-lasting snow cover, where the growth period starts as late as mid-summer in some years. The nival species pool is not as sharply restricted to specific habitats as the alpine species pools. These generalised niche schemes are shown in Figure 17.4.

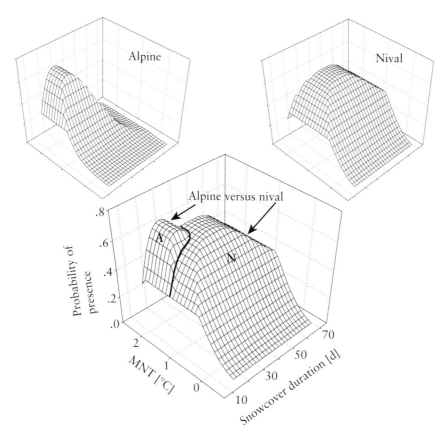

Figure 17.5 The ecological niches of alpine and nival species at the alpine–nival ecotone and in the nival zone with respect to mean nighttime temperature (MNT) and snow cover duration under the current climate (diagrams at top). Bottom: combination of these two niches, separated by a solid line. A: habitats where alpine species currently dominate. N: habitats where nival species currently dominate. Axis scales and labels are equal for all three diagrams

Whether these schemes reflect the fundamental niche or the realised niche of the species remains open to debate. It is argued that, for the alpine species, which are more or less absent in the nival zone, it is more a fundamental niche shaped by temperature and snow cover thresholds which restricts growth processes in terms of metabolic constraints and available time. In nival species, the key factor seems to be competition. While the nival species reach their maximum abundance in the nival zone, they are also present in alpine grassland communities, though mostly only in gaps occurring (Pauli *et al.*, 1999, and unpublished data). Nevertheless, shifts in these climatic niche schemes are likely to have effects on the species distribution patterns.

CLIMATE CHANGE SCENARIOS

By plotting the generalised niche schemes (above) in three dimensions (Figure 17.5, top left, top right), and by then combining these two schemes (Figure 17.5, bottom), we can deduce the ecological conditions under which the optima of alpine species prevail over those of nival species, and vice versa. Following this scheme, climatic changes would induce the following effects. If temperature rises and snow cover duration remains as today, the alpine

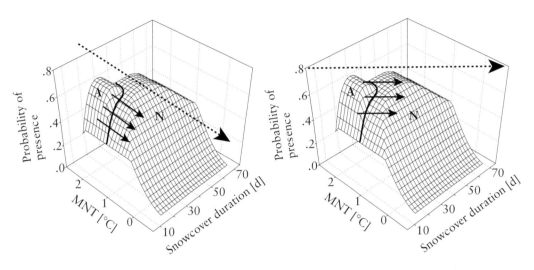

Figure 17.6 Climate change scenarios. *Left*: mean nighttime temperature (MNT) increasing, no change in snow cover duration. *Right*: MNT increasing, snow cover duration decreasing. Arrows: directions of climatic shifts (dashed) and the resulting pressure (solid) of the alpine habitat (A) on the nival habitat (N). Compare also Figure 17.5 and explanations in the text

niche would be enhanced along the temperature axis in a negative direction, producing moderate pressure on the habitats of nival species (Figure 17.6, left). Note that nival species are restricted in their ability to avoid this competition by moving away because the nival zone is altitudinally limited by the summits. This scenario is somewhat unlikely (though possible) because increased snowy precipitation would occur but would also be subject to accelerated melting due to increased temperature. The future development of precipitation in the Alps is currently uncertain (Gyalistras *et al.*, 1998). If solid precipitation does not increase markedly, however, a second scenario with decreased snow cover duration due to increased temperature (Martin *et al.*, 1994) seems likely. In this case, all nival plants without exception would be affected. Their habitat would shrink considerably (Figure 17.6, right) and summits would become nival 'species traps' (Grabherr *et al.*, 1995).

'NIVAL' HABITATS IN A WORLD-WIDE PERSPECTIVE

Life zones with nival conditions are found in high mountain systems all over the globe.

Whether they are labelled superparamo in the tropical Andes or East African mountains (Díaz *et al.*, 1997; Schröder, 1998), cryoromediterranean belt in Mediterranean mountains (Peinado Lorca and Rivas-Martinez, 1997), free gelifluction belt in the Himalaya (Miehe, 1997), subnival in the New Zealand Alps (Mark and Dickinson, 1997), Golec in Russian mountain systems (Schröder, 1998), or simply as the upper alpine belt in the Scandes (Dierssen, 1996), is less important in the present context, especially as these names, to some degree, reflect historical aspects of terminology. What most of these mountain zones have in common, however, are their two main climatic determinants, low temperatures and snow; together, these allow only scattered vegetation patterns, composed of highly specialised species. Subtropical and tropical mountains, where other factors were reported to constitute the plant limit (Halloy, 1989), may be an exception. Although a comparative account is still lacking, high numbers of endemic taxa restricted to the uppermost zones have been reported for mountain systems, e.g. for the Caucasus, where 300 plant species occur around the alpine/nival ecotone, 109 of them being restricted to this zone (Kharadze, 1965;

Nakhutsrishvili, 1998). Another good example is the Sierra Nevada in Southern Spain. Of about 180 higher plants occurring in the uppermost – the cryoromediterranean – belt, 30% are endemics of this mountain system (Molero Mesa and Pérez-Raya, 1987). As this narrow belt is today restricted to a few summit areas, massive biodiversity threats would be the likely result with a drastic change in climate.

Since high mountain systems are generally biodiversity hot spots (Barthlott *et al.*, 1996; Messerli, 1999), one can assume that many of them harbour distinct species groups with nival habitat requirements. Depending on the respective situation, they are more or less threatened by climate warming. This is important for any global account of mountain biodiversity and its possible changes. This prompts us to suggest the use of the terms alpine and nival, or any other pair of terms, in high mountain ecology whenever applicable. Areas above the treeline which are intersected by another distinct ecotone should not be treated as a single unit.

CONCLUSIONS

Whether the high mountain area above the treeline should be viewed as a single unit or divided into an alpine zone – between the treeline and the upper boundary of closed grasslands – and a nival zone – the area with scattered vegetation on scree and rock above the grasslands – has always been a matter of dispute. There are some good reasons to distinguish between alpine and nival with respect to altitudinal zones and to plant species, as demonstrated by a case study from the European Alps:

(1) The nival zone hosts a number of characteristic species which are only exceptionally found in the grasslands below.

(2) This pool of nival species reacts quite differently to microclimatic factors compared to species typical for alpine grasslands: We correlated the presence/absence of species to nighttime temperatures and snowcover duration at 32 locations on a typical high mountain slope system. Whereas alpine species clearly reach their physiological limits at the alpine/nival ecotone, nival species here, or even higher, show optima with respect to these two factors.

(3) Climate change scenarios suggest that, as a consequence of earlier snowmelt and/or warming, alpine species may fill the nival niche, putting strong pressure on the less competitive nival species.

(4) We suggest that these observations can be extrapolated to many other mountain systems of the world and conclude that this may lead to remarkable biodiversity losses, especially because the uppermost zones or the uppermost ecotones often host unique species.

(5) In the light of this, we suggest that areas above the treeline which are intersected by another distinct ecotone should not be treated as a single unit.

ACKNOWLEDGEMENTS

The research for this paper would not have been possible without the generous assistance of many people who supported us at Schrankogel. Special thanks in this respect to Hermann Stockinger and Markus Gottfried, to Reinhard Lentner of Tiroler Landesregierung and to the team of the Ambergerhütte. The study was financed by the Austrian Academy of Sciences (IGBP 17/1997–99, IGBP 22/2000).

References

Anchisi E (1986). Quatrième contribution à l'étude de la Flore Valaisanne. *Bulletin Murithienne*, **102**:115–26

Arnone JA and Körner C (1997). Temperature adaptation and acclimation potential of leaf dark respiration in two species of *Ranunculus* from warm and cold habitats. *Arctic and Alpine Research*, **29**:122–5

Austin MP, Nicholls AO, Doherty MD and Meyers JA (1994). Determining species response functions to an environmental gradient by means of a ß-function. *Journal of Vegetation Science*, **5**:215–28

Barry RG (1981). *Mountain Weather and Climate*. Methuen, London

Barthlott W, Lauer W and Placke A (1996). Global distribution of species diversity in vascular plants: towards a world map of phytodiversity. *Erdkunde*, **50**:317–27

Billings WD and Bliss LC (1959). An alpine snow-bank environment and its effects on vegetation, plant development, and productivity. *Ecology*, **40**:388–97

Braun-Blanquet J (1954). *La Vegetation alpin et nivale des Alpes francaises*. Station Internationale de Géobotanique Méditerranéenne et Alpine, Montpellier, Communication No **125**:27–96

Braun-Blanquet J (1964). *Pflanzensoziologie*, 3rd edn. Springer, Wien

Brzoska W (1971). Energiegehalte verschiedener Organe von nivalen Sproßpflanzen im Laufe einer Vegetationsperiode. *Photosynthetica*, **5**:183–9

Díaz A, Péfaur JE and Durant P (1997). Ecology of the South American Páramos with emphasis on the fauna of the Venezuelan Páramos. In Wielgolaski FE (ed.), *Polar and Alpine Tundra Ecosystems of the World*, vol. 3. Elsevier, Amsterdam, pp 263–310

Dierssen K (1996). *Vegetation Nordeuroas*. Ulmer, Stuttgart

Dullinger S (1998). *Vegetation des Schrankogel, Stubaier Alpen*. MSc Thesis, University of Vienna, Austria, 189 pp

Ehrendorfer F (1973). *Liste der Gefäßpflanzen Mitteleuropas*, 2nd edn. Fischer, Stuttgart

Ellenberg H (1996). *Vegetation Mitteleuropas mit den Alpen*. Ulmer, Stuttgart

Friedel H (1961). Schneedeckendauer und Vegetationsverteilungen im Gelände. *Mitt Forst Bundesversuchsanst Mariabrunn* (Wien), **59**:317–69

Galen C and Stanton ML (1995). Responses of snowbed plant species to changes in growing-season length. *Ecology*, **76**:1546–57

Gottfried M, Pauli H and Grabherr G (1994). Die Alpen im 'Treibhaus': Nachweise für das erwärmungsbedingte Höhersteigen der alpinen und nivalen Vegetation. *Jahrb Ver Schutz Bergwelt*, **59**:13–27

Gottfried M, Pauli H and Grabherr G (1998). Prediction of vegetation patterns at the limits of plant life: A new view of the alpine–nival ecotone. *Arctic and Alpine Research*, **30**:207–21

Gottfried M, Pauli H, Reiter K and Grabherr G (1999). A fine-scaled predictive model for changes in species distribution patterns of high mountain plants induced by climate warming. *Diversity and Distributions*, **5**:241–51

Grabherr G (1997). The high-mountain ecosystems of the Alps. In Wielgolaski FE (ed.), *Polar and Alpine Tundra Ecosystems of the World*, vol. 3. Elsevier, Amsterdam, pp 97–121

Grabherr G, Brzoska W, Hofer H and Reisigl H (1980). Energiebindung und Wirkungsgrad der Nettoprimärproduktion in einem Krummseggenrasen (Caricetum curvulae) der Ötztaler Alpen, Tirol. *Oecol Plant*, **1**(15):307–16

Grabherr G, Gottfried M, Gruber A and Pauli H (1995). Patterns and Current Changes in Alpine Plant Diversity. In Chapin III FS and Körner C (eds), *Arctic and Alpine Biodiversity: Patterns, Causes and Ecosystem Consequences*. Ecological Studies, 113. Springer, Berlin, pp 167–81

Grabherr G, Gottfried M and Pauli H (1994). Climate effects on mountain plants. *Nature*, **369**:448

Grabherr G, Gottfried M and Pauli H (2001). Long-term monitoring of mountain peaks in the Alps. In Burga C and Kratochwil A (eds), *Vegetation Monitoring/Global Change*. Kluwer, Dordrecht

Gyalistras D, Schär C, Davies HC and Wanner H (1998). Future Alpine Climate. In Cebon P, Dahinden U, Davies H, Imboden DM and Jaeger CC (eds), *Views from the Alps Regional Perspectives on Climate Change*. MIT Press, Cambridge, pp 171–223

Halloy S (1989). Altitudinal limits of life in subtropical mountains: What do we know? *Pacific Science*, **43**:170–84

Hill MO (1979). *TWINSPAN – a FORTRAN program for arranging multivariate data in an ordered two-way table by classification of individuals and attributes*. Cornell University, Ithaca, N.Y.

Kharadze AL (1965). O subnival'nom pojase Bol'shogo Kavkaza [On the subnival belt of the Greater Caucasus]. *Zametki po sistematike i geografii rastenii Instituta Botaniki AN GSSR*, **25**:103–14

Körner C (1999). *Alpine plant life: functional plant ecology of high mountain ecosystems*. Springer, Berlin

Körner C and Larcher W (1988). Plant life in cold climates. In Long SF and Woodward FI (eds), *Plant and Temperature*. Symp Soc Exp Biol, 42. The Company of Biologists, Cambridge, pp 25–57

Körner C and Pelaez Menendez-Riedl S (1989). The significance of developmental aspects in plant growth analysis. In Lambers H (ed.), *Causes and consequences of variation in growth rate and productivity of higher plants*. SPB Acad. Publ., The Hague, pp 141–57

Landolt E (1983). Probleme der Höhenstufen in den Alpen. *Botanica Helvetica*, **93**:255–86

Larcher W, Wagner J and Lütz C (1997). The effect of heat on photosynthesis, dark respiration and cellular ultrastructure of the arctic-alpine psychrophyte *Ranunculus glacialis*. *Photosynthetica*, **34**:219–32

Mark AF and Dickinson KJM (1997). New Zealand Alpine Ecosystems. In Wielgolaski FE (ed.), *Polar and Alpine Tundra Ecosystems of the World*, vol. 3. Elsevier, Amsterdam, pp 311–46

Martin E, Brun E and Durand Y (1994). Sensitivity of the French Alps snow cover to the variation of climatic variables. *Annals Geophysicae*, **12**: 469–77

Messerli B (1999). The global mountain problematique. In Price M *et al.* (eds), *Global Change in the Mountains*. The Parthenon Publishing Group, New York, pp 1–3

Miehe G (1997). Alpine vegetation types of the Central Himalaya. In Wielgolaski FE (ed.), *Polar and Alpine Tundra Ecosystems of the World*, vol. 3. Elsevier, Amsterdam, pp 161–97

Moiseev P (2002). A comparison of contemporary and old landscape photographs of the upper treeline in the Southern Urals. Submitted

Molero Mesa J and Pérez-Raya F (1987). *La flora de la Sierra Nevada – Avance sobre el catálogo florístico nevadense*. Universidad de Granada, Diputacion Provincial de Granada

Moser W, Brzoska W, Zachhuber K and Larcher W (1977). Ergebnisse des IBP-Projekts 'Hoher Nebelkogel 3184 m'. *Sitzungsber Oesterr Akad Wiss, Wien, Math Naturwiss Kl Abt I*, **186**:387–419

Nakhutsrishvili GS (1998). The vegetation of the subnival belt of the Caucasus Mountains. *Arctic and Alpine Research*, 30:207–21

Ozenda P (1985). *La végétation de la Chaine alpine dans l'ensemble montagnard européen*. Masson, Paris

Pauli H, Gottfried M and Grabherr G (1996). Effects of Climate Change on Mountain Ecosystems – Upward Shifting of Alpine Plants. *World Resource Review*, 8:382–90

Pauli H, Gottfried M and Grabherr G (1999). Vascular plant distribution patterns at the low-temperature limits of plant life – the alpine–nival ecotone of Mount Schrankogel (Tyrol, Austria). *Phytocoenologia*, 29:297–325

Peinado Lorca M and Rivas-Martinez S (eds) (1997). *La vegetation de Espana*. Universidad de Alcala de Henares

Reisigl H and Pitschmann H (1958). Obere Grenzen von Flora und Vegetation in der Nivalstufe der zentralen Ötztaler Alpen (Tirol). *Vegetatio*, 8:93–129

Schröder FG (1998). *Lehrbuch der Pflanzengeographie*. Quelle & Meyer, Wiesbaden

Schröter C (1926). *Das Pflanzenleben der Alpen. Eine Schilderung der Hochgebirgsflora*. Raustein, Zürich

ter Braak CJF and Looman CWN (1995). Regression. In Jongman RHG, ter Braak CJF and van Tongeren OFR (eds), *Data analysis in community and landscape ecology*. Cambridge University Press, pp 29–72

Tranquillini W (1979). *Physiological Ecology of the Alpine Timberline*. Springer, Berlin

Walter H (1985). *Vegetation of the earth and ecological systems of the geo-biosphere*, 3rd edn. Springer, Berlin

Variations in Community Structure and Growth Rates of High-Andean Plants with Climatic Fluctuations

18

Stephan R.P. Halloy

INTRODUCTION

Species richness and community structure have long been debated as emergent properties of biological communities affected by environmental conditions (Brown *et al.*, 2001) and potentially linked to properties of resilience and stability (McGradySteed *et al.*, 1997; Tilman, 1999). Given concerns over losses of biodiversity due to climate change (IPCC, 2000), it is appropriate to examine the sensitivity of community biodiversity traits to climatic variability (Wall *et al.*, 2001). The high Andes are relatively poorly studied in this respect, yet there are indications of major climatic changes in progress.

Patterns of primary productivity have been studied intensively at some high mountain sites (Bliss, 1966; Larcher *et al.*, 1975; Meurk, 1978; Webber, 1974; Williams, 1977), but relatively little is known of individual plant growth patterns, survival and longevity during full life cycles. There are not many studies of natural plant growth and productivity in the Andes although there are a good number of physiological and phenological studies, as well as productivity of agricultural plants (Arroyo *et al.*, 1981; Halloy, 1983; Hermann and Heller, 1997; Mann, 1966). One growth study on *Azorella* cushion plants of the high Andes in Peru estimated a remarkable longevity of over 3000 years (Ralph, 1978). Can such measurements of growth be replicated in other regions and over many years? What are the growth rates of other characteristic high-Andean plants, and how do they relate to community structure changes and climatic variability? The answers to such questions become important, given the urgent need for environmental indicators in the face of major global environmental changes.

This paper explores dynamic changes in the structure of diversity and functional responses of key indicator species of high-Andean vegetation over 22 years and how it correlates to environmental factors in the Cumbres Calchaquíes mountains of NW Argentina between 3750 and 4650 m altitude. In doing so, I test the above questions as operating hypotheses. The site is well suited to the purpose of understanding sensitivity of biodiversity to climate change. It represents a high altitude area sensitive to global warming incorporating vegetation boundaries, is climatically and biologically relevant to a wide section of the Central Andes, and is relatively high in biodiversity.

METHODS

Location

The Cumbres Calchaquíes are an isolated mountain group belonging to the Pampean system and separated from the Andes by a deep valley. The mountain forms a broad plateau above 4000 m, occupied by over 20 shallow glacial lakes, the centre of which is referred to as Huaca-Huasi (26° 40′ S, 65° 44′ W).

Lake levels

The lake level of Laguna Nostra, one of the most stable of the central Huaca-Huasi lakes, was measured along the horizontal slope on the edge. These measures were transformed to depths through a detailed survey of the variations in slope when the lake was empty. The lake level acts as a proxy for hydrological balance, integrating precipitation and evapotranspiration.

Temperatures

Soil temperatures were measured in open-ended plastic tubes installed from −10 to −190 cm depth. A collection of such tubes has been permanently installed on the flat top of a moraine with rocky–silt–clay soils with cryptofruticetum vegetation (a species-rich vegetation dominated by plants completely flattened on the ground, including flat rosettes, plaques and cespitose herbs) characteristic of the Huaca-Huasi plateau at 4250 m. Daily maximum and minimum of hourly to two-hourly interval measurements (from dawn till dusk) during a few days each month provided representative values for each month. The anomalies in standard deviations from the monthly means over several years were calculated.

Biodiversity

Because of isolation, the Cumbres Calchaquíes are considered a unique biogeographic unit, the Calchaquí District of the high-Andean biogeographic province (Halloy, 1985b; Halloy, 1997). This district has evolved a unique and diverse flora and fauna, with general high-Andean connections but a high degree of endemism. The flora and fauna of the area have been inventoried and quantitatively sampled from 1977 to 1985. Prior to that, scientific collecting expeditions have left important reference collections since the late 1800s (deposited at the Fundación Miguel Lillo, Tucumán), providing benchmark material for species presence and morphology.

Communities can be grouped into seven major vegetation types. Permanent plots were established within these associations from 4240 to 4640 m. Plots ranged in size from 1 × 1 m (22 plots in denser vegetation) to 10 × 10 m (sparse vegetation at the limit, 8 plots). The distribution of cover among species in a quadrat allows the determination of total plant cover, species richness, the Shannon–Weaver diversity index ($H = -\Sigma\, p_i \ln p_i$, where p_i = cover of species i divided in total vascular plant cover), and the chi^2/n index of distance from the lognormal (a measure of disturbance, Halloy and Barratt, 1999). Plots were resampled four times between 1977 and 1985.

Linear regression analysis was used to explore the correlation between lake level, soil temperature and growth rates of *Azorella* for the 12 years from 1977 to 1988. For a chronology of measurement dates see Halloy (1985a). Because small year-to-year variations constitute noise that can hide a correlation with extreme events, a second analysis was done using only four extreme year events (1977, 1978, 1983, 1984). This analysis included the diversity parameters of the cryptofruticetum vegetation, chosen as a focus because of its species richness (it contains over half of the vascular plant species censused in relevés) and representativeness (it covers around 40% of the Huaca-Huasi basin).

Growth

Total diversity measurements are time consuming and hence very rarely done. In addition to the quadrat diversity data, and as part of a larger study, growth parameters of several dozen species of plants were measured at seven main locations along an altitudinal transect from 2850 m to 4650 m. This Chapter reports on six key species and four of these locations. Entrada Ciénaga Grande (or Entrada for short) is a SW facing area at 3750 m on sloping ground close to the lower end of a high-Andean bog (ciénaga). The dominant vegetation is tussock grassland with shrubs among rocks and sand. Piedra Grande is on a rocky and sandy N facing slope beside the upper reaches of Ciénaga Grande at 4050 m. Vegetation consists of dwarf shrubs, spiny tussocks and rosettes. The Huaca-Huasi site is in the centre of the plateau on gently undulating glacial till terrain of silt–clay and stone, dominated by cryptofruticetum. The Negrito site is on the north side of the summit of the Negrito mountain, a rocky slope above the limit of closed vegetation at 4430 m with sparse shrubs, cushions and rosettes.

Species

Of the six key species, the first four are listed as endangered by the World Conservation Union (Jenkins, 1995). The *Azorella* cushion plant (*Azorella compacta*), is a dense massive cushion growing up to 5–6 m in diameter (more frequently 1–2 m). The species is sparsely distributed in high mountains from south Peru through Bolivia to north Argentina and Chile. Outside of its peculiar adaptations, *Azorella* is of interest for its multiple uses as a high caloric, low smoke fuel, as well as medicinal value for treating diabetes, asthma, colds, bronchitis, kidney and womb complaints (Wickens, 1995). The following co-existing species bracket the *Azorella* data: a woody central-rooted cryptofrutex plaque (flat cushion, Halloy, 1990) called cuernito (*Adesmia crassicaulis*, endemic), a herbaceous central-rooted plaque (*Geranium planum*, restricted endemic), a central-rooted rosette herb (*Nototriche caesia*, endemic), a loose adventitious-rooted plaque called yaretilla (*Pycnophyllum convexum*, Puna distribution), and a wave-forming grass (*Festuca nardifolia* var. *calchaquiensis*, endemic subspecies). For convenience, these species will be referred to by the genus name.

Growth measurements

Plants ranging in size from seedlings to large adults were marked by distances and directions from a metal stake or by position with reference to rocks. Measurements included:

– crosswise diameters to extreme green parts of plant (in curved cushions following the curvature), providing a horizontal radial growth rate.
– for cushions, marks were engraved at 1 cm intervals in bedrock schist touching the plant edge. Distance to these marks indicated seasonal contraction and increase. Depending on the slope of the rock, this also provides a horizontal to vertical growth rate.

Over 40 *Azorella* were measured but the data reported here relate to 19 plants with consistent measures since 1977 (some died, some were marked later and will be used in future studies). Photographic records, annotations of shape and phenological state, comparing total diameters with rock measurements, and in *Adesmia* comparison of estimates with growth rings provided validation of growth rates.

RESULTS

Water and temperature

Lake levels showed remarkable multi-annual variations, characterised by rapid filling then a slow decline which lasts for 10–20 years (Figure 18.1). There seems to be a moderate inverse relation between this hydrological cycle and the global El Niño cycle. Photographic records and old lakeshores provide evidence of high lake levels in the recent past, but the lake has not reached overflow within the last 70 years. Present levels seem to be the lowest in hundreds, if not thousands, of years. This is consistent with a general drying trend observed in the whole Puna and Altiplano region from here to Lake Titicaca, associated with glacier retreat (Cabrera, 1968; Hastenrath, 1979).

The mean annual air temperature is 1.5 °C for 1977–1979 (Table 18.1; all climatic measurements are in the centre of the plateau at 4250 m). Annual precipitation is estimated around 385 mm (1977–1978), falling mostly as snow during the summer months, but year to year variation is substantial. A detailed description of the environment and soils is found in Halloy (1985a).

Soil temperatures are within ranges similar to those found in other alpine areas, with a mean of 3.7 °C at − 30 cm, and an absolute maximum of 13.7 °C and minimum of −4.2 °C. At −150 cm, the absolute maximum was 8.1 °C and the minimum 0.8 °C. Subtracting 2.2 °C from the monthly values of the −150 cm temperatures gives a close (SE ± 0.4) approximation to the monthly air temperature at 1.5 m aboveground, providing a practical climatic indicator with only a few measurements at that depth. Because of the

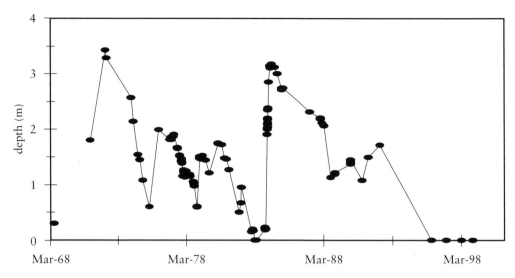

Figure 18.1 Lake level fluctuations in Laguna Nostra, Huaca-Huasi, 4250 m

Table 18.1 Mean monthly values of temperatures (air and soil), and precipitation, at Huaca Huasi. From Halloy (1985a) and subsequent data

	Jan	Feb	Mar	Apr	May	Jun	Jul	Aug	Sep	Oct	Nov	Dec	Mean or total
Temperature °C													
Air at 1.5 m	4.5	3.4	5.5	–1.2	–2.0	–2.7	–3.5	–0.8	0.7	3.5	5.6	4.5	1.5
Soil at – 30 cm	8.6	7.4	6.7	4.6	0.7	–1.7	–1.9	–0.4	0.3	4.0	8.6	8.0	3.7
Soil at –150 cm	6.9	7.0	6.8	–	–	–	–	1.0	–	5.7	–	6.2	–
Precipitation mm													
	58	80	37	11	2	10	1	2	4	24	46	74	349

low short-term variations in temperature at this depth, long-term temperature changes become much more obvious.

Species richness and diversity

Some 201 species (in 103 genera and 43 families) of vascular plants have been recorded in the 150-km² area above 4000 m, representing the high-Andean or high-alpine vegetation. This is around 12% of the flora recorded in the Anconquija centre of plant diversity, including lowland subtropical forest down to 400 m. In addition, 53 species of vertebrates have been recorded above 4000 m. The high-Andean vegetation transitions on the eastern slopes at around 3900 m with grassland páramo, while on the drier western slopes it transitions around 3750 m with the shrub páramo and, further down, prepuna vegetation. Vegetation becomes sparse above 4400–4500 m and the regional maximum altitudes recorded for vascular plants is between 5100 and 5500 m.

Short-term climatic variations have little effect on the vegetation of mostly long-lived plants, but long-term variations may have important effects. Since the 1940s, at least four species of plants and three of vertebrates have disappeared from the area above 4000 m, while three new species of plants are now found which

Table 18.2 Diversity changes with time in a typical cryptofruticetum vegetation plot (C1)

Vascular plants	Summer of			
	1977	1978	1984	1985
Total cover	18.9%	18.1%	17.5%	20.2%
n species	16	18	12	20
Species lost		1	6	0
Species added		3	0	8
Shannon–Weaver index	2.10	2.02	1.73	2.33
chi²/n	1.11	0.64	1.23	0.64

were not found before (Halloy, 1985a). Although the plants may have been affected by increasing grazing pressure and taruca deer may have been hunted to extinction, it seems likely that the drying climate is implicated in the local disappearance of some of the plants and of two flamingo species. As a result of the major 1983 and 1995 onward droughts, many species of aquatic birds have entirely disappeared from the area. The endemic lizard, *Liolaemus huacahuasicus*, has had its population decline to less than a tenth of its pre-drought condition. *Liolaemus* relies on flowers for food and although its food plants have not disappeared, they do not flower abundantly during droughts. Invertebrate prey are probably also affected.

It is known that even when species richness and diversity remain stable, there is a continuous turnover of species within equilibrium systems (e.g. Brown *et al.*, 2001; Halloy, 2000; MacArthur and Wilson, 1963; Simberloff, 1974). Thus the appearance of new species, or the disappearance of others as part of the normal dynamics of the system has to be distinguished from similar events which may be due to major disturbances to the system.

The measurements of the cryptofruticetum vegetation show that, in addition to species turnover, the species richness and diversity fluctuated substantially, indicating events affecting the system as a whole. Species number and the Shannon–Weaver diversity index plummeted, while the chi²/n disturbance index rose following the 1983 drought (Table 18.2), indicating a substantial stress to the community.

Growth rates: species variations

Growth against rocks was, on average, about half the radial growth rate (overall mean for *Azorella* of 1.64 mm a^{-1} vs 3.39 mm a^{-1}). The only other available growth rates for *Azorella compacta*, published by Ralph (1978) for Peru, fit well with the rock-based measurements made here (mean of 1.35 to 1.69 mm a^{-1}, Table 18.3). The related *Azorella selago* growing close to sea level in the subantarctic islands shows growth rates which are higher than the high-Andean *Azorella compacta*, both in horizontal and vertical measurements. Growth rates decrease with altitude in *Azorella*. Although the two lower sites have the same mean radial growth rate of 4.17 mm a^{-1}, the maximum is lower at the 4050 m site. At the top site of 4430 m, mean radial growth is down to 1.55 mm a^{-1}, lower than both the other sites ($p < 0.05$).

Adesmia showed a minimum growth rate (4.9 mm a^{-1}) more than double the mean for *Azorella* at the same site, but the mean and maximum were around three times larger. The lower growth rate seems more compatible with the rate indicated by growth rings. *Pycnophyllum* also showed more than double the growth rate of *Azorella*, with a mean of 4.27 mm a^{-1} at the highest site. The circular *Festuca* grasses spread at a relatively rapid average rate of 18.7 mm a^{-1}. This is achieved by a strategy of an advancing front leaving no live tissues behind to consume resources (Halloy, 1985a).

The interspecific differences in growth rates are all significantly different at $p < 0.05$. Although these rates may seem low compared to lowland plant growth rates, they are relatively high compared to some alpine plant growth rate estimates (e.g. *Carex curvula* rhizome growth rates in the Alps between 0.4 and 0.9 mm a^{-1}, Steinger *et al.*, 1996).

Temporal variations in diversity and growth

Like plants in most seasonal climates, high-Andean plants exhibit a diverse range of phenological strategies. *Azorella* remains green

Table 18.3 Growth rates of Andean plants. *p* values are in comparison to *Azorella* at 4430 m

Species	Location	Type	Growth rate, mm a⁻¹			Longevity y	Source
			Mean	Max	Min		
Azorella compacta	Cumbres Calchaquíes, 3750 m	Radial	4.17 (p = 0.002)	16.4	-3.0	>400-1000	this Chapter
	Cumbres Calchaquíes, 4050 m	Radial	4.17 (p = 0.04)	10.1	-2.0	>400-1000	
	Cumbres Calchaquíes, 4430 m	Radial	1.55	12.3	-8.9	>400-1000	
Adesmia crassicaulis	Cumbres Calchaquíes, 4430 m	Radial	14.0 (p < 0.001)	15.1	4.9	~50-140	
Geranium planum	Cumbres Calchaquíes, 4250 m	Small plaques with large tap root (Halloy, 1998)				~50	Halloy (1983)
Nototriche caesia	Cumbres Calchaquíes, 4250 m	Small rosettes with thick tap root				~50	
Pycnophyllum convexum	Cumbres Calchaquíes, 4430 m	Radial	4.27 (p = 0.01)	15.9	2.1	~25-190	this Chapter
Festuca nardifolia calchaquiensis	Cumbres Calchaquíes, 4250 m	Radial	18.7 (p < 0.001)	23.9	15.3	Indefinite (clonal)	
Azorella compacta	Southern Peru, 3960 m	Small, 30-75 mm plants, vertical	1.69	3.75			Ralph (1978)
	Southern Peru, 4500 m	Large, > 300 mm plants, vertical	1.35	3.25		850-3000	
Azorella selago	Marion Island	Radial	< 5	~10 or more		83 to > 100	Huntley (1972)
		Vertical, fjaeldmark	1.8	2.4			
		Vertical, herbfield	6				
	Macquarie Island	Radial	6.35	7.6			Taylor (1955)

Figure 18.2 *Azorella compacta* cushion at Entrada Ciénaga Grande, 3750 m on 5 October 1980 just after sunrise. From Halloy (1985a)

throughout the year (Figure 18.2). *Adesmia* and *Geranium* lose their leaves in winter, while *Nototriche* leaves die but remain attached to form a stypochaetium (thick sheath of dead leaves around the stem). The growth and decrease of aboveground biomass and productivity of *Nototriche caesia* has been reported elsewhere (Halloy, 1983). *Pycnophyllum* turns yellow-green and stops growing in winter but does not shed leaves. *Festuca nardifolia* turns yellow with only a few purple-green leaves remaining alive.

Azorella not only stops growing in winter, it exhibits an annual cycle of growth and retraction (Figure 18.3). Although most plants retract during a dormant season, this retraction is achieved by shedding or dying of plant parts. *Azorella*'s contraction is more like that of a cactus (although all other characters are entirely different), where the whole green body remains intact yet contracts.

From 1977 to 1981 the −30 cm soil temperatures were moderately stable (mean anomaly no larger than 0.3 standard deviations away from the mean temperature), but then show a drop in 1982 and 1984 and a strong rise in 1983 and 1985. The warming of 1983 corresponds to the extreme drought event (Figure 18.4). As one would expect the soil temperature is cooler in wetter years and warmer in the drier years. The 12-year data set gives no

significant relation between temperature and lake level; but growth rates gave a significant positive relation both with lake levels and temperature anomalies. However in the 4-year sample, the extreme drought of 1983 linked low water to higher temperatures, creating a non-significant inverse relation between lake level and temperature (Table 18.4).

Correlations do not indicate cause and effect, but within a set of similar conditions they may provide a practical capability to predict with some confidence margin one set of parameters from another. Higher lake levels meant a higher species richness and diversity, with less disturbance effect. Growth rates of *Azorella* provide a reasonable proxy for community effects on species richness and stability (low disturbance). Although significant correlations are found (Table 18.4), the large noise of environmental variation, combined with complex issues of time lag (do we correlate present species richness with lake levels, this year, one year back, two years back, average the two previous years, use only the minimum, only the max, etc.?) place restrictions on the capability of statistical tools to produce meaningful results.

Longevity

Mean radial growth rates over all sites for *Azorella* rapidly decrease with size from 4.93 mm a^{-1} for plants below 11.2 cm diameter, 3.03 for 11.2 to 25.2 cm, and 2.43 for plants over 25.2 cm. Assuming growth rate remains constant above 25.2 cm diameter, a 3-m individual of *Azorella* would attain over 500 years of age. Because growth rates decrease more with age, some *Azorella* reach diameters of 5 m or more, and individual plants grow slower than the mean, it is likely that individual *Azorella* may reach above 1000 years of age. Given that *Azorella* seems to frequently go through years without any growth at all or even some contraction, these estimates are conservative but still imply lower ages than the great longevity estimated by Ralph (1978) for Peruvian plants. Like many long-lived plants (Harper, 1977), *Azorella*

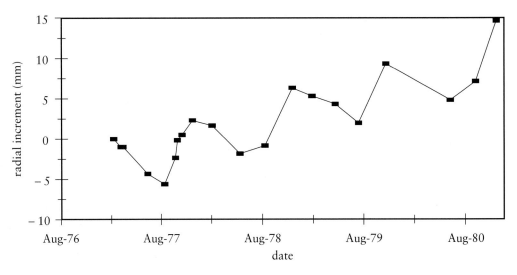

Figure 18.3 Cyclic growth and retraction of a typical *Azorella* (n1 Entrada) from initial marking point of 0 in March 1977

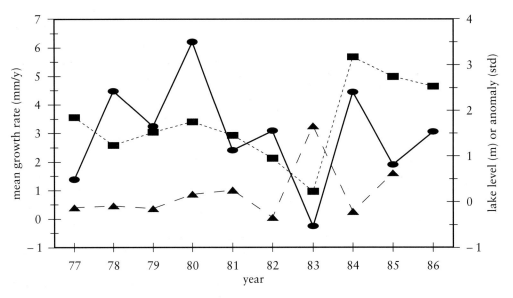

Figure 18.4 Yearly variations of mean growth rates of 19 *Azorella* (bold line, circles) at three sites, compared to lake level variations (squares) and soil temperature anomalies (triangles) at Huaca-Huasi

Table 18.4 Correlation coefficients of environmental and biodiversity factors for extreme years at Huaca-Huasi. The sign indicates the direction of the relation. *significant at $p <= 0.05$

	Soil temperature	Growth rates	Species richness	Diversity (Shannon)	Disturbance (Chi²/n)
Lake levels	−0.91*	+0.67	+0.91*	+0.99*	−0.86*
Soil temperature		−0.63	−0.80*	−0.77*	+0.51
Growth rates			+0.89*	+0.56	−0.99*
Species richness				+0.86*	−0.84*
Diversity (Shannon)					−0.50

takes some years to first flower and fruit, the youngest individual yet seen in flower was about 16 years old.

The faster-growing *Adesmia* and *Pycnophyllum* may reach close to 150–200 years of age, while the smaller plaques of *Geranium planum* will reach around 50 years for a final diameter of 16 cm. Even the small rosettes of *Nototriche caesia* are only beginning to flower at 4 years at the earliest and have a good number of flowers only by 15–20 years of age (being 3–4 cm in diameter at that stage). This plant is also estimated to reach an age of around 50 years (Halloy, 1983).

Assuming that growth rates remain constant in *Festuca nardifolia* var. *calchaquiensis*, as there is no accumulation of body mass, the waves of advancing *Festuca* seen in the field probably reach several hundreds or thousands of years of age. There is no indication that such clones 'die' and many may be as old as the last glaciation (~12 000 years).

Productivity and biomass

Total net primary productivity for the Huaca-Huasi cryptofruticetum vegetation has been estimated at 0.3 to 1.2 t ha^{-1} a^{-1} based on growth rates and dry matter for *Nototriche caesia* and *Calycera pulvinata* (Halloy, 1981; Halloy, 1983). Total biomass estimates based on the former plant were 0.27 t ha^{-1} on actual cover basis, and 5.79 t ha^{-1} based on a total community extrapolation (60% cover).

Based on the reported dry matter density of 0.35 g cm^{-3} for *Azorella compacta* reported by Ralph (1978), and assuming a medium growth over the whole plant surface of 2 mm a^{-1} (between the mean growth against rock of 1.64 and horizontal growth of 3.39), the plant would produce 700 g m^{-2} a^{-1} of aboveground dry matter. Assuming a 150-day production period, this translates to 4.67 g m^{-2} d^{-1} aboveground production. This compares to 0.2–0.8 g m^{-2} d^{-1} of total above- and belowground production for *Nototriche caesia* (Halloy, 1983). If *Azorella* covered 100% of the ground, this would mean 7 t ha^{-1} a^{-1} aboveground production, but it rarely covers more than 1% in this area (Ralph reported 12% for the Peruvian study site).

Although total productivity on a per area basis is low due to low plant cover, individual *Azorella* plants show a total aboveground dry mass accumulation which is as high or higher than most alpine communities (compare for example data in Meurk, 1978; Scott and Billings, 1964; and that reviewed in Körner, 1999). The biomass of *Azorella* estimated at 110 t ha^{-1} in a dense Peruvian stand is high by alpine standards and comparable to that of temperate forests (Ralph, 1978). With a 1.8% cover, as at a particularly dense area of the Negrito site, aboveground biomass would be 4.2 t ha^{-1}. Bliss (1966) and Körner (1999) noted that alpine plants of temperate zones were as efficient in dry matter production as temperate lowland plants. The above numbers show that in the subtropical Andes, species like *Calycera pulvinata* and *Nototriche caesia* may be considerably less efficient, while species like *Azorella compacta* are at the upper range of published estimates.

DISCUSSION

Species richness in the Cumbres Calchaquíes is high compared to other high mountains of comparable size, and particularly to mountains in the Central Andes. Compilations from around the world suggested that whole alpine floras of single mountain ranges are typically in a range of 200–280 species belonging to around 40 families (Körner, 1999). However the Cumbres Calchaquíes possess this number of species for the high-Andean zone only (i.e. there will be another few hundred species below this level in the tussock grassland zone). In comparison, for the neighbouring Andes of northwestern Argentina, on a total area almost ten times as large as that of the Cumbres Calchaquíes, Ruthsatz (1977, in Simpson and Todzia, 1990) obtained 273 species in 171 genera above 3700 m, or 182 species for the high-Andean region alone (Ruthsatz and Movia, 1975). Cabrera (1958) recorded 221 species of vascular plants for the high-Andean province of the whole northwestern Argentina region. In a comparison of high-alpine sites of New Zealand (four sites), the Andes (three sites) and Europe (one site), Halloy and Mark (1996) found well-studied areas of 110 to 170 km^2 had 74 to 156 species of vascular plants. The alpha diversity per square metre of 20 species and above in the cryptofruticetum is also high.

Adaptive strategy in *Azorella*

Azorella cushions have achieved a unique level of condensation of the plant body as only few other cushions have. The plant body is so firmly compact that the photosynthetic surface is a single large dome, made up of numerous small leaves in intimate contact with each other. Through the effect of mass and reduced surface to volume ratio, this compact body exhibits smaller thermal and water fluctuations than its environment (Halloy, 1985a; Ruthsatz, 1978). *Azorella* cushions are so compact that a cushion of *Azorella madreporica* of Chile turned off a revolver bullet according to Reiche (1907) quoted by Huntley (1972). As a result, no other plants can grow in their midst, the perfect aboveground competitive strategy (Grime, 1979). As *Azorella*s grow they slowly but relentlessly smother any competing plant. *Azorella* is also unique in reaching some of the highest dry matter accumulations per unit time and area of any high altitude plant. This dense cushion is likely to be particularly efficient in recirculating carbon dioxide from the decomposition of its older leaves back to newer leaves, a subject which deserves research (Halloy, 1985a; Körner, 1993).

Longevity and environmental recording

Although clonal plants are estimated to reach impressive ages, individual plants do not do so as easily. Trees rarely surpass 500 years, with the longest-lived believed to be the bristlecone pine, at over 4600 years. As single individual plants, *Welwitschia mirabilis* of the Namib desert is often regarded as the longest-lived non-tree plant, claimed to reach over 1500 years (Klötzli, 1991). In high mountains, the cushion *Acantholimon diapensoides* of the Pamir is claimed to reach 400 years (Agakhanyantz and Lopatin, 1978). In contrast, clonal plants have been estimated at many thousands or even tens of thousands of years (e.g. *Populus tremuloides* > 10 000 years old, Kemperman and Barnes, 1976; *Lomatia tasmanica* 43 000 years old, Gray, 1997; and the alpine *Carex curvula* reaching 2000 years of age, Steinger *et al.*, 1996). Körner (1999: p 289) used the term 'functionally immortal' for such clonal plants. This paper confirms the multi-centenary age of individual *Azorella compacta*, centenary age of several other high-Andean plants, and probable millennial age of high-Andean clonal plants. This places *Azorella* among a restricted number of the most long-living plants, and makes it probably older than most non-tree species known until now.

Long-lived plants act as recorders of environmental history and it seems that there would be interesting opportunities for plants like *Azorella* and *Adesmia* to provide alternative environmental proxies with innovative methods. Even at the lowest estimated ages, many *Azorella* now living have lived through major climatic changes, including the Little Ice Age. *Azorella*'s compact growth retains all leaf remains, as well as falling dust and pollen particles, suggesting the possibility of finding a chronological sequence of leaf (to explore leaf anatomical changes for example) and pollen (to explore vegetation changes) over several hundred years. It is as yet unknown to what degree decomposition and movement (percolation) within the plant mass may have degraded these messages, but the compact, resin-bound interior suggests these phenomena may be minimal.

Because growth rates vary with time and between individuals, establishing longevity from growth rates is at best an approximation which becomes more precise the longer the measurement period.

Vulnerability

The slow growth of high-Andean plants represents a risk for their survival when confronted with increasing human pressure. Dense populations of *Azorella* have been heavily exploited for fuel in the treeless Andes of Chile, Bolivia and Argentina (Wickens, 1995), and large, older specimens are becoming increasingly rare. During the period from 1915 to 1958 the mine of Chuquicamata alone consumed half a million tonnes of *Azorella*. Although the plant is now legally protected to some degree in Chile and Peru, large numbers are still being burnt locally there as well as in Bolivia and Argentina. Assuming a hemispherical shape, a 1 m diameter *Azorella* growing to 20 cm high would weigh around 37 kg (dry weight), while a 2 m × 40 cm *Azorella* would weigh close to 300 kg (not counting roots). Thus the numbers above may imply the extraction of well over a million multi-centenary plants. *Azorella* is concentrated on restricted high altitude rocky slopes, and is absent from the rest of its large range, implying that a very large area has been exploited. *Azorella* has become an iconic tourist attraction representing the unique strangeness and vulnerability of high altitude environments in the Andes (e.g. CONAF, 2001).

Adesmia is also used for fuel, but a 100-year-old *Adesmia* will provide only a small cooking fire for maybe an hour or less. Unlike *Azorella* which is harvested live and left to dry, most *Adesmia* are harvested as dead specimens. Other threats to these and most flat-growing Andean plants, such as *Pycnophyllum* plaques and *Nototriche* rosettes, are trampling by stock, people and, increasingly, by all-terrain vehicles. Long-term climate changes may affect many of these and other species, as suggested by the diversity data presented here. *Azorella* has been classified as vulnerable by the World Conservation Monitoring Centre (1994 listing) and *Geranium planum* is an extremely rare and vulnerable plant (Halloy, 1998). Unfortunately, despite efforts since 1913 to protect the Cumbres Calchaquíes and Anconquija from increasing external pressures, the area remains virtually unprotected (Halloy, 1997).

CONCLUSION

The results of measurements in the high-Andean region of the Cumbres Calchaquíes show broad patterns which are relevant to a large area of the Central Andes and to the whole high-Andean biogeographic province, given the similarities in flora and environment. High-Andean plant communities, as exemplified by the Cumbres Calchaquíes, are vulnerable to climatic fluctuations and direct human impacts. It has been suggested that global warming will place high mountain species at risk as they 'fall off the top' when rising temperatures raise vegetation limits. However, in the Central Andes, the most direct effect of climate change seems to be through desiccation. The drought events of the last decades have been related to significant losses of diversity and decreases in growth rates.

The long-term study of plots and marked individuals has provided valuable information on growth rates, longevity and environmental sensitivity of various high-Andean plant species. Growth rate variations between species constitute a direct functional aspect of biodiversity inasmuch as it represents the capture and allocation among species (biomass distribution) of energy, carbon and mineral resources. Plant growth, being at the base of the ecosystem pyramid, also implies knock-on effects throughout the food chain, as evidenced in the drought period with the reduction of plankton-feeding birds and flower-eating lizards. Expensive and time consuming climatic monitoring is not often a practical option in areas like the Andes. Cheap and simple measures, including lake levels, deep soil temperatures and plant size measures of key species, can be widely utilised as indicators of change.

ACKNOWLEDGEMENTS

Although they are too numerous to be named, I am deeply grateful to all those who have helped in the arduous field work in often difficult conditions, as well as those who have supplied lake level observations, filling in gaps. During the period of this study my successive employers have supported the vision of valuable results to be obtained at some future date. These include the Universidad Nacional de Tucumán and Fundación Miguel Lillo (Argentina) and Ministry of Agriculture and

Fisheries and Crop and Food Research (New Zealand). The final writing up of this data would not have been possible without the stimulus provided by the Global Mountain Biodiversity Assessment through C Körner and G Grabherr. My thanks for the detailed comments from reviews by C Körner, E Spehn and an anonymous referee.

References

Agakhanyantz OE and Lopatin IK (1978). Main characteristics of the ecosystems of the Pamirs, USSR. *Arctic and Alpine Research*, 10:397–407

Arroyo MTK, Armesto JJ and Villagrán C (1981). Plant phenological patterns in the high Andean cordillera in central Chile. *Journal of Ecology*, 69:205–23

Bliss LC (1966). Plant productivity in alpine microenvironments on Mt. Washington, New Hampshire. *Ecological Monographs*, 36:125–55

Brown JH, Morgan Ernest SK, Parody JM and Haskell JP (2001). Regulation of diversity: maintenance of species richness in changing environments. *Oecologia*, 126:321–32

Cabrera AL (1958). La vegetación de la Puna argentina. *Revista de Investigaciones Agrícolas*, 11:317–412, 16 Lám

Cabrera AL (1968). Ecología vegetal de la Puna. In Troll C (ed.), *Geo-Ecology of the Mountainous Regions of the Tropical Americas*. Colloquium Geographicum, Bonn, pp 91–116

CONAF (2001). *Parque Nacional Lauca*. CONAF

Gray A (1997). Sterile shrub is world's oldest plant. *Geneflow*, 1997:27

Grime JP (1979). *Plant Strategies and Vegetation Processes*. John Wiley & Sons, Chichester

Halloy SRP (1981). La presión de anhidrido carbónico como limitante altitudinal de las plantas. *Lilloa*, 35:159–67

Halloy SRP (1983). Datos ecológicos sobre *Nototriche caesia* Hill, Malvacea altoandina en las Cumbres Calchaquíes. *Lilloa*, 36:85–104

Halloy SRP (1985a). *Climatología y Edafología de Alta Montaña en Relación con la Composición y Adaptación de las Comunidades Bióticas (con especial referencia a las Cumbres Calchaquíes, Tucumán)*. University Microfilms International publ, Ann Arbor, MI

Halloy SRP (1985b). Reencuentro de *Azorella biloba* (Schlecht.) Wedd. en Tucumán. *Lilloa*, 36:267–9

Halloy SRP (1990). A morphological classification of plants, with special reference to the New Zealand alpine flora. *Journal of Vegetation Science*, 1:291–304

Halloy SRP (1997). Anconquija Region, Northwestern Argentina. In Davis SD, Heywood VH, Herrera-MacBryde O, Villa-Lobos J and Hamilton AC (eds), *Centres of Plant Diversity – A guide and strategy for their conservation*. WWF, IUCN, Cambridge, UK, pp 478–85

Halloy SRP (1998). A new and rare plate-shaped Geranium from the Cumbres Calchaquíes, Tucumán, Argentina. *Brittonia*, 50:467–72

Halloy SRP (2000). Effects on system structure, diversity and stability of the distance exponent in a resource attraction model. In Halloy SRP and Williams T (eds), *Applied Complexity – from neural nets to managed landscapes*. New Zealand Institute for Crop & Food Research, Christchurch, NZ, pp 328–46

Halloy SRP and Barratt B (1999). Patterns of species abundance as indicators of ecosystem status. In Holt A, Dickinson K and Kearsley GW (eds), *Environmental Indicators*. Environmental Policy and Management Research Centre, University of Otago, Dunedin, NZ, p 87

Halloy SRP and Mark AF (1996). Comparative leaf morphology spectra of plant communities in New Zealand, the Andes and the Alps. *Journal of the Royal Society of New Zealand*, 26:41–78

Harper JL (1977). *Population Biology of Plants*. Academic Press, London

Hastenrath S (1979). Clima y sistemas glaciales tropicales. In Salgado-Labouriau ML (ed), *El Medio Ambiente Páramo*. Centro de Estudios Avanzados, Mérida, Venezuela, pp 47–53

Hermann M and Heller J (eds) (1997). *Andean roots and tubers: Ahipa, arracacha, maca, yacon*. Institute of Plant Genetics and Crop Plant Research and Gatersleben/International Plant Genetic Resources Institute, Rome

Huntley BJ (1972). Notes on the ecology of *Azorella selago* Hook. f. *Journal of South African Botany*, 38:103–13

IPCC (2000). *Climate Change: Impacts, Adaptation, and Vulnerability*. Intergovernmental Panel on Climate Change (IPCC)

Jenkins C (1995). *List of threatened plants from WCMC database*. World Conservation Monitoring Centre (WCMC), Cambridge, UK

Kemperman JA and Barnes BV (1976). Clone size in American aspens. *Canadian Journal of Botany*, 54:2603–07

Klötzli F (1991). Niches of longevity and stress. In Esser G and Overdieck D (eds), *Modern Ecology:*

basic and applied aspects. Elsevier, Amsterdam, pp 97–110

Körner C (1993). Das 'Ökosystem Polsterflanze': Recycling und Air condition. *Biologie in unserer Zeit*, 23:353–5

Körner C (1999). *Alpine Plant Life – Functional Plant Ecology of High Mountain Ecosystems*. Springer, Berlin

Larcher W, Cernusca A, Schmidt L, Grabherr G, Nötzel E and Smeets N (1975). Mt. Patscherkofel, Austria. *Ecological Bulletin*, 20:125–39

MacArthur RH and Wilson EO (1963). An equilibrium theory of insular biogeography. *Evolution*, 17:373–87

McGradySteed J, Harris PM and Morin PJ (1997). Biodiversity regulates ecosystem predictability. *Nature*, 390:162–5

Mann FG (1966). *Bases ecológicas de la explotación agropecuaria en la América Latina*. Serie de Biología, OEA, Monografías 2, Washington DC

Meurk CD (1978). Alpine phytomass and primary productivity in central Otago, New Zealand. *New Zealand Journal of Ecology*, 1:27–50

Ralph CP (1978). Observations on *Azorella compacta* (Umbelliferae), a tropical andean cushion plant. *Biotropica*, 10:62–7

Ruthsatz B (1977). Pflanzengesellschaften und ihre Lebensbedingungen in den Andinen Halbwüsten Nordwest-Argentiniens. Quoted in Simpson BB and Todzia CA (1990). Patterns and processes in the development of the high Andean flora. *American Journal of Botany*, 77:1419–32

Ruthsatz B (1978). Las plantas en cojín de los semi-desiertos andinos del Noroeste Argentino. *Darwiniana*, 21:491–521

Ruthsatz B and Movia CP (1975). *Relevamiento de las Estepas Andinas del Noreste de la Provincia de Jujuy, República Argentina*. Fundación para la

Educación, la Ciencia y la Cultura (FECYC), Buenos Aires

Scott D and Billings WD (1964). Effects of environmental factors on standing crop and productivity of an alpine tundra. *Ecological Monographs*, 34:243–70

Simberloff DS (1974). Equilibrium theory of island biogeography and ecology. *Annual Review of Ecology and Systematics*, 5:161–82

Simpson BB and Todzia CA (1990). Patterns and processes in the development of the high Andean flora. *American Journal of Botany*, 77:1419–32

Steinger T, Körner C and Schmid B (1996). Long-term persistence in a changing climate: DNA analysis suggests very old ages of clones of alpine *Carex curvula*. *Oecologia*, 105:94–9

Taylor BW (1955). The flora, vegetation and soils of Macquarie Island. *ANARE Reports, Botany*, 1:1–192

Tilman D (1999). The ecological consequences of changes in biodiversity: a search for general principles. *Ecology*, 80:1455–74

Wall D, Mooney H, Adams G, Boxshall G, Dobson A, Nakashizuka T, Seyani J, Samper C and Sarukhán J (2001). An International Biodiversity Observation Year. *Trends in Ecology and Evolution*, 16:52–4

Webber PJ (1974). Tundra primary productivity. In Ives JD and Barry RG (eds), *Arctic and Alpine Environments*. Methuen, London, pp 446–73

Wickens GE (1995). Llareta (*Azorella compacta*, Umbelliferae): A review. *Economic Botany*, 49:207–12

Williams PA (1977). Growth, biomass, and net productivity of tall-tussock (*Chionchloa*) grasslands, Canterbury, New Zealand. *New Zealand Journal of Botany*, 15:399–442

A Scenario for Mammal and Bird Diversity in the Snowy Mountains of Australia in Relation to Climate Change

Ken Green and Catherine M. Pickering

19

INTRODUCTION

The Snowy Mountains (Figure 19.1) contain the largest contiguous area of subalpine and alpine habitat in Australia. Other subalpine and alpine habitats occur as relatively isolated areas in the mainland Victorian Alps and in the Central Highlands and higher peaks of the island of Tasmania (Costin, 1989). Altogether 11 500 km², or 0.15% of the continent is subject to winter snow cover (Green and Osborne, 1994). Predicted changes to the climate are likely to result in a dramatic decline in the total area receiving snow. The Intergovernmental Panel on Climate Change assessment (IPCC, 1996) presented scenarios for global warming for 1990–2100 with predicted warming of 0.7–2.1 °C by 2070. From these, the Australian CSIRO Climate Impact Group estimated regional warming and precipitation values for Australia (CIG, 1996). Under these scenarios the 'best case' scenario for snow is the least increase in temperature and the greatest increase in winter precipitation (see Whetton *et al.*, 1996 for a full discussion). For example, even a modest warming ('best case scenario' of only + 0.6 °C by 2070; Table 19.1) will result in a 39% reduction in the area that receives 30 days of snow per year in the Snowy Mountains and Victorian Alps (Figure 19.2; Whetton, 1998). Under the 'worst case scenario' (Table 19.1) a reduction of 96% is forecast by 2070 (Figure 19.2). A reduction of this scale in snow cover will affect the unique and biologically important flora and fauna of the

region (Green, 1998). Changes in the diversity and abundance of plants and animals may be particularly severe in Australia because the extent of the true alpine habitat is minimal, with limited high altitude refuges. For example, the highest mountain in Australia, Mt Kosciuszko (2228 m), is 500–600 m lower than the theoretical nival zone (Slatyer *et al.*, 1984), so that there is no opportunity for an altitudinal shift in the alpine zone.

Seasonal cover of snow is regarded as a major determinant of the faunal composition of the subalpine and alpine areas of the Snowy Mountains above 1500 m (Green and Osborne, 1994, 1998). Within the latitudinal band of southeastern mainland Australia which encompasses the Snowy Mountains, 25 species of mammals found between the coast and the western slopes occur in areas of winter snow cover (Green and Osborne, 1998). Most of these are also common at low altitudes with two exceptions, the mountain pygmy-possum (*Burramys parvus*) and the broad-toothed rat (*Mastacomys fuscus*). In the Snowy Mountains, the mountain pygmy-possum is found only above the level of the winter snowline and the broad-toothed rat only above 1000 m (Green and Osborne, 1994). In addition to native species, feral mammals including foxes, hares and horses are found in the mountains (Green and Osborne, 1994). Among the birds there are no species confined to the mountains, however, it is the composition of the avifauna and particularly the absence of some species, which are found nearby at lower

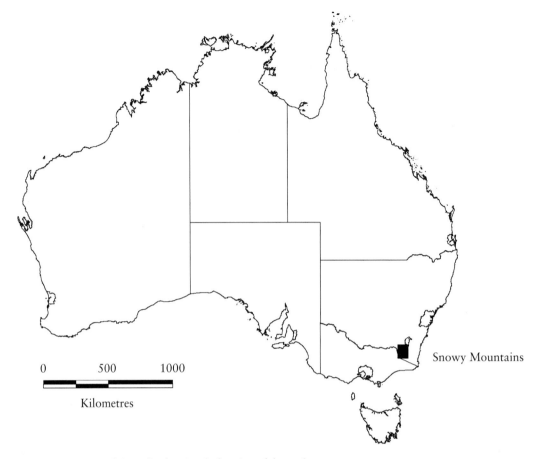

Figure 19.1   Map of Australia showing the location of the study area

Table 19.1   Climate change scenarios for the Snowy Mountains and Victorian Alps from Whetton (1998)

Changes in	Best case 2030	Best case 2070	Worst case 2030	Worst case 2070
Temperature	+0.3 °C	+0.6 °C	+1.3 °C	+3.4 °C
Precipitation	0%	0%	−8%	−20%

altitudes which is a characteristic of the snow-country avifauna (Osborne and Green, 1992). Of 271 native species of birds occurring at sea level, only 66 species are found above 1500 m, with only 13 of these being winter residents (Green and Osborne, 1994). In the present paper only the two faunal groups with the best data (mammals and birds) are examined, although there is greater alpine endemism among other groups such as the reptiles and the invertebrates (in which the majority of the endemic species and the alpine adapted forms are to be found) (Green and Osborne, 1994).

To determine whether there is any evidence for existing changes in animal distribution with climate, data on snow cover and animal distribution for the last 45 years are examined. This provides information on the potential impact of the dramatic climatic change predicted for the next 70 years for the Australian snow country.

240

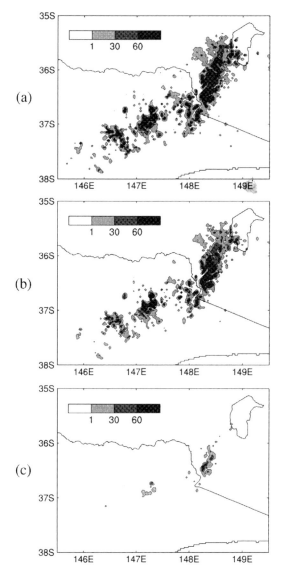

Figure 19.2 Simulated average duration of snow cover in days: a) in the current climate, b) the best case scenario for 2070, and c) for the worst case scenario in 2070 (figures from Whetton, 1998, based on Intergovernmental Panel on Climate Change scenarios for global warming)

METHODS

Site description

Areas of the Australian mainland subject to persistent winter snow are restricted to the southeast of the continent where the Snowy Mountains and the Victorian Alps contain about 2350 km^2 of land subject to a minimum of 60 days of snow cover in winter (Green and Osborne, 1994). The Snowy Mountains (Figure 19.1) lie on a north–south orientation centred on Mt Kosciuszko (36° 27′ S, 148° 16′ E). The alpine zone, extending from around 1830 m at the treeline to the top of Mt Kosciuszko, is characterised by continuous snow cover for at least four months per year and six to eight months with minimum temperatures below freezing. Precipitation is in the range 1800–3100 mm per year with about 60% of this falling as snow (Costin, 1957). The subalpine zone, extending from the winter snowline (around 1500 m) to the treeline is characterised by continuous snow cover for one to four months per year and minimum temperatures below freezing for about six months per year. Precipitation is in the range 770–2000 mm per year (Costin, 1975).

Trends in snow cover for the last 45 years

Snow data were obtained from a Snowy Mountains Hydro-electric Authority snow course at Spencers Creek in the Snowy Mountains that has been visited weekly through the snow season since 1954. To reflect depth and duration of snow cover, these data were transformed into metre-days of snow per year. This was done by multiplying the depth of snow by the numbers of days at that depth and summing the figures for each year. The data are presented by year with the running 5-year mean (Figure 19.3).

Impact of snow cover on animal distribution

The *Atlas of New South Wales Wildlife* which contains wildlife records for the past 30 years was examined to determine how many mammal species occurred in the southeastern corner of New South Wales, between 35° S and the Victorian State border and 147° E and the eastern seaboard. A Geographic Information System was used to allocate altitude to the nearest 100 m for each record. For the species

241

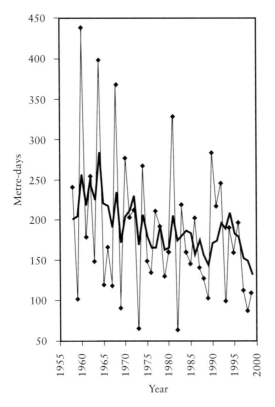

Figure 19.3 Annual snow cover in metre-days at Spencers Creek in the Snowy Mountains, with running five-year mean. Data courtesy of Snowy Mountains Hydro-electric Authority

of terrestrial mammals that declined in number of records with increasing altitude, the effect that snow cover had on their distribution was rated as: no observed effect (group 1), effect through competition (group 2) or direct effect (group 3) (see Green and Osborne, 1998).

To determine the possible impact of a reduction in snow cover on these species, the mean of the three highest records for each species was calculated from the same data set for the three decades 1970–1979, 1980–1989 and 1990–1999. This was only done for those animals found at or about the winter snowline and which did not inhabit the full altitudinal range, in order to determine whether and in what direction there was a shift in the highest altitudinal distribution of species. The search effort over this time is difficult to quantify, but

if the number of records placed on the data base per decade is taken as an index of effort, then this appears not to have changed greatly; with the number of records ≥ 1500 m altitude in the three decades being 586, 470 and 597. Sources of information independent of this data set were also recorded, such as observations by long-term residents at higher altitudes where a species had not previously been seen, indirect evidence and data from pest animal control studies.

For all bird species occurring seasonally above 1500 m, the earliest records above that altitude for the three decades were obtained from the *Atlas of New South Wales Wildlife*.

RESULTS AND DISCUSSION

Osborne *et al.* (1998) recorded a significant decline in mean snow cover at Spencers Creek snow course in the two 15-year periods 1960–74 and 1975–89, corresponding to a 27% reduction. Examination of the data from the snow course record show a total of 2283 metre-days of snow in the 1960s, with a 20% reduction to the 1970s (1843 metre-days) and a further 10% reduction in the 1980s and 1990s (1655 and 1706 metre-days, respectively). The last five years, occurring in the warmest decade of the century (Australian Bureau of Meteorology), had the lowest five-year average of the series; 7.5% less than the previous lowest 5 years and 53% less than the highest 5 years (Figure 19.3). If the fauna were sensitive to depth and extent of snow cover, this decline might be expected to be reflected in changes in their distribution.

Based on the pooled data from the *Wildlife Atlas*, 35 species of mammals, six of which are feral, generally decline in number of records with increasing altitude (Table 19.2). For 20 of the species, including all the bats, the decline in number of records with altitude is likely to be unrelated to snow (Green and Osborne, 1998; Group 1 in Table 19.2). For example, the seven possum species are dependent upon trees, and the house mouse *Mus musculus* might be excluded above the winter snowline by lack of suitable food (Green and Osborne, 1998).

Table 19.2 Species of mammal that decline in numbers of records in the *New South Wales Wildlife Atlas* with increasing elevation, together with probable reasons for decline and relatedness to presence of snow

Declining species	Reason
Group 1: Decline not snow-related in the short term	
Bats (11 species)	Reduction in flying insects
Possums (7 species)	Require trees
Koala	Absence of food tree species
House Mouse[a]	Lack of food for a specialist granivore
Group 2: Possible changes in the longer term	
Dog[a]	Absence of large prey/competition with fox[a]
Agile antechinus	Competition with dusky antechinus[c]
Swamp rat	Competition with broad-toothed rat[c]
Eastern pygmy-possum	Competition with mountain pygmy-possum[c]
Spotted-tailed quoll	Competition with fox[a]
Group 3: Decline likely to be snow-related	
Kangaroos and wallabies (3 species)	Mobility in snow/access to food
Wombat	Access to ground-based food
Echidna	Access to ground-based food
Bandicoot	Access to ground-based food
Cat[a]	Hunting method
Rabbit[a] Pig[a] Horse[ab]	Access to ground-based food

[a]Feral animals. [b]Within the study area, feral horses are not common below 1000 m. [c]These species increase in numbers with increased altitude

Table 19.3 Mean of three highest records (metres) for mammals in the *New South Wales Wildlife Atlas* (sample size in parenthesis), together with other sources of evidence for an altitudinal shift

Decade	1970–1979	1980–1989	1990–1999	Other evidence
Number of records ≥ 1500 m	586	470	597	
Grey kangaroo	1500(10)	1400(39)	1700(28)	
Swamp wallaby	1233(3)	1267(16)	1700(32)	1, 2, 4
Red-necked wallaby	1533(17)	1333(41)	1567(17)	
Cat[a]	1267(3)		1600(3)	1, 3
Horse[a]	1233(3)	1000(2)	1900(17)	1, 2
Pig[a]	1400(22)		1533(12)	1, 2
Rabbit[a]	1670(17)	1300(7)	1800(26)	1, 3, 5

[a]Feral animals. 1. Anecdotal records by long-term residents, 2. indirect evidence (dung or rooting), 3. increased presence in predator scats, 4. occurrence of tracks on regular snow transects, 5. rabbit control required for first time at ski resort at 1800 m

For a second group of species, the excluded or numerically reduced species may be out-competed by the more common higher altitude species which may derive a competitive advantage from the presence of snow (Dickman *et al.,* 1983). The three successfully competing native species, dusky antechinus *Antechinus swainsonii,* broad-toothed rat and mountain pygmy-possum, are the only ones that increased in numbers of records with altitude. Finally, there is a third group of mammals, six native and four feral species, where snow appears to be the major factor in the reduced numbers of records at higher altitude (Green and Osborne, 1998; Table 19.2).

There is some evidence that there is an increasing altitudinal distribution of animals over the 30-year period to 1999. *Wildlife Atlas* records indicate a higher maximum altitudinal distribution for all macropod species and for the four species of feral mammals (Table 19.3). Other evidence for increasing activity by feral mammals at higher altitudes supports this trend (Table 19.3). In the 1970s, Snowy Plains (1370 m) was regarded as climatically marginal for rabbits (Dunsmore, 1974), yet during the summer of 1998/99 the National Parks and Wildlife Service was forced to institute a rabbit control programme at Perisher Valley (1800 m) (Sanecki and Knutson, 1998). During the period 1980–1988, Green (1988) conducted regular small mammal trapping near the tree-line on the South Ramshead with access up a spur from Dead Horse Gap (1580 m) without once recording evidence of horses above the gap. Currently, this route is heavily used by horses for access to the alpine zone where sightings are now common (Green, pers obs.).

Among the macropods there is little evidence of altitudinal movements of red-necked walla-bies *Macropus rufogriseus* and grey kangaroos *M. giganteus* outside of the data set, and these records may be of aberrant individuals. For example, the grey kangaroo is a social species (Bennett, 1995) yet the high altitude records were all of individuals. The situation with the swamp wallaby *Wallabia bicolor* is, however, different. While the other two species of macropods are predominantly grazers, the swamp wallaby is a browser and therefore seasonal snow cover will not severely reduce its access to food. Despite an extensive winter fauna survey (Osborne *et al.*, 1978) and further winter work through 1979 (Green and Osborne, 1981), no tracks of swamp wallabies were observed in winter along the main access trail for these studies, whereas in recent winters (Green, unpublished) swamp wallaby tracks have been observed on this trail on vir-tually all weekly winter visits.

Additional support for the impact of snow cover on animal numbers comes from the response of species to years of shallow snow cover. In these years there is evidence for a reduction in populations of the three species of native mammal in the second group that increased in number of records with altitude (Table 19.2). Populations of dusky antechinus (Green, 1988) and broad-toothed rats (Green, unpublished) declined in poor snow years, while in the mountain pygmy-possum there

was also lowered recruitment (L. Broome, pers. comm. 2000). The latter species depends upon snow cover for stable, low temperatures for hibernation (Walter and Broome, 1998), whereas the two former species are active under the snow throughout winter (Green, 1998) and are therefore subject to predation by foxes (Green and Osborne, 1981). Using a bio-climatic analysis and prediction system to examine present and predicted future animal distribution, Brereton *et al.* (1995) suggested a potential decrease in areas of suitable habitat for broad-toothed rats with global warming. Already the broad-toothed rat is under pressure at Barrington Tops, 600 km north of the Snowy Mountains. The population studied there in the 1980s (Dickman and McKechnie, 1985) was in serious decline by 1999 in the face of invasion by swamp rats *Rattus lutreolus* (Green, 2000). A similar process may occur in the Snowy Mountains with a reduction in snow cover. This is likely to reduce the competitive advan-tage of broad-toothed rats and may act syner-gistically with increased incursions of feral animals, particularly foxes, whose hunting is made easier by shallow snow (Halpin and Bissonette, 1988).

In addition to the apparent changes in the distribution of some mammal species associa-ted with changes in snow cover, migratory birds may also be affected. Among the migra-tory birds, the only observable change in timing of arrival, for those species for which there was a sufficient sample size, was one of earlier arrival in the 1980s and/or 1990s compared to the 1970s. For the 11 bird species for which there were sufficient data (Table 19.4), the earliest record was in the 1990s in five species (four of these occurring in an earlier month) and the 1980s (generally differing by only a few days from the earliest date for the 1990s) in four. For two species (grey fantail *Rhipidura fuliginosa* and silvereye *Zosterops lateralis*) there was virtually no difference across the three decades. Whilst there has been a greater search effort in the 1990s by one of us (KG), the search effort in the period 1971–1980 was boosted by a large fauna survey (CSIRO, unpublished) and two studies of the avifauna

Table 19.4 Time of first records of migratory birds at or above 1500 m after winter, from the *New South Wales Wildlife Atlas* to the end of 1997. The earliest record is given in each decade and all other dates (maximum one per year) earlier than the first record in 1970–1979

Species	1970–1979	1980–1989	1990–1999
Number of records from August to October	124	65	152
Crescent honeyeater	19 Oct	26 Oct	12 Sep, 17 Sep, 30 Sep
Olive whistler	15 Sep	22 Sep	21 Aug
Flame robin	2 Sep	17 Aug	21 Aug
Grey fantail	19 Oct	26 Oct	23 Oct
Striated pardalote	16 Sep	24 Aug, 15 Sep	30 Aug
Yellow-faced honeyeater	18 Sep	26 Oct	12 Sep
Australian kestrel	5 Nov	20 Sep, 26 Oct	30 Aug, 8 Sep, 23 Sep, 28 Sep
Fantail cuckoo	25 Nov	21 Sep, 20 Oct	23 Oct
Red wattlebird	14 Oct	13 Sep, 20 Sep	20 Sep
Richards pipit	16 Sep	5 Sep	28 Aug
Silvereye	19 Oct	18 Oct, 20 Oct	22 Oct, 23 Oct

(Longmore, 1973; Gall and Longmore, 1978). While the earlier records in the 1990s might be ascribed to a greater search effort (152 records entered compared to 124 from August to October in the 1970s), the same explanation cannot be used for the 1980s. There were only half the number of records in the same period in the 1980s and yet six of the nine species that varied in time of immigration were recorded earlier in the 1980s than the 1970s.

The response of the birds is variable depending upon their foraging techniques. The birds that are recorded as arriving earlier include three species of honeyeaters that depend on the flowering of shrubs (see also Osborne and Green, 1992). The Australian kestrel *Falco cenchroides* is largely dependent upon snow-free ground for foraging. The ground-feeding flame robins *Petroica phoenicea* and Richards pipit *Anthus novaeseelandiae* arrive early in spring and feed on insects immobilised on snow, but the very fact of the earlier presence of these insects is associated with sufficient warmth in their point of origin for metamorphosis and flight. Olive whistlers *Pachycephala olivacea* and striated pardalotes *Pardalotus striatus* glean active insects off shrubs and trees and movements of fan-tailed cuckoos *Cuculus flabelliformis* must be attuned to the breeding timetable of their hosts. The two species that appear not to arrive earlier,

despite changes in snow cover over the three decades, are the grey fantail which catches insects in flight and the silvereye which is involved in long migratory flights, the timing of which may be independent of local events.

There are obvious problems with the data sets examining altitudinal distribution of mammals and time of first arrival of migratory bird species, such as sample size, sampling effort, sampling timing, and other provisos. The use of the numbers of records ≥ 1500 m entered per decade and the number of records in the three-month period of influx of migratory birds as indices, while not perfect, does suggest that there is no great bias in search effort. This paper does not state unequivocally that all of the changes in animal distribution (both spatially and temporally) are a direct result of observed changes in snow cover. For many of the changes there are plausible alternative explanations. Changing land use, however, is not one of those. During the course of this study period there have been no major changes in land-use patterns at the higher altitudes to explain differences in animal distribution. The study area was declared a State Park in 1944 and summer grazing was withdrawn from the alpine area of the Main Range in 1946 and banned above 1360 m in 1958 (Good, 1992). Any increased use by humans has largely been confined to the few ski resorts

245

and access routes. However, the changes are those that might reasonably be hypothesised to result from reduced snow cover and the observed change in snow cover is the only explanation that can account for all data. Therefore, the patterns documented here could be a model for likely changes to animal distribution, and hence biodiversity, with predicted changes in snow cover. Further global warming resulting in declining snow cover might, therefore, have a major impact upon the faunal composition of the alpine/subalpine areas of the Snowy Mountains, allowing greater access by feral animals and reducing the competitive advantage of the higher altitude species. As such, while possibly increasing the numbers of species in the Snowy Mountains, this process might reduce the regional biodiversity by the loss or serious reduction of populations of endemic species.

CONCLUSION

The only observed impacts of a loss of snow cover on biodiversity in the Snowy Mountains has been on mammals. Of the three classifications used here, there has been no observed altitudinal shift in the group of species whose distribution is unrelated to snow. For other species that increase in numbers of records

with increased altitudes, there have been documented impacts of shallow snow in some years, but because of the high fecundity of these species, populations have recovered in subsequent good years and no trend is apparent. It may take a greater loss of snow cover to reduce the competitive advantage of these species more than has so far occurred. In a third group, animals that are constrained by the presence of snow, there is a trend towards higher altitude and/or winter occupancy by species normally excluded by snow cover. Further loss of snow cover may see a greater upward movement of these species. If sufficient snow is lost, an altitudinal increase by the grazing macropods may follow the upward move by the browsing swamp wallaby.

ACKNOWLEDGEMENTS

We thank Cate Gillies for manipulating the *Atlas* data set and drawing Figure 19.1. Peter Whetton allowed us to use Figure 19.2. The Snowy Mountains Hydro-electric Authority allowed access to data to construct Figure 19.3 which was drawn by Mike Young. Professor JB Kirkpatrick (University of Tasmania) and Dr. WS Osborne (University of Canberra) commented on an earlier draft of the manuscript.

References

Bennett AF (1995). Eastern Grey Kangaroo. In Menkhorst PW (ed.), *Mammals of Victoria*. Oxford University Press, Melbourne, pp 138–40

Brereton R, Bennett S and Mansergh I (1995). Enhanced greenhouse climate change and its potential effect on selected fauna of south-eastern Australia: a trend analysis. *Biological Conservation*, 72:339–54

CIG (1996). *Climate Change Scenarios for the Australian Region*. Climate Impact Group, CSIRO Division of Atmospheric Research, Melbourne

Costin AB (1957). The high mountain vegetation of Australia. *Australian Journal of Botany*, 5:173–89

Costin AB (1975). Sub-alpine and alpine communities. In Moore RM (ed.), *Australian Grasslands*.

Australian National University Press, Canberra, pp 191–8

Costin AB (1989). The Alps in a global perspective. In Good R (ed.), *The Scientific Significance of the Australian Alps*. Australian Alps Liaison Committee, Canberra, pp 7–19

Dickman CR, Green K, Carron PL, Happold DCD and Osborne WS (1983). Coexistence, convergence and competition among *Antechinus* (Marsupialia) in the Australian high country. *Proceedings of the Ecological Society of Australia*, 12:79–99

Dickman CR and McKechnie CA (1985). A survey of the mammals of Mount Royal and Barrington Tops, New South Wales. *Australian Zoologist*, 21:513–43

Dunsmore JD (1974). The rabbit in subalpine south-eastern Australia. 1. Population structure and productivity. *Australian Wildlife Research*, **1**:17–26

Gall BC and Longmore NW (1978). Avifauna of the Thredbo valley, Kosciusko National Park. *Emu*, **78**:189–96

Good RB (1992). *Kosciusko Heritage*. New South Wales National Parks and Wildlife Service, Hurstville

Green K (1988). *A Study of Antechinus swainsonii and Antechinus stuartii and their prey in the Snowy Mountains*. Ph.D. thesis, Zoology Department, Australian National University, Canberra

Green K (1998). Introduction. In Green K (ed.), *Snow: A Natural History; an Uncertain Future*. Australian Alps Liaison Committee, Canberra/Surrey Beatty & Sons, Sydney, pp xiii–xix

Green K (2000). *A Survey of the Broad-toothed Rat at Barrington Tops*. Unpublished Report to New South Wales National Parks and Wildlife Service

Green K and Osborne WS (1981). The diet of foxes, *Vulpes vulpes* (L.) in relation to abundance of prey above the winter snowline in New South Wales. *Australian Wildlife Research*, 8:349–60

Green K and Osborne WS (1994). *Wildlife of the Australian Snow-Country*. Reed, Sydney

Green K and Osborne WS (1998). Snow as a selecting force on the alpine fauna. In Green K (ed.), *Snow: A Natural History; an Uncertain Future*. Australian Alps Liaison Committee, Canberra/ Surrey Beatty & Sons, Sydney, pp 141–64

Halpin MA and Bissonette JA (1988). Influence of snow depth on prey availability and habitat use by red fox. *Canadian Journal of Zoology*, **66**:587–92

IPCC (1996). *Climate Change 1995: Contribution of Working Group 1 to the Second Assessment Report of the Intergovernmental Panel on Climate Change*. (eds), Houghton JT, Meira Filho LG, Callander BA, Harris N, Kattenberg A and Varney SK. Cambridge University Press, Cambridge

Longmore W (1973). Birds of the alpine region, Kosciusko National Park. *Australian Birds*, **8**:33–5

Osborne WS and Green K (1992). Seasonal changes in composition, abundance and foraging behaviour of birds in the Snowy Mountains. *Emu*, **92**:93–105

Osborne WS, Davis MS and Green K (1998). Temporal and spatial variation in snow cover. In Green K (ed.), *Snow: A Natural History; an Uncertain Future*. Australian Alps Liaison Committee, Canberra/Surrey Beatty & Sons, Sydney, pp 56–68

Osborne WS, Preece M, Green K and Green M (1978). Gungartan: A winter fauna survey above 1500 metres. *Victorian Naturalist*, **95**:226–35

Sanecki GM and Knutson R (1998). *Perisher Range ski resorts rabbit control program*. Unpublished report, NSW National Parks and Wildlife Service, Jindabyne

Slatyer RO, Cochrane PM and Galloway RW (1984). Duration and extent of snow cover in the Snowy Mountains and a comparison with Switzerland. *Search*, **15**:327–31

Walter M and Broome L (1998). Snow as a factor in animal hibernation and dormancy. In Green K (ed.), *Snow: A Natural History; an Uncertain Future*. Australian Alps Liaison Committee, Canberra/Surrey Beatty & Sons, Sydney, pp 165–91

Whetton P (1998). Climate change impacts on the spatial extent of snow-cover in the Australian Alps. In Green K (ed.), *Snow: A Natural History; an Uncertain Future*. Australian Alps Liaison Committee, Canberra/Surrey Beatty & Sons, Sydney, pp 195–206

Whetton PH, Haylock MR and Galloway RW (1996). Climate change and snow-cover duration in the Australian Alps. *Climatic Change*, **32**:447–79

247

Modelling and Monitoring Ecosystem Responses to Climate Change in Three North American Mountain Ranges

20

Daniel B. Fagre and David L. Peterson

INTRODUCTION

Mountains play a significant role in human activities, providing critical resources such as minerals, forest products, and 50% of the freshwater consumed by people (Liniger *et al.*, 1998). Recently, mountains assumed a different strategic role because they are now recognised as important reserves of biodiversity, with potential value that we are just beginning to comprehend (Messerli and Ives, 1997).

Mountain protected areas and biodiversity

Mountains have greater biodiversity than their lowland counterparts because steep environmental gradients compress life zones, resulting in large species turnover over short distances (Beniston and Fox, 1996). Rugged mountain topography leads to higher rates of endemism through genetic isolation, and many mountain areas have been refugia for species during past climate changes. Lastly, mountains historically have been less altered by humans because of the harsher climate and logistical difficulties inherent in mountain living. Thus, the loss of biodiversity in agricultural and urban areas has not been mirrored to the same extent in the mountains. Mountain environments are better represented than lowlands in the world's inventory of parks and protected areas and, in the United States, many national parks were established *before* increasing human populations significantly modified the region. Although mountain ecosystems have retained their inherently greater biodiversity more than human-dominated environments, new pressures have led to mountain ecosystems being declared one of the Earth's most threatened resources. Stress caused by climate change may be more profound and occur earlier in mountains because of their steep climatic gradients (Oerlemans, 1994). Climate change will interact more with direct human activities, such as deforestation and habitat fragmentation, to severely threaten extant biodiversity of mountain ecosystems.

Climate change research at US mountain National Parks

We describe in this paper a multi-scale, multi-disciplinary approach examining how climatic change and variability drive ecosystem dynamics in three North American mountain protected areas – Olympic, North Cascades, and Glacier National Parks (Figure 20.1). Olympic and Glacier National Parks have been part of the US National Park Service (NPS) and US Geological Survey Global Change Research Program (GCRP), established by Congress in 1990 to look at impacts of global change on park natural resources. National Parks are excellent sites for climate change detection and research because it is easier to attribute changes in relatively pristine natural systems to regional and global-scale climatic shifts rather than local, human factors. National Parks represent an increasingly unique condition in which the original species and relationships are largely unaltered and available for study. Lastly, park management needs to know how

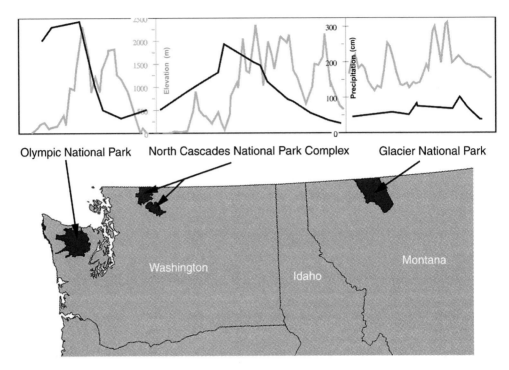

Olympic National Park   North Cascades National Park Complex   Glacier National Park

Figure 20.1 National parks dominated by mountains along the United States and Canadian border (49°N). A transect from Olympic National Park to Glacier National Park represents gradients in climatic variability and surrounding land use

climate change will affect the biodiversity they are charged with protecting. Because park boundaries are fixed and species responses to environmental gradients are spatially dynamic, under future climatic conditions many parks may contain different flora and fauna.

Although these protected areas are relatively pristine, they already have been subjected to climate change. For example, at Glacier National Park, glaciers have been reduced to mere remnants of their previous sizes and over two-thirds of the estimated 150 glaciers have disappeared since 1850 (Carrara, 1989), leaving approximately 37 today (Key et al., 1997). Because glacial meltwater can be critical for stream baseflow in late summer, there are significant ecological implications of glacier disappearance in park watersheds. Temperature-sensitive stream organisms such as net-spinning caddisflies (e.g. Hydropsyche spp.) may become less abundant, shift distribution or disappear as summer water temperatures increase in the

future, without glacial meltwater to keep streams flowing with cold water (Fagre et al., 1997). Major changes in glacial extent strongly indicate other fundamental changes in mountain ecosystems are occurring at Glacier and Olympic National Parks. Since 1990, different approaches were taken at each park to identify climatically-driven shifts in species distribution, species growth patterns, and ecosystem processes. As studies incorporated broader scales and regional landscape connections, the separate research groups coalesced. We summarise here the achievements of this diverse group, whose individual contributions are cited throughout this paper and who are listed in the Acknowledgements.

STUDY AREAS

Glacier, North Cascades and Olympic National Parks are large, wilderness-dominated parks near the United States–Canada border in the northern Rocky Mountains, the Cascade Mountains and on the Olympic Peninsula, respectively (Figure 20.1). Each park encompasses mountains with similar topographic relief, numerous glaciers and expansive conifer forests; each is snow-dominated, acts as the headwaters for its region and contains relatively intact floral and faunal assemblages. Climate is controlled by dominant air masses, providing Olympic with a maritime climate, North Cascades a transitional climate, and Glacier with a more continental climate. Thus, winter temperatures are moderate in the Olympics and cold in the northern Rockies. Summer precipitation as a proportion of annual precipitation is greater in the northern Rockies than in Olympics. Precipitation varies dramatically between westside and eastside locations within each park. For example, precipitation in the Olympic Mountains ranges from > 600 cm yr^{-1} on Mt Olympus to only 40 cm yr^{-1} in the northeastern rainshadow. Precipitation in the northern Rockies varies from 350 cm yr^{-1} (westside, high elevation) to 40 cm yr^{-1} (eastside, low elevation). This contrast in precipitation over relatively small distances has a profound impact on microclimate, vegetation distribution and disturbance regimes. Vegetation is dominated by coniferous forest, with species distribution and abundance varying along elevational gradients (extending to alpine vegetation) and from westside to eastside (including grassland). The western Olympics are dominated at low elevations by temperate rain forests with high biomass and abundant woody debris. Biomass and productivity generally are lower in the northern Rockies. The parks have 10 coniferous species and several plant communities in common, which allows comparisons of biotic responses to climatic shifts.

Lake McDonald and St Mary watersheds in Glacier National Park

The modelling approaches were developed initially in the Lake McDonald watershed and later applied to St Mary watershed (Figure 20.2). Lake McDonald watershed is a 462-km^2 forested watershed on the western slopes of the

Figure 20.2 Map of Glacier National Park with McDonald and St Mary drainages. Much of the model development and validation took place in McDonald drainage area

continental divide in the park. Elevation ranges from 948 m at the outlet to 2895 m in the alpine areas of the watershed. Approximately 75% of the watershed is forested, predominantly by conifers. Seven second-order streams drain the basin and flow into Lake McDonald. The watershed encompasses more than a dozen small lakes and approximately 250 km of streams and rivers. The basin comprises a mix of argillaceous limestones and tertiary sediments deposited by repeated glaciations. The St Mary watershed and St Mary Lake are similar in most respects to Lake McDonald, except that there is a distinct shift to more arid-adapted species, less biomass, more open forest canopies, reduced extent of forests and more expanses of grass.

RESULTS

Ecosystem modelling and monitoring in large landscapes

To address how climate change would alter western mountain National Parks, we chose a combined approach of ecosystem modelling and long-term monitoring. Few natural

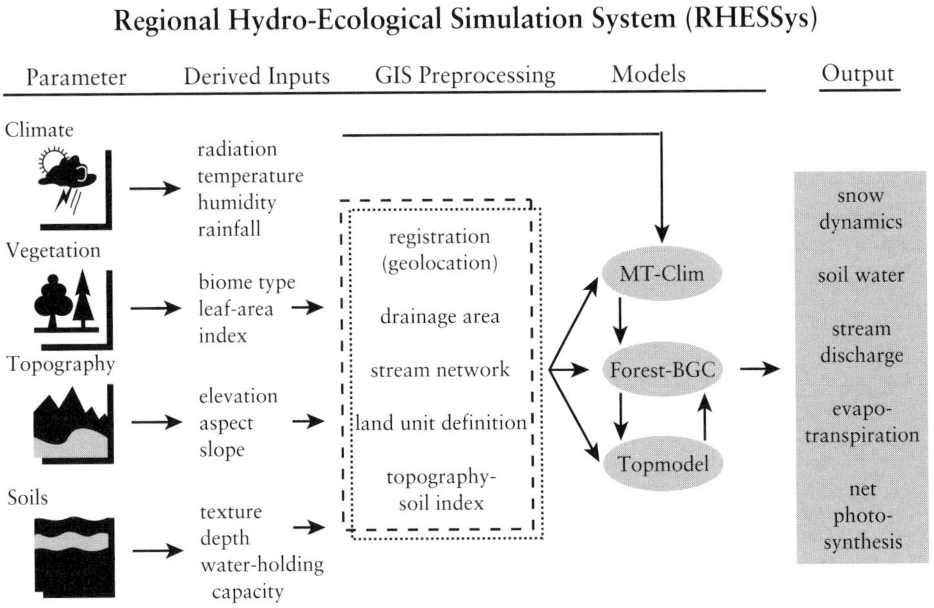

Regional Hydro-Ecological Simulation System (RHESSys)

Figure 20.3 Schematic of RHESSys organization depicting data inputs and ecosystem outputs. See text for details

resources have been monitored in any National Park for more than two decades, so long-term studies were initiated in the Olympic, North Cascades, and Glacier National Parks to document changes with the knowledge that their value might be many years in the future. These monitoring efforts, however, also provided parameterisation and validation data for ecosystem models. To predict potential climate change effects, we initially established an ecosystem modelling framework for Glacier National Park under current climatic conditions. Ecosystem modelling is a necessary tool for organising complex relationships in the biophysical environment into a structured logic that can produce quantitative predictions with which field data can be integrated and compared. Ecosystem models, once tested and validated, can help us to understand responses to climate change that at first seem counter-intuitive. An example is found in Running and Nemani's (1991) simulation of the response of a northwestern Montana forested watershed to a doubling of atmospheric CO_2. Despite a 10% increase in annual precipitation, streamflow decreased by 30%, especially in late summer

because snowpack duration was diminished by 2 months and, with a longer growing season increasing annual photosynthetic production up to 30%, greater evapotranspiration losses from the watershed occurred. These results, if true, would profoundly affect the watershed ecosystems of the Glacier National Park area.

Ecological models

We further developed the Regional Hydro-Ecological Simulation System (RHESSys) and FIREBGC (Fire–BioGeoChemical) models and applied them to Glacier National Park to answer questions about the state of natural resources and biodiversity in the future using several climate change scenarios.

RHESSys is a collection of evolving tools and interacting models that can be customised for specific needs (Band *et al.*, 1993) (Figure 20.3). RHESSys utilises remotely-sensed data from satellite platforms, such as Landsat Thematic Mapper data, to calculate metrics such as the Normalised Vegetation Difference Index (NDVI) and Leaf Area Index (LAI) for specific landscapes (Running and Gower, 1991). NDVI and LAI are coarse indices of vegetation function and structure that are applied to, or 'draped', over a Digital Elevation Model of Glacier Park within a Geographic Information System (GIS) to provide a three dimensional view of a cyber-ecosystem. A mountain climate simulator (MTCLIM) uses existing base station daily climate data to calculate daily variables (e.g. maximum and minimum temperatures, solar radiation, daily precipitation) for all slope, aspect, and elevation combinations (Thornton et al., 1997). These daily simulated climate data, coupled with a soils map, are used by a core model component, FOREST-BGC, to estimate ecosystem processes, such as gross photosynthesis, and calculate ecosystem outputs, such as net primary productivity and hydrologic discharge. FOREST-BGC is a forest growth model that calculates tree responses to nutrients, moisture and energy, and incorporates algorithms that describe details like carbon allocation to roots and stems (Running and Gower, 1991). Responses to daily climate variables are summed over the year to estimate annual amount of carbon fixed, available nitrogen and other ecosystem attributes. RHESSys is a flexible and modular system. Early in the project, a hydrologic routing model (TOPMODEL) was incorporated to account for subsurface water flow through topographically complex terrain and to estimate the daily flux in basin outflow (Band et al., 1993). This proved to be effective, estimating both daily and annual stream discharge accurately when compared to actual stream discharge measurements (Fagre et al., 1997). Recently improved routing models can easily be incorporated, underscoring the point that RHESSys is continually evolving. The key products from RHESSys for managers are spatially-explicit estimates of ecosystem responses to daily climate, displayed in a three-dimensional format.

FIREBGC is a companion model that is driven from the same forest process model, FOREST-BGC, as RHESSys but has a focus on tree stand dynamics and responses to fire as a disturbance process (Keane et al., 1996). FIREBGC is an individual tree model created by merging the gap-phase process-based model

FIRESUM with the mechanistic ecosystem biogeochemical model FOREST-BGC. It has mixed spatial and temporal resolution such that ecological processes that act at a landscape level, such as fire and seed dispersal, are simulated annually from stand and topographic information. Stand-level processes, such as tree establishment, growth and mortality; organic matter accumulation and decomposition, and undergrowth plant dynamics are simulated daily and annually. Tree growth is mechanistically modelled based on daily carbon fixed by forest canopy photosynthesis at the stand level. Carbon allocated to the tree stem at the end of the year generates the corresponding diameter and height growth. Outputs from FIREBGC include spatially-explicit estimates of tree species dominance, stem density, and forest floor biomass. These forest attributes help to determine the fuel loading and continuity necessary to generate stand-replacing forest fires. FIREBGC explicitly simulates long-term changes in fuels, fire hazard, fire behaviour and consequent effects on ecosystem characteristics of standing crop biomass, nitrogen cycling and leaf area index. FARSITE (Fire Area Simulator) is a model integrated with FIREBGC that maps fire extent and intensity for different climate conditions and successional trajectories in forest response. Thus, as the Glacier National Park cyber-ecosystem runs through daily cycles of climate and ecosystem response, these models allow us to 'see' — over a three-dimensional landscape — the changing structure and distribution of the forest, as well as the invisible ecosystem processes that underlie them.

How well do the models work?

Since 1992 we tracked daily, monthly, and annual dynamics of the Lake McDonald watershed and, less intensively, of the St Mary watershed. A key test was conducted by comparing daily estimates of temperature and precipitation with observed data from seven automated climate stations situated at various points in the watershed. These stations were placed to maximise topographic variability to

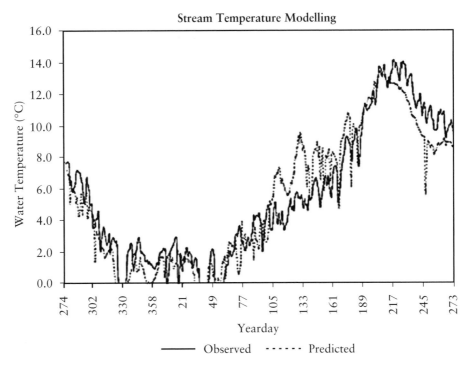

Figure 20.4 Relationship between predicted (simulated) and observed (measured) stream temperatures for McDonald drainage, Glacier National Park

test the full range of conditions within the watershed, i.e. different combinations of slope, aspect, and elevation. Snowpack characteristics, such as the snow water content, were generated by RHESSys from these climate station data and over 5 000 snow measurements were taken throughout the watershed for validation. Individual site snow measurements were well-correlated ($R^2 = 0.78$) with model estimates, but it was concluded that better estimates of snow sublimation in tree canopies depended upon accurate leaf area index (LAI) estimates (Fagre *et al.*, 1997). Because LAI was derived from Landsat TM data at 30 m pixel resolution, further improvements in satellite-based remote sensing will be necessary to improve snow estimates. However, at the watershed scale, the spatially aggregated snow water estimates were better correlated with observed values ($R^2 = 0.95$) (White *et al.*, 1998) and proved to be sufficiently accurate to provide close agreement between observed and simulated stream discharges

during spring run-off (Fagre *et al.*, 1997). Additionally, modelled estimates of stream temperatures throughout the watershed closely matched the annual temperature pattern from seven monitored streams (Figure 20.4) (Fagre *et al.*, 1997).

White *et al.* (1998) concluded that reasonable estimates of ecosystem processes were generated for Lake McDonald watershed by RHESSys. These included net primary productivity, evapotranspiration, available nitrogen and other outputs that characterise the major forest processes driving ecosystem change. Estimates of forest stand characteristics from FIREBGC were assessed with 100 circular plots (0.05-ha) located in stands typical of each plant community and distributed throughout the watershed by slope, aspect and elevation (Keane *et al.*, 1996). Tree structure, age, forest floor biomass, soil depth and texture and undergrowth canopy cover were measured on each plot. FIREBGC estimates provided reasonable agreement with field data. For instance, trends in tree ring widths for the past 40 years generally agreed with FIREBGC predictions ($R^2 = 0.89$) (Keane *et al.*, 1996).

What do we do with the models?

With functional ecosystem models in hand, we addressed potential ecosystem responses to future environmental conditions rather than to current conditions. We chose an initial climate scenario that increased annual precipitation by 10% and annual temperature by 4 °C and found that the LAI increased about 40% as trees responded to better growing conditions. Ferguson (1997) evaluated four general circulation models and several downscaling approaches to provide a 'most likely' climate change scenario for the Columbia River Basin (of which Lake McDonald watershed is a part). This scenario projects a 30% annual precipitation increase and a 0.5 °C annual mean temperature increase by 2050. Applying these changes to Glacier National Park, we found a distinct shift toward more mesic tree species at lower elevations near Lake McDonald. Western red cedar (*Thuja plicata*) and hemlock (*Tsuga heterophylla*) expand in valley bottoms by 2050. This shift in tree species composition has significance for park management because these cedar–hemlock groves are currently the easternmost distribution of these tree species and receive special consideration during decision making. The cedar–hemlock plant association includes devil's club (*Oplopanax horridum*) and a variety of other species more typical of the humid Pacific northwest forests that contribute to the biodiversity of the park. Other dominance shifts include a reduction in subalpine fir (*Abies lasiocarpa*) as treelines rise and a significant expansion of Engelmann spruce (*Picea engelmannii*) at the expense of lodgepole pine (*Pinus contorta* var. *latifolia*). Although these are only estimates of future conditions, the results underscore the potential for changes in the species composition at Glacier National Park.

We also examined the effects on Glacier National Park ecosystems of an extremely variable climate but without long-term increases in temperature or precipitation. Overall conifer net primary productivity decreased between 4% on the western side of the continental divide and 13% on the eastern side (White *et al.*, 1998) but broad-leaved shrubs and alpine vegetation increased 2–7%. However, the interannual variation in productivity for all vegetation types except conifers increased as much as 110%. Adding disturbance factors such as fire, a 120-year simulation indicated net primary productivity decreases for the eastern side of the park in the St Mary watershed but the western side of the park became more productive by 3000 kg of carbon per hectare annually. These potential changes would cause significant shifts in vegetation composition because the lowered productivity of the St Mary watershed changes the long-term dynamic to favour grasses. In fact, the lower treeline (the forest–grassland ecotone) rises under the variable scenario, permanently reducing the amount of forest cover in the St Mary watershed.

The models can spatially identify areas of greatest potential biodiversity change, essentially mapping vulnerable parts of mountain ecosystems. We compared the difference between current and future water temperatures in Lake McDonald watershed under a climate change scenario (Figure 20.5). The areas with greatest temperature change, rising + 4.5 °C, would seem to be the most sensitive to climate change and likely have significant shifts in alpine stream insect populations. This would also be a logical place to focus monitoring efforts, alerting managers to more pervasive changes going on throughout the park.

Wildland fire is the primary disturbance process in northern Rocky Mountain forests and greatly influences carbon cycles in mountain ecosystems. Fire infrequently burns upper elevation forests but fire suppression is one cause for the decline of whitebark pine (*Pinus albicaulis*). Under the future climate scenario, FIREBGC clearly indicates that the resulting more productive forest landscapes will generate more frequent and severe fires than the same landscapes experienced historically, even with the increase in annual precipitation

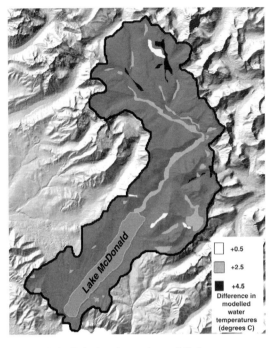

☐	+0.5
▦	+2.5
■	+4.5

Difference in modelled water temperatures (degrees C)

Figure 20.5 Relative change in modelled stream water temperatures from current climate to a future climate scenario. Note greater change in upper McDonald drainage relative to the rest of drainage area. Future climate scenario used a 30% increase in annual precipitation and + 0.5 °C warming

(Keane *et al.*, 1997). Keane *et al.* (1997) examined the interplay of different fire management policies, coupled with climate change, and asserted that fire-maintained early successional communities create overall landscapes that release less carbon to the atmosphere and are more diverse than landscapes without fire.

Scaling up to regional and global scales

The collection of modelling tools embodied in RHESSys and FIREBGC performed well for Glacier National Park and, in initial applications, expanded our understanding of mountain ecosystem responses to climate change and fire. Although designed as models that incorporate universal principles, much of the development was centred on the northern Rocky Mountains. It seemed important to apply these modelling capabilities to other mountain ecosystems and regional landscapes to ensure that they performed equally well. We aimed to quantitatively examine how ecosystem processes differed within mountain protected areas and the larger, altered landscapes surrounding them. Regional scale forecasts of changes in ecosystem processes will suggest the future role of mountains in safeguarding biodiversity and providing ecosystem 'services' to society. These needs were addressed by establishing the mountain ecosystem transect from the Pacific Ocean to the Great Plains (Fagre and Peterson, 2000).

Regional scale

A simulation run for the entire region was completed using the model DAYMET (Thornton *et al.*, 1997). DAYMET builds on the logic and proven capabilities of MTCLIM (Running *et al.*, 1987) but extrapolates climate data three-dimensionally for entire landscapes (Thornton *et al.*, 1997), providing daily measures of temperature, precipitation, shortwave radiation and relative humidity. DAYMET maps deliver detailed, spatially-explicit estimates of daily climatology in remote and rugged mountainous areas that have no instrumentation and, hence, no local climate information. This allows us to compare patterns of alpine vegetation with climatic drivers over broad areas. For instance, the frost-free days that partially delimit the growing season for alpine plants can be mapped and compared to various plant community distributions. In the Bitterroot Mountains in Montana, we found that the variance in frost-free days is highest for the upper elevation sites that have the fewest frost-free days. Although these patterns have been documented for specific sites before, the DAYMET model's capabilities make it possible to examine regional patterns for the first time.

Global scale

Trends in global mountain biodiversity should be contrasted with global-scale patterns in vegetation cover and primary productivity. The dynamics of terrestrial landscapes are an integral part of global change processes and, for the first time, such large-scale dynamics are becoming available in near real-time. The TERRA satellite, launched in December 1999, has on board the MODerate resolution Imaging Spectroradiometer (MODIS), a sensor that can track the distribution, state and productivity of global vegetation each day at a 1-km^2 resolution (150 million pixel cells). These include cover type and vegetation density (LAI), net primary productivity and seasonality. This allows mountain ecologists to track seasonal trends in vegetation growth across entire mountain ranges by downloading weekly information from the Internet. Monitoring long-term changes in vegetation cover type by this means provides measures of landscape fragmentation within and around mountain environments, even in places where no field data exist. These perspectives will be a valuable adjunct to direct studies of mountain biodiversity at both regional and global scales.

Scaling down to specific communities in the landscape

In contrast to the regional scale picture of ecological dynamics across the mountain transect, our species-scale studies address interactions and biodiversity for specific components of the three mountain ecosystems. The examples below are integrated with the ecological modelling approaches previously described.

High-elevation forests

At all three parks, high-elevation tree species have responded where temperature and permanent snow coverage previously limited tree establishment and growth. In the Pacific northwest mountains, subalpine fir have been displacing subalpine meadows, particularly since the 1930s (Rochefort and Peterson, 1996). This rapid regeneration of subalpine fir is most pronounced on the (wet) west side of the mountains during periods of warmer, drier climate and on the (dry) east side of the mountains during periods of cooler, wetter climate. Precipitation is more critical than temperature where duration of snowpack limits length of the growing season (west side) and summer soil moisture limits seedling survival (east side). Analysis of repeat photographs in Glacier National Park has documented similar invasions of meadows by subalpine fir. If climate becomes warmer and drier during the next century, continued regeneration of trees may continue to displace meadows within wetter regions of the subalpine forest–meadow mosaic. More vigorous establishment and growth of high-elevation forests is also evident at the treeline. At Logan Pass in Glacier National Park, krummholz patches have expanded to fill inter-patch spaces, and krummholz have shifted to upright tree forms (Klasner and Fagre, in press). Although treelines generally have not moved upslope, the treeline ecotone throughout the park has become more abrupt as the density of krummholz increased (Butler *et al.*, 1994).

High-elevation forests at several locations in western North America have experienced increased growth rates, controlled by snowpack duration but presumably also related to increased atmospheric CO_2 or other factors (Peterson, 1998; McKenzie *et al.*, 2001). The response of tree growth to climatic variability is spatially and temporally variable with aspect, elevation, landform, soil and other site characteristics being important. Finally, a chronology of treeline expansion into alpine tundra has been documented in a few locations at Glacier National Park (Bekker *et al.*, 2000). Post Little Ice Age (ca. 1850) advances of subalpine fir krummholz occurred until the early 1900s in fingers extending into the alpine tundra, less than 100 m from the forest. However, numerous isolated trees and fingers of trees less than 80 years old have established as upright tree forms and up to several hundred metres from the forest. This indicates that limitations to seedling establishment and subsequent tree growth had been ameliorated, beginning about 1920. This corresponds to the rapid rise in summer average temperature that is also related to the period of most rapid glacial recession (Fagre and Peterson, 2000).

Meadows

Biodiversity shifts in subalpine meadows and a high-elevation wetland in Glacier National Park have been studied by Lesica and Steele (1996) because they support the most diverse plant assemblages in the park and include numerous species at the edges of their ranges. These areas will act as climate change indicators because slight changes in moisture will eliminate marginal species. After only ten years, Lesica (2000) found four plant species declining but the trend was not statistically significant.

Non-native plants

At all three parks numerous non-native plants have established populations and are considered a management problem. At Olympic, there are nearly 1400 vascular plant species of which about 12% are non-native. At Glacier, there are 1150 vascular plants of which 10.5% are non-native. A distinct elevation gradient exists with most occurrences of non-native plants remaining at low elevations and relatively few in alpine areas. The extreme variability scenario (White *et al.*, 1998) indicates rapid transitions to grasses in lowland areas of the park where most roads, camp grounds, and buildings exist. This suggests a greater vulnerability to establishment of non-native species.

CONCLUSIONS

The multi-scale, multidisciplinary research and monitoring programme described here is a nested approach that provides scientific insights for ecologists, park managers, and policy makers. It provides spatial and temporal continuity by using plot-level data as inputs to ecosystem simulation in broader landscapes and under possible future climates. It provides a bridge between fine-grained data at small spatial scales and coarse-grained data at sub-continental scales. Regional ecosystem models can play an important role in synthesising disparate existing information into an integrated assessment of the state of mountains and of the biodiversity they support.

ACKNOWLEDGEMENTS

The authors are indebted to other members of the research teams whose work is summarised in this paper. At Glacier National Park, these members include (1) Steven Running, Joseph White, and Peter Thornton at the Numerical Terradynamics Simulation Group, School of Forestry, University of Montana, (2) Robert Keane and Kevin Ryan, Intermountain Fire Sciences Laboratory, US Forest Service, Missoula, Montana, and (3) F. Richard Hauer and Jack Stanford, Flathead Lake Biological Station, University of Montana. At Olympic and North Cascades National Parks, these include Don McKenzie, Amy Hessl, and David W. Peterson at the University of Washington. Staff of the National Park Service have cooperated in long-term studies at all three parks. US Geological Survey support staff have made field studies possible. Lisa McKeon provided expert assistance on graphics. Most of the research described herein was supported by the USGS Global Change Research Program.

References

Band LE, Patterson P, Nemani RR and Running SW (1993). Forest ecosystem processes at the watershed scale: incorporating hillslope hydrology. *Agricultural and Forest Meteorology*, **63**:93–126

Bekker MF, Alftine KJ and Malanson GP (2000). Effects of biotic feedback on the rate and pattern of edge migration. *Ecological Society of America*, Abstracts 85:247

Beniston M and Fox DG (1996). Impacts of climate change on mountain regions. In Watson RT, Zinyowera M, Moss RH and Dokken DJ (eds), *Climate Change 1995 — Impacts, Adaptations* and Mitigation of Climate Change: Scientific-Technical Analyses. Cambridge University Press, Cambridge, pp 191–213

Butler DR, Malanson GP and Cairns DM (1994). Stability of alpine treeline in northern Montana, USA. Phytocoenologia, 22:485–500

Carrara PE (1989). Late Quaternary glacial and vegetative history of the Glacier National Park region, Montana. US Geological Survey Bulletin, 1902:1–64

Fagre DB, Comanor PL, White JD, Hauer FR and Running SW (1997). Watershed responses to climate change at Glacier National Park. Journal of the American Water Resources Association, 33:755–65

Fagre DB and Peterson DL (2000). Ecosystem dynamics and disturbance in mountain wildernesses: assessing vulnerability of natural resources to change. In McCool SF, Cole DN, Borrie WT and O'Loughlin J (eds), Wilderness science in a time of change. US Department of Agriculture, Ogden UT, pp 74–81

Ferguson S (1997). A climate-change scenario for the Columbia River Basin. In Quigley TM (ed.), Interior Columbia Basin Ecosystem Management Project: Scientific Assessment. Pacific Northwest Research Station, Portland, OR

Keane RE, Hardy CC, Ryan KC and Finney MA (1997). Simulating effects of fire on gaseous emissions and atmospheric carbon fluxes from coniferous forest landscapes. World Resource Review, 9:177–205

Keane RE, Ryan KC and Running SW (1996). Simulating effects of fire on northern Rocky Mountain landscapes with the ecological process model FIRE-BGC. Tree Physiology, 16:319–31

Key CH, Johnson S, Fagre DB and Menicke RK (1997). Glacier recession and ecological implications at Glacier National Park, Montana. In Williams RS and Ferrigno JG (eds), Final report of the workshop on long-term monitoring of glaciers of North America and northwestern Europe. USGS, Woods Hole MA, pp 88–90

Klasner FL and Fagre DB (2002, in press). Spatial changes in alpine treeline vegetation patterns along hiking trails in Glacier National Park, Montana. Journal of Arctic, Antarctic and Alpine Research

Lesica P (2000). Monitoring the effects of global warming using peripheral rare plants in wet alpine tundra in Glacier National Park, Montana. Progress report, 2000

Lesica P and Steele BM (1996). A method for monitoring long-term population trends: an example

using rare arctic–alpine plants. Ecological Applications, 6:879–87

Liniger H, Weingartner R and Grosjean M (1998). Mountains of the world: water towers for the 21st century. University of Bern, Switzerland

McKenzie D, Hessl AE and Peterson DL (2001). Recent growth in conifer species of western North America: assessing spatial patterns of radial growth trends. Canadian Journal of Forest Research, 31:526–38

Messerli B and Ives JD (eds) (1997). Mountains of the World: A Global Priority. Parthenon Publishing, New York

Oerlemans J (1994). Quantifying global warming from the retreat of glaciers. Science, 264:243–5

Peterson DL (1998). Climate, limiting factors and environmental change in high-altitude forests of Western North America. In Beniston M and Innes JL (eds), The impacts of climate variability on forests. Springer, New York, pp 191–208

Rochefort RM and Peterson DL (1996). Temporal and spatial distribution of trees in subalpine meadows of Mount Rainier National Park, Washington, USA. Journal of Arctic and Alpine Research, 28:52–9

Running SW, Nemani RR and Hungerford RD (1987). Extrapolation of synoptic meteorological data in mountainous terrain and its use for simulating forest evapotranspiration and photosynthesis. Canadian Journal of Forest Research, 17(6):472–83

Running SW and Gower ST (1991). FOREST-BGC, a general model of forest ecosystem processes for regional applications. II. Dynamic carbon allocation and nitrogen budgets. Tree Physiology, 9:147–60

Running SW and Nemani RR (1991). Regional hydrologic and carbon balance responses of forests resulting from potential climate change. Climatic Change, 19:349–68

Thornton PE, Running SW and White MA (1997). Generating surfaces of daily meteorological variables over large regions of complex terrain. Journal of Hydrology, 190:214–51

White JD, Running SW, Thornton PE, Keane RE, Ryan KC, Fagre DB and Key CH (1998). Assessing regional ecosystem simulations of carbon and water budgets for climate change research at Glacier National Park, USA. Ecological Applications, 8:805–23

Scenarios of Plant Diversity in South African Mountain Ranges in Relation to Climate Change

21

David J. McDonald, Guy F. Midgley and Les Powrie

INTRODUCTION

Africa has a long and impressive geological history that has resulted in the mountains we see today. The most significant mountains of South Africa are the Drakensberg Range and the Cape Fold Mountains. Compared with mountainous regions of the world, including the mountains of East and Central Africa, the mountains of southern Africa are relatively low. Only a few peaks in the Cape Fold Mountains exceed 2000 m asl and the highest peak in southern Africa, Thabana Ntlenyana, found in Lesotho and part of the Drakensberg–Maloti (Lesotho) Range, reaches only 3482 m. Also, in contrast to other mountains on the African continent, the mountain ranges in southern Africa are not continental but lie relatively close to the coast. The mountain ranges in South Africa extend in a wide arc from the Cape Fold Mountains in the west to the Drakensberg and other ranges in the east and northeast. These mountains form a boundary between the coastal zone to the south and east and the high inland plateau, having a marked influence on the climate of both the coastal and inland regions. In general the mountains lie within a temperate climatic region with a distinctly mediterranean-type climate in the west (winter rainfall) that influences the Cape Fold Mountains, whereas the Drakensberg Range is situated in the summer rainfall zone (Killick, 1963, 1978a,b,c, 1979). We briefly consider the Cape Fold Mountains, their altitudinal and phytogeographical zonation, and then focus on the Drakensberg and Maloti Mountains and the high altitude zone of these ranges.

WHERE IS THE ALPINE ZONE IN SOUTHERN AFRICA?

The Cape Fold Mountains

The Cape Fold Mountains fall within one of the worlds 'hottest hotspots' of plant diversity, the Cape Floristic Region. The sclerophyllous, shrubby vegetation with a herbaceous understory is commonly referred to as 'fynbos'. About 68% of the 8550 plant species are endemic and a large number of these are confined to the mountains. There is growing concern that increased global temperatures will prolong the already long, hot, dry summers and that drought conditions will increase. Although most fynbos species have means to overcome long periods of drought stress, excessively long dry periods could lead to the demise of many species.

Fynbos is a pyrophytic vegetation type where plants are adapted to regenerating after fires. If fires are too frequent, sensitive species are negatively affected and are lost, leading to local extinctions in some cases. If species are already rare and locally restricted, such species may be lost altogether. With desiccation of fynbos habitats, it is likely that mean annual temperature and summer maximum temperature will increase, there is also a strong chance that fires will increase in frequency and intensity. This will greatly threaten fynbos plant communities and even those species well adapted to resisting fire will be compromised. Fynbos at moist highland sites and marginal arid localities would most likely be worst affected (Cowling *et al.*, 1997).

Although the Cape mountains are not high, there is nevertheless some evidence for the altitudinal zonation of the vegetation in these mountains. At the beginning of the twentieth century, Rudolf Marloth (1902) made some observations on the growth form changes in Cape mountain vegetation with increasing altitude. Most later studies attributed observed species richness in the Cape flora to the effects of rainfall and substrate rather than altitude. To remedy this, a study was conducted in the Klein Swartberg mountains in 1992 by a group of Cape botanists (Linder *et al.*, 1993; McDonald *et al.*, 1993) to investigate whether an alpine zone exists in the Cape Fold Mountains. Details of this study are available in Linder *et al.* (1993) but the general conclusion was that a high-altitude zone is distinguishable in the Cape mountains. Evidence for this is found in the distinction of vegetation structure between this zone and those of the middle and lower slopes. In the high altitude zone there is a high dominance of the families Restionaceae (Cape reeds) and Poaceae and the dicotyledonous shrubs are mostly prostrate (Linder *et al.*, 1993).

The sub-alpine zone in the Cape Fold Mountains, therefore, does not in reality represent alpine conditions although some of the vegetation structure suggests this. Further research is required to investigate the zonation of these mountains in a Cape Floristic Region context but in a global sense of alpine ecosystems no further attention needs be given to these areas.

The Drakensberg—Maloti Mountains

The true alpine zone in southern Africa is found in the complex of mountains known as the Drakensberg or Ukhahlamba (Zulu: Barrier of Spears), essentially a mountain escarpment on the boundary between KwaZulu-Natal and Lesotho and the Maloti Mountains, forming the highlands in the Kingdom of Lesotho. The alpine zone covers about 40 000 km² with a range in elevation from 1800–3482 m. This zone consists of a subalpine belt that ranges in altitude from 1800–2800 m and the alpine belt that ranges from 2800–3482 m, the elevation of the summit of Thabana Ntlenyana, the highest peak in southern Africa.

The alpine zone in southern Africa, given the name of 'Austro-afroalpine Belt' by Coetzee (1967), is now known as the Drakensberg Alpine Region (Killick, 1978, 1994). Ecological conditions of this zone can be compared with, but are somewhat different from, those in the Afroalpine zone of east and northeast Africa where there is a high diurnal temperature range: summer every day and winter every night (Hedberg, 1964). In the Drakensberg alpine region, where the climate is seasonal, there are also obvious diurnal changes in temperature but they are not as pronounced as in the East African Mountains. Seldom do the maximum temperatures in summer exceed 16 °C and, at soil level, low minimum temperatures may occur at any time during the year. The region falls within the summer rainfall area of southern Africa with most of the rain falling as a result of thunder showers in the cool to hot summers. Rainfall varies with locality; in excess of 1600 mm has been measured on the Drakensberg escarpment summit in one season. Very little if any rain falls in winter, leading to dry, cold winters often with frost. Snowfalls may occur a few times in the winter and, depending on the intensity of the falls, snow may persist on the ground for up to six months.

The Drakensberg–Maloti mountain complex is overlain by amygdaloidal basalt that was deposited in the Early Jurassic (van Zinderen-Bakker and Werger, 1974). In places it is up to 1500 m thick. The basalt has been deeply dissected over time, resulting in the present-day dramatic relief of the Lesotho Highlands. The soils formed in the alpine zone in open areas are shallow and subject to frost heaving in the winter. Needle-ice forms and soil material is sorted as a result of repeated freezing and thawing. 'Frost terracettes', narrow parallel soil ridges, caused by simultaneous frost and wind action can also be observed on the mountain slopes (van Zinderen-Bakker, 1965; Troll, 1944 in van Zinderen-Bakker and

Werger, 1974). In depressions, bogs are encountered where the soils are deep accumulations of organic material or peat. Many of these bogs have a characteristic microtopography with hummocks or thufur formed by successive freezing and thawing of the peaty soils.

The vegetation of the subalpine and alpine belt is treeless. The subalpine belt is covered by grasslands of two main types, the *Themeda triandra* grassland (Sebokv grassland) and the *Festuca-Merxmuellera* grassland (Letsiri grassland) (van Zinderen-Bakker and Werger, 1974). The first type is found on the drier slopes above the lowest basalt cliffs to about 2600 m and on the cooler, moister slopes, extending down to 2140 m (Killick, 1990). The Letsiri grassland equates with what Killick (1990) refers to as Alpine Grassland, found on the southern slopes above 2100 m and on the northern slopes above 2700 m. Killick (1990) provides a more detailed classification of the sub-units of these grasslands.

An analysis of the composition of the grasslands of the Drakensberg and Maloti mountains shows that they are dominated by C4 grasses. This is not what would be expected and is not in line with the current understanding of temperature: C4 grass relationships. The observed pattern is probably the result of a carry-over of C4 grass species dominance from when there were lower levels of CO_2 during glacial times.

The vegetation of the alpine belt of the Drakensberg alpine region consists of a mosaic of communities. It is mainly an alpine heathland or dwarf shrubland dominated by species of *Erica* and *Helichrysum* and referred to by Killick (1990) as Alpine Heath; it consists of five distinct heath communities which he recognised. Interspersed with the heathland is the alpine grassland. Also found in the alpine belt are aquatic and hygrophilous communities. These can be classified into (i) bog communities and (ii) streambank communities (Killick, 1990). The bog communities, in turn, are of two types, those found as seepage areas on mountain slopes and those found at riverheads and in depressions where extensive swampy areas are formed (van Zinderen-Bakker and Werger, 1974; Killick, 1990). These bogs may vary in size from a few hundred square metres to several km^2 .

Humans actively inhabit the alpine zone of the Drakensberg–Maloti mountains. The parts with highest elevation have been mainly grazed by cattle, sheep, goats and horses in the summer months in the past. However, usage patterns of the alpine zone are changing and these areas are being increasingly grazed in the winter. There is, consequently, very little opportunity for the vegetation to recover, with the result that overgrazing is becoming prevalent.

Potential effects of climate change on the subalpine and alpine zones of South African mountains: The Drakensberg–Maloti Mountains

The climatic regime and immediate impression given by the Drakensberg and Maloti Mountains suggests that the grasses should be predominantly C3 grasses, i.e. those grasses favouring cool, moist climates on nutrient-rich substrates. However, this is not the case. The grasslands are dominated by C4 grasses. What can explain this?

Fire has been an important abiotic driving force over a long period of time, greatly influencing the evolution of the subalpine grasslands of the Drakensberg and Maloti mountains. These fires have been frequent and, coupled with high plant nutrient use efficiencies, have allowed the C4 grasses to reduce nutrient levels to a point that excludes the C3 species. This explanation may be somewhat simplistic as expressed here, suggesting that careful research is necessary to tease out the complexities of the system.

Despite the level of our understanding of the complexities of the interaction of the C3 and C4 grasses of the high-altitude grasslands, we can speculate as to the possible effects of global warming on these ecosystems. An increase in global temperatures will predictably result in aridification of the mountain grasslands thus, in turn, increasing the fire frequency to levels even greater than those presently experienced. Levels of CO_2 are likely to increase. This would usually favour C3 grasses but the prediction is that C4 grasses will continue to

Table 21.1 The species selected to represent the 'Alpine Community'

Species	Family	Growth form
Agrostis subulifolia Stapf	Poaceae	Herb
Aloe polyphylla Schönland ex Pillans	Aloaceae	Succulent rosette herb
Aster erucifolius (Thell.) Lippert	Asteraceae	Forb
Craterocapsa congesta Hilliard & B.L. Burtt	Campanulaceae	Forb
Dierama dracomontanum Hilliard	Iridaceae	Herb
Disa fragrans Schltr.	Orchidaceae	Herb
Erica algida Bolus	Ericaceae	Dwarf shrub
Erica dominans Killick	Ericaceae	Dwarf shrub
Eumorphia sericea J.M. Wood & M.S. Evans ssp. *robustior* Hilliard & B.L. Burtt	Asteraceae	Low shrub
Helichrysum sessilioides Hilliard	Asteraceae	Forb
Helichrysum trilineatum DC.	Asteraceae	Low shrub
Hirpicium armerioides (DC.) Roessler	Asteraceae	Forb
Merxmuellera stereophylla (J.G. Anderson) Conert	Poaceae	Herb
Pentaschistis oreodoxa Schweick.	Poaceae	Herb
Rhodohypoxis rubella (Baker) Nel	Hypoxidaceae	Herb
Zaluzianskya microsiphon (Kuntze) K. Schum.	Scrophulariaceae	Forb

dominate even though levels of CO_2 are predicted to increase.

MODELLING THE EFFECTS OF CLIMATE CHANGE IN THE ALPINE ZONE IN LESOTHO

Methods

Relatively few studies of the vegetation of the alpine zone in Lesotho have been done. Consequently, there is not a substantial amount of information upon which to base conclusions about changes in vegetation, both in respect of species composition and structure. To examine the possible effects of changing climate in the alpine zone, a selection of 36 species known to occur in the alpine zone was made, based on collections in the National Botanical Institute Herbarium in Pretoria (PRE), South Africa. Locality information was extracted from the PRECIS taxonomic database of the National Botanical Institute. From the initial set of 36 species, a subset of 16 species, collected from a variety of localities within the alpine zone and including a number of different growth forms from grasses and other herbaceous plants to woody dwarf shrubs, was chosen for modelling

Table 21.2 Climate model: Envelope with boundary parameters

Factor	Range
Altitude	2800–3484 m
Potential evaporation/annum	976–1675 mm
Mean annual precipitation	362–1625 mm
Heat units	0–3 °C
Soil moisture days in winter	12–25 days
Soil moisture days in summer	30–81 days
Mean minimum cold	–11–2 °C

purposes. This set of 16 species was termed the 'Alpine Community' (Table 21.1).

A climatic envelope was then modelled for the 'Alpine Community'. The modelling criteria were set to examine the distribution of species in only those areas above 2800 m, i.e. the alpine zone. Although the selected species may not occur over the whole of the area of the present alpine zone, if all ecological factors were suitable for them, climate would presently not be limiting and they could occur there. If, however, the climate changes, the area they could potentially inhabit would predictably reduce dramatically. A model was constructed using the climatic envelope given in Table 21.2

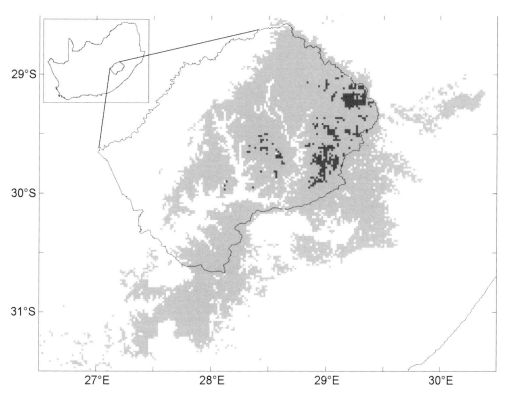

Figure 21.1 The present extent of the alpine zone in southern Africa (light grey) with the possible reduction in the alpine zone as a consequence of global warming in the future (dark grey). The internal boundary represents the border of the Kingdom of Lesotho (with its position in South Africa, as shown in the small map)

to investigate the extent to which the potential distribution area of the selected species would diminish.

Results

The climate envelope developed for the 'Alpine Community' covers a large part of the present alpine belt. From the model, it was predicted that the Drakensberg Alpine Region will dramatically diminish in size, contracting from its present 40 000 km², (Figure 21.1) to less than half of its present extent. The alpine belt will be found only at the highest elevations, mainly in central Lesotho and around Thabana Ntlenyana, the highest peak. It will also become very fragmented, with the result that populations of species will become isolated. The effects of this

isolation on plant species could result in either speciation or extinction.

The predictable physical effects include reduced snow cover in the winter, and what snow may fall would probably melt more rapidly, with a resultant loss of available water to the vegetation of the alpine belt and increased run-off. Greater run-off may add to the already high degree of soil erosion found in the mountain catchments of Lesotho.

Anthropogenic effects in the alpine belt of the Drakensberg–Maloti mountains are significant and these, coupled with the effects of changing climate, are predicted to hasten the pace of environmental degradation in these highlands. Invasive shrubs such as *Chrysocoma ciliata*, indicative of disturbance through activities such as overgrazing, are evident in many parts of the Lesotho highlands. Arguably these effects will have the most impact on the landscape and the biota in the first place, with the effect of these impacts predicted to be made worse by the effects of climate change.

by climate change will most likely be outstripped and exacerbated by heavy impacts, such as overgrazing and inappropriate land management, in the short term.

CONCLUSIONS

The investigation presented is preliminary and requires more rigorous testing. The signs are that global climate changes will dramatically affect the distribution of plant species in the alpine zone of the Drakensberg–Maloti mountain complex, but that these changes will most likely be superseded by anthropogenic influences. Changes in the alpine ecosystem caused

ACKNOWLEDGEMENTS

The authors acknowledge the contribution to this work of Prof Roland E Schulze, Department of Agricultural Engineering, University of Natal, Pietermaritzburg, who developed the climate change models. David McDonald also wishes to thank the organisers of the GMBA Conference for sponsorship to attend the conference at Rigi Kaltbad, Switzerland, September 2000.

References

Coetzee JA (1967). Pollen analytical studies in East and Southern Africa. In van Zinderen EM (ed.), *Palaeoecology of Africa* 3. Balkema, Bakker, Cape Town, pp 146

Cowling RM, Richardson DM and Mustart, PJ (1997). Fynbos. In Cowling, RM, Richardson DM and Pierce SM (eds), *Vegetation of Southern Africa*. Cambridge University Press, Cambridge, pp 99–130

Hedberg O (1964). Features of Afroalpine plant ecology. *Acta Phytogeographica Suecica*, **49**:1–144

Killick DJB (1963). An account of the plant ecology of the Cathedral Peak area of the Natal Drakensberg. *Memoires of the Botanical Survey of South Africa*, **34**:178

Killick DJB (1979). African mountain heathlands. In Specht RL (ed.), *Heathlands and related shrublands*. Ecosystems of the World 9A, Elsevier, Heidelberg, pp 97–116

Killick DJB (1978a). Notes on the vegetation of the Sani Pass area of the southern Drakensberg. *Bothalia*, **12**:537–42

Killick DJB (1978b). The Afro-Alpine region. In Werger MJA (ed.), *Biogeography and Ecology of Southern Africa*. Junk, The Hague, pp 515–60

Killick DJB (1978c). Further data on the climate of the alpine vegetation belt of eastern Lesotho. *Bothalia*, **12**:567–72

Killick DJB (1990). *A field guide to the flora of the Natal Drakensberg*. Jonathan Ball & Ad. Donker, Johannesburg.

Linder HP, Vlok JH, McDonald DJ, Oliver EGH, Boucher C, Van Wyk B-E and Schutte A (1993). The high altitude flora and vegetation of the Cape Floristic Region, South Africa. *Opera Botanica*, **121**:247–61

Marloth R (1902). Notes on the occurrence of alpine types in the vegetation of the higher peaks of the south-western region of the Cape. *Transactions of the South African Philosophical Society*, **11**:161–8

McDonald DJ, Oliver EGH, Linder HP and Boucher C (1993). Is there alpine vegetation on the mountains of the south-western Cape? A botanical expedition to the Klein Swartberg. *Veld & Flora*, **79**:17–19

Troll C (1944). Strukturböden, Solifluktion und Frostklimate der Erde. *Geol. Rundschau*, **34**:545–694

Van Zinderen-Bakker EM (1965). Ueber Moorevegetation und den Aufbau der Moore in Süd- und Ostafrika. *Botanischer Jahrbucher*, **84**: 215–312

Van Zinderen-Bakker EM and Werger MJA (1974). Environment, vegetation and phytogeography of the high-altitude bogs of Lesotho. *Vegetatio*, **29**:37–49

Biodiversity in Mountain Medicinal Plants and Possible Impacts of Climatic Change

22

Aditya N. Purohit

INTRODUCTION

The earliest recorded evidence of use of herbal medicine dates to about 5000 years ago. Even today, Ayurvedic, Unani, Tibetan and Chinese systems of medicine mostly use herbal plants, and some products of animal and mineral origin. Although these medical systems have undergone many changes in the course of their long history, they remain the mainstays of medical relief for the majority of the people in these countries. Since it is believed that drugs from plant sources have fewer side effects than synthetic drugs, recent past herbal medicare has gained considerable public favour, even in developed counties. Researchers are investigating many of the plants used as folk remedies in developing countries. Many multinational pharmaceutical companies are now in the race to discover new drugs from natural sources. Therefore, the traditions of developing countries are gradually leading to a large trade at national and international levels. The trend in the global market size (Table 22.1) is an index of this development.

While on one hand all these facts indicate a process of globalisation of herbal drugs, on the other hand, importers are in search of patentable products derived from these plants, with the aim of ultimately commercialising such compounds. Therefore, the herbal resources of the continent are caught in a precarious situation, with respect to their use and financial value of processes and products. Since these resources have now attained international commercial importance, every country is examining their databases to ascertain:

(1) which plants are in traditional usage,
(2) which medicinal plants are available and in what quantities;
(3) which taxa are most important in terms of their marketable quantity;
(4) the extent of their taxonomic accuracy;
(5) knowledge of their bioactivity;
(6) gaps in knowledge bases;
(7) access to screening technologies for cost-effective phytomedicinal research;
(8) possibilities for immediate cultivation, conservation and commercialisation;
(9) the implications of the biodiversity treaty for medicinal plant trading.

INDIAN SCENARIO

There are about 250 000 higher plant species in the world. Of these 15 000 species are found in India, of which 2500 are in ethnomedicinal use (Jain, 1991). Out of these, 1500 are used in Indian traditional systems of medicine, 600 species in Ayurvedic medicine, 500 species in Unani medicine and 550 species in phytopharmaceutical industries (Bhakuni, 1998). A general account of known medicinal plants of India and their uses is well documented (Kirtikar and Basu, 1918; Chopra *et al.*, 1956; Jain, 1991; Biswas, 1956; Gaur *et al.*, 1984; Singh and Aswal, 1992; Joshi *et al.*, 1993; Kaul, 1997). The central Drug Research Institute, Lucknow, launched a biological screening programme in 1963. A small fraction of flowering plants that have, so far, been investigated have yielded about 120 therapeutic agents of known structure from about

Table 22.1 Global market size of herbal medicines (Brower, 1998)

Country	Herbal drug sale 1996 (US$ billion)	Herbal drug sale (guestimate) 2000 (US$ billion)
Europe	~10.0	~20.0
USA	4.0	~8.0
India	1.0	4.0
Other countries	5.0	~10.0
Total	~20.0	~30.0–60.0

Table 22.2 Major medicinal plants being exported from India (Based on BCIL, 1996)

Botanical name	Part used	Distribution (m asl)
Aconitum sp.	Root	2800–4500
Podophyllum emodi	Rhizome	2800–4000
Rheum emodi	Rhizome	2800–3500
Picrorhiza kurrooa	Root	2800–4500
Juniperus communis	Fruit	3000–4800
Juniperus macropoda	Fruit	3000–5000
Inula racemosa	Rhizome	2000–2800
Saussurea lappa	Root	1800–2500
Valeriana jatamansi	Rhizome	1800–3000
Swertia chirayita	Whole plant	1800–2600
Berberis aristata	Root	800–1800
Hedychium spicatum	Rhizome	1800–2800
Heracleum candicans	Rhizome	1500–3400
Colchicum luteum	Rhizome and seed	800–2400
Plantago ovata	Seed and husk	500–2200
Punica granatum	Flower, root and bark	1000–1800
Rawolfia serpentina	Root	500–800
Juglans regia	Bark	1000–2200
Zingiber officinale	Rhizome	400–1200
Acorus calamus	Rhizome	1000–2500
Adhatoda vasica	Whole plant	600–1000
Cassia angustifolia	Leaf and pod	1000–1500

90 species of plants. Some examples of useful high altitude plant-derived drugs include taxol, podophyllotoxin, chiratin, allicin, valerian, artemisinin and ephedrine. A number of CSIR laboratories and academic institutions in the country are involved in developing new drugs from plant sources. Two of the prominent new leads obtained from high altitude plants are an antihepatotic in *Picrorhiza kurrooa* and a tranquilliser in *Valeriana wallichii*.

The trade in medicinal plants in India is largely unorganised and uncertain, both in terms of demand and price structure. Actual available figures of the plant-based drugs exported from India between 1985 and 1996 indicate that export is increasing at an exponential rate. Twenty-three high-priced medicinal plants exported from India are listed in Table 22.2. In addition, species like *Angelica glauca*, *Taxus baccata* and *Nardostachys jatamansi* that are not mentioned in this list, are also exported to a very large extent. Likewise, owing to large internal demand and export potential, several aromatic species are exploited for their essential oils.

From the above data, one can see that two-thirds of the exported species are of mountain origin, which constitutes only one tenth of the

Table 22.3 Proportional distribution pattern of medicinal plants in subalpine and alpine zone in Indian Himalaya (based on Dhar *et al.*, 2000)

Distribution	Subtropical (< 1800 m)	Temperate (1801–2800 m)	Subalpine (2801–3500 m)	Alpine (> 3500 m)
Exclusive	1217	281	3	47
Subtropical–temperate	200	200	–	–
Subalpine–alpine	–	–	152	152
Temperate–subalpine–alpine	–	128	128	128
Temperate–subalpine	–	154	154	–
Total	1417	763	437	327

total geographical area of the country. Therefore, mountain medicinal plants form one of the major resource bases for economic transformation in mountains in the future.

MEDICINAL PLANT DIVERSITY IN THE INDIAN HIMALAYAN REGION

Diversity of medicinal plants can be analysed from three viewpoints, phytotaxonomy, ecophysiology, and chemotaxonomical diversity. A detailed survey of the literature conducted by Samant *et al.* (1998) indicates that more than 50% of the ethnomedicinally important species are from the Himalayan region. Among the families, Asteraceae contains the highest number of ethnomedicinal plants (129 spp.). In terms of altitudinal distribution, out of 1748 about 1217 species of medicinal plants are found below 1800 m asl. Of these, 962 species are found only between the temperate to alpine zone (Table 22.3). Many of the species show overlapping distribution patterns within different altitudinal zones.

According to a survey conducted by Dhar *et al.* (2000), a total of 280 plant species are being utilised in 316 formulations by herbal drug companies. One in hundred and seventy-five of these species are from the Indian Himalayan region, and 24 are endemic to this region. The region is rich in herbs as compared to shrubs and trees. Across altitudinal zones, the percentage of medicinal plants increases considerably from the subtropical (7.5%) to the alpine (76.9%) zone.

The above analysis does not include diversity based on bioactive constituents in plants. This type of diversity is of greater significance in the case of medicinal plants because the quality as well as the quantity of chemical constituents depend on conditions in which the plants grow. For example, in the subalpine to alpine species, *Podophyllum hexandrum*, resin and bioactive constituents seem to increase with increasing altitude (Purohit *et al.*, 1999). However, some populations of this plant from lower elevations also show high quantities of bioactive constituents, which might be of interest in multiplication and cultivation, particularly if the bioactive production potential can be genetically controlled. A survey of different populations of *Podophyllum hexandrum* revealed that there are four morphotypes, with one, two, three and four leaves, as compared to the common two-leaved type. The four morphotypes showed distinct differences in their bioactive constituents and resin profiles. Analysis of the rhizomes revealed that the highest levels of resin and bioactive constituents were found in the plants with one leaf and lowest in those with four leaves (Purohit *et al.*, 1998). The polypeptide profiles of the seeds indicate that these four morphotypes are genetically different. Generally, it is argued that species of restricted distribution exhibit low genetic diversity (Kress *et al.*, 1994). Apart from the considerable variation found in *Podophyllum hexandrum*, variations have also been found in several species of another important genus, *Aconitum* (Bahuguna, 2000). Therefore, the magnitude of diversity seems to

Table 22.4 Number of species listed as rare, endangered and vulnerable from the Indian Himalaya in the Red Data Book of India (Nayar and Sastry, 1987–1990)

Status	Total Himalayan region	Subalpine–alpine zone
Rare	90	21
Endangered	50	08
Vulnerable	70	18
	214	47

be much higher than expected, as exemplified by *Podophyllum hexandrum*.

Over-exploitation of medicinal plants has created very serious problems in terms of conservation of such species. Some of the medicinal plants are now on the verge of extinction. As many as 214 species from the Indian Himalayan region have been listed in Red Data Book (Table 22.4). Out of these, nearly 36 are in commercial use as medicinal plants and most can only be found in the subalpine and alpine zone. The main cause of the extinction risk is harvesting of wild plant populations in a destructive manner (whole plant, rhizome, root and stem). However, the impact of habitat loss and degradation of natural ecosystems, as well as climatic shifts, might also play a role, especially as these high altitude species are very specific in terms of their adaptation to climate.

Medicinal plants that are critically endangered in India include *Aconitum dedinorrhizum* and *A. heterophyllum*. Endangered high-altitude medicinal plants include: *Angelica glauca*, *Arnebia benthamii*, *Atropa acuminata*, *Berberis aristata*, *Colchicum luteum*, *Dioscorea deltoidea*, *Ferula jaeschkeana*, *Gentiana kurroo*, *Nardostachys jatamansi*, *Orchis latifolia*, *Rheum emodi* and *Swertia chirata*. The most vulnerable medicinal plants are *Artemisia brevifolia*, *A. maritima*, *Corydalis govaniana*, *Ephedra gerardiana*, *Hedychium spicatum*, *Jurinea dodlomiaea*, *Picrorhiza kurroa*, *Valeriana wallichii* and *Zanthoxylum alatum*. All of these species are highly priced in the market but few are being commercially cultivated.

CLIMATIC REQUIREMENTS AND POSSIBLE IMPACT OF CLIMATE CHANGE ON HIGH ALTITUDE MEDICINAL PLANTS

In general, under extreme stress, such as that found at high altitudes, excessive restriction in the diversity of alleles for genes that regulate stress tolerance may occur and this can potentially result in fixed alleles (only one type of allele for the gene). Moreover, most mountain plants species are habitat-specific and flourish only within a narrow range of environments. For example, *Acorus calamus*, *Hedychium* spp. and *Dactylorhiza atagirea* are restricted to moist areas; *Bergenia* prefers wet, rocky sites; some species of *Valeriana* and *Viola* are essentially shade plants; while *Aconitum* spp., *Jurinea macrocephala*, *Picrorhiza kurrooa* and *Rheum emodi* grow only on open slopes. Many of these species require more than one growing season to develop sufficient biomass to allow harvesting for the desired secondary metabolites. Therefore, even slight changes in microclimate may have a great impact on a species distribution and survival. The higher sensitivity of alpine plants is argued on the basis of their ecophysiological response to various types of environmental change (Körner and Diemer, 1987; Nautiyal, 1996; Pandey and Purohit, 1984; Mooney and Billings, 1961). Although we cannot know how vegetation patterns will change over time with altitude in the Himalayan region, it is likely that the impact of climatic change will be more prominent in the alpine and subalpine zones than at lower altitudes.

A study of vegetation patterns made at ten-year intervals in a fully protected area in Tungnath (alpine area) by Nautiyal and colleagues (unpublished data, Table 22.5) indicates that overall species richness has increased during the last ten years. Late-seasonal species are invading at a faster rate than early-season species, indicating that microtemperature changes are probably taking place, leading to warmer winters. Similarly, a considerable increase in fern abundance indicates that soil moisture is increasing in these areas.

Table 22.5 Changes in species richness (per m^2) according to growth initiation period during 10 years in Tungnath (Nautiyal, unpublished data)

Growth form	Total species		Early (May 31st)		Late (May 31st)	
	1988	1998	1988	1998	1988	1998
Shrubs & undershrubs	0	4	0	0	0	4
Tall forbs	6	11	2	2	4	9
Medium forbs	5	10	2	4	3	6
Short forbs	14	22	8	12	6	10
Grasses, sedges and other monocots	2	4	0	0	2	5
Total species number	27	52	12	18	15	34

Socioeconomic climate, however, is changing worldwide at a faster rate than the global physical climate, especially in relation to use and trade of medicinal plants, and is directly affecting biodiversity of such medicinal plants. Globalisation of use of herbal medicines and privatisation of processes and products involving these plants are the two basic dilemmas. Diversity of these plants will be affected more by these factors than by climate changes.

PROPOSALS TO USE MEDICINAL PLANT RESOURCES FOR ECONOMIC DEVELOPMENT IN MOUNTAIN REGIONS

The situation for medicinal plants described above indicates that Himalayan mountains in particular and all mountain regions in general, which were once regarded as hostile and economically non-viable regions, are now likely to attract major economic investments for herb-based industries. However, while mountains are often rich centres of plant diversity, the capacity to convert this diversity into economic returns through science and technology resides predominantly outside the mountains, and is mainly in the hands of private industry. It is clear that medicinal plant culture could offer opportunities for improving socioeconomical conditions for mountain people, but the question is how mountain people can take advantage of these bioresources. Clearly, random harvesting of wild plants, as presently practised in mountains, is not the key to socioeconomic growth or conservation. Mountain people of the Himalaya are not currently benefiting financially from such harvesting methods. Recently, the Indian government has enacted export restrictions on some plants to prevent their unsustainable utilisation and has listed 44 species whose export is banned. This is the result of overexploitation due to heavy international demand by pharmaceutical firms. However, conservation cannot be achieved merely by promulgating a ban on export or by fencing areas rich in medicinal species.

Conservation follows a sequence of changes, from demand → collection → consumption → cultivation → commercialisation → conservation. To achieve conservation and maintain biodiversity, it is necessary to incorporate incentives for cultivation so that the species banned from export become a priority for cultivation. To obtain lasting benefits from the national and international biotrade, mountain people need to follow one of two basic strategies, production or processing of the medicinal plants. The people can either become low-cost suppliers, providing substantial quantities of materials at low relative cost (the production-based approach) or they can aim to become the value-added supplier, providing valuable, screened plant material with known characteristics and high assurance of re-supply (processing-based approach).

To follow the production strategy, one needs to go for mass production of high quality material of selected species. This requires development of agrotechnology for these species. The processing-based strategy proposed above is only possible if there is a will to invest in mountain infrastructure and to allow retention in the community of some of the added value. Both strategies are based on market intelligence. While for the production strategy ecophysiological and phytochemical studies of plant populations are essential, the processing-based strategy requires state of art facilities to process the raw materials.

Subalpine and alpine areas are the store-houses of high-value medicinal plants and the conservation of this resource is essential. However, the approach should be science- and technology-based so that mountain people are able to derive benefit from this biodiversity. Simply banning the collection or use of species and fencing areas rich in rare medicinal species is not going to lead to their conservation.

References

Bahuguna R (2000). *Study of variation in medicinally active constituents of Aconitum species of different populations of Garhwal Himalaya.* DPhil Thesis, H.N.B. Garhwal University, Srinagar Garhwal, India

BCIL (1996). *Sectoral Study of Indian Medicinal Plants – Status, Perspective and Strategy for Growth.* Biotech Consortium India Ltd., New Delhi

Bhakuni DS (1998). Medicinal plants of commerce of Uttarakhand and problems of phytopharmaceutical industries. In Nautiyal AR, Nautiyal MC and Purohit AN (eds), *Harvesting Herbs 2000.* Bishan Singh Mahindra Pal Singh, Dehra Dun. pp 35–44

Billings WD and Mooney HA (1968). The ecology of arctic and alpine plants. *Biological Review*, 43:481–529

Biswas K (1956). *Common medicinal plants of Darjeeling and Sikkim Himalaya.* Ms Bengal Govt. Press, West Bengal

Brower V (1998). Nutraceuticals: Poised for a healthy slice of the healthcare market ? *Natural Biotechnology*, 16:728–31

CHEMEXCIL (1996). *Medicinal plants and essential oil export data: 1985–1995.* Basic Chemicals, Pharmaceuticals & Cosmetics Export Promotion Council, Bombay, India

Chopra RN, Nayar SL and Chopra IC (1956). *Glossary of Indian Medicinal Plants.* CSIR, New Delhi

Dhar U, Rawal RS and Upreti J (2000). Setting priorities for conservation of medicinal plants – a case study in the Indian Himalaya. *Biological Conservation*, 95:57–65

Gaur RD, Semwal JK and Tewari JK (1984). A survey of high altitude medicinal plants of Garhwal Himalaya. *Bulletin of Medicoethnobotanical Research*, 4:102–16

Grabherr G, Gottfried M, Graber A and Pauli H (1995). Patterns and current changes in alpine plant diversity. In Chapin FS III and Körner CH (eds), *Arctic and Alpine Biodiversity: Patterns, causes and ecosystem consequences.* Ecological studies, vol. 113, Springer, Berlin, pp 167–81

Jain SK (1991). *Dictionary of Indian Folk Medicine and Ethnobotany.* Deep, New Delhi

Joshi GC, Tiwari KC, Tewari RN, Pandey NK and Pandey G (1993). Resource survey of the pharmaceutically important plants of Uttarapradesh Himalaya. In Dhar U (ed.), *Himalayan Biodiversity–Conservation Strategies.* Gyanodya Prakshan, Naini Tal pp 279–91

Kaul MK (1997). *Medicinal plants of Kashmir and Ladakh – temperate and arid Himalaya.* Indus Publishing Co., New Delhi

Kirtikar KR and Basu BD (1918). *Indian Medicinal Plants.* (IIed., 1984). Bishen Singh Mahendra Pal Singh, Dehra Dun

Körner CH and Dimer M (1987). *In situ* photosynthetic responses to light, temperature and carbon dioxide in herbaceous plants from low and high altitude. *Functional Ecology*, 1:179–94

Kress WJ, Maddox GD and Roesel CS (1994). Genetic variation and protection priorities in *Ptilimnum nodosum* (Apiaceae), an endangered plant of the Eastern United States. *Conservation Biology*, 8:271–6

Mooney HA and Billings WD (1961). Comparative physiological ecology of arctic and alpine populations of *Oxyria digyna*. *Ecological Monographs*, 31:1–29

Nautiyal MC (1996). Cultivation of medicinal plants and biosphere reserve management in Alpine zone. In Ramakrishnan *et al.* (eds), *Conservation and management of bioresources in Himalaya.* Oxford and IBH Co. Ltd., New Delhi, pp 569–82

Nayar MP and Sastry ARK (1987–1990). *Red data book of Indian plants*, vol. 1–3. Botanical Survey of India, Howrah

Pandey OP and Purohit AN (1984). Activity of PEP-carboxylase and two glycolate pathway enzymes in C_3 and C_4 plants grown at two altitudes. *Current Science*, 49:263–5

Purohit AN, Lata H, Nautiyal S and Purohit MC (1998). Some characteristics of four morphological variants of *Podophyllum hexandrum* Royle. *Plant Genetics Research Newsletter*, 114:51–2

Purohit MC, Bahuguna R, Maithani UC, Purohit AN and Rawat MSM (1999). Variation in podophylloresin and podophyllotoxin contents in different populations of *Podophyllum hexandrum*. *Current Science*, 77(8):1078–80

Samant SS, Dhar U and Palni LMS (1998). *Medicinal Plants of Indian Himalaya*. Himavikas, 13. GBPIHED, Kosi, Almora, India

Silander JA Jr (1985). Microevolution in clonal plants. In Jackson JBC, Buss LW and Cook RE (eds), *Population biology and evolution in clonal plants*. Yale University Press, New Haven CN, pp 107–52

Singh PB and Aswal BS (1992). Medicinal plants of Himachal Pradesh used in Indian pharmaceutical industry. *Bulletin of Medicoethnobotanical Research*, 13:172–208

Part IV

Mountain Biodiversity, Land Use and Conservation

Land Use and Biodiversity in the Upland Pastures in Ethiopia

23

M.A. Mohamed-Saleem and Zerihun Woldu

INTRODUCTION

Land clearing and sedentary agriculture have changed the mountain vegetation in the East African region for over 3000 years (Hancock, 1985). For example, the natural vegetation of Ethiopian mountains has disappeared under mounting demographic pressure, except for a few patches around the holy places and in inaccessible terrain. Historically, the vegetation on the mountains of Ethiopia has oscillated with climatic episodes. The climate in Ethiopia is controlled by macro-scale pressure changes, and monsoon flows with elevation and topography contributing to the seasonal and inter-annual variations. The extremes of cold and warmth that the earth has experienced over the past 20,000 years have significantly affected the vegetative cover of Ethiopia. In Ethiopia, both climatic changes and human subsistence activities have caused changes in the vegetation types, making it difficult to apportion the relative importance of each of these factors.

According to the National Conservation Strategy (1997), four broad 'pristine' vegetation types occur above 1500 m asl in Ethiopia, namely:

(1) Afroalpine and sub-afroalpine vegetation (>3600 m),
(2) Dry evergreen montane vegetation (1900–3600 m, N + NW),
(3) Moist evergreen montane forest (1900–3600 m, SW),
(4) Evergreen scrub (1500–1900 m).

The natural vegetation types vary from afroalpine and sub-afroalpine at above 3600 m asl to dry evergreen montane forest in the northern and northwestern part and moist evergreen montane forest in southwestern parts of the country down to ca. 1900 m asl. Dry evergreen scrub is the dominant vegetation between 1500 and 1900 m asl.

The natural evergreen afro-montane forests on the northwestern plateau are dominated by *Olea africana* ssp. *cuspidata*, *Juniperus procera* and *Podocarpus falcatus*. This vegetation must have developed in the late Tertiary Period, because there is palynological evidence that *Podocarpus* and *Juniperus procera* were absent around 8 million years BP; instead Pteridophytes were abundant, indicating wetter and more closed forests (Kedamavit *et al.*, 1985). A study further east of Ethiopia, indicates the presence of thick vegetation cover which lasted until ca. 5000 BP. The Intertropical Convergence Zone, which determines the climate in this region, must have reached to the north of the Sahara, closer to the Mediterranean Sea around 8000 BP, and appears to have been much further south around 18,000 BP (Messerli and Winiger, 1992). This caused considerable changes in moisture regimes and glacial activities and shifted the treeline up and down the mountains at different times in history.

The last ice age terminated around 12,000 BP and had a marked effect on the treeline and zonation of the vegetation. The lower limit of trees may have moved down to 3600 m during Neoglacial times (Messerli *et al.*, 1977). Apparently, the confluence of many water bodies formed a single large lake in the Rift Valley of Ethiopia between 9600 and 9400 BP, evidencing much wetter conditions, and smaller disjointed lakes appeared between 5600 and

5300 BP as the water receded (Grove and Goudie, 1971). The climate became more humid again around 2000 BP, leading to a dominance of *Juniperus procera* and *Podocarpus falcatus* north and south of the lake region of Ethiopia (Mohammed and Bonnefille, 1991). The period between (e.g. around 4000 BP) appears to have been much more arid.

More recent descriptions of the vegetation are based on the accounts of European travellers in Ethiopia (Pankrhust, 1961, 1992; Pankrhust and Ingrams, 1988) and it appears that the present-day open, impoverished vegetation in Ethiopia was already evident in the seventeenth century. Over most parts of Ethiopia, grass dominated mountains, with only small remnant forests in isolated areas, demonstrate the effects of thousands of years of intensive human intervention. Beginning around 3500 BP into 3000 BP, the northern Tigray region had sustained settlements with intense livestock grazing and plough agriculture with irrigation (Anfray, 1967; Fattovich, 1988).

Land use practice has continued to modify the vegetation until today. For instance, in the western Shewa region large areas were cleared for cereal cultivation for several centuries, often leading to degraded dry evergreen shrubland. In the sixteenth century, however, pastoralists from the south moved into this area and replaced plough-based cereal cultivation by grazing. As a result, these shrublands regenerated to tall forest, with tree species apparently resembling those that were found in the pristine state, thus demonstrating vegetation changes solely due to varying land-use practices (Tamrat, 1994). In an adjacent area, a progression towards forest was arrested by a periodical reversion of the land to cereal cultivation. Hence, the degree of shrub cover reflects cultivation intensity (Zerihun and Backéus, 1991). In a recent study, livestock exclosures in the Tigray region resulted in rapid recovery of shrublands from almost bare land within only 15 years (Feoli *et al.*, 1995).

As population increased in montane Ethiopia, forests were cleared and grassland and shrublands became the dominant vegetation. The central plateau of Ethiopia is now characterised by mixed cereal and livestock agriculture, and 80% of the country's livestock population are found 1500 m asl. However, the altitude at which herded livestock graze depends on the season, reaching 4000 m asl and on steeper slopes during the rainy period when most areas at lower altitudes and with less steep slopes are under cereal cultivation. After crop harvest, grazing extends to below 1000 m asl. The species composition of the vegetation and the productivity of grazing land are, therefore, mainly influenced by livestock type and stocking rate, but climatic and edaphic factors are also significant co-factors (Zerihun, 1986). In general, defoliation, uprooting, trampling and desiccation are important stress factors in such pastures. Through their feeding behaviour or foraging choice, livestock determine the ways seeds disperse and vegetative cover regenerates (Zerihun and Mohamed Saleem, 2000). An experiment was designed to study the species composition of the natural upland grasslands under different grazing pressures across the landscape in one of the International Livestock Research Institute's (ILRI) mountain case study areas.

METHODOLOGY

A long-term experiment is in progress at Ghinchi Research Station, where ILRI is working with a consortium of institutions on natural resource management issues. The Ghinchi watershed is located in the Ethiopian Highlands at 2300–3000 m asl, 80 km west of Addis Ababa (9°02′ N; 38°07′ E). Although this study area only covers a 700-m range in elevation, it is part of a larger transect study on biodiversity changes between 500 and 4500 m asl, and similar monitoring sites such as that at Ghinchi are planned at 500-m intervals in the future.

The study site is a communally-grazed area. It receives average annual rainfall of 1150 mm, the main rainy season starts in June, peaks in August and tapers off in September. The soils vary from vertisols in the lower part to cambisols and regosols in the upper part of the watershed. Six sites with slopes varying from 0 to 8% were selected. Each site has nine 30 × 30-m plots, randomly assigned at the beginning of the experiment, to accommodate the following treatments: (a) no grazing (NOG), (b) moderate grazing (MDG at a stocking rate of 1.8 Tropical Livestock Unit per Month = TLUM/ha), and (c) very heavy grazing (VHG at a stocking rate of 4.2 TLUM/ha). The VHG treatment represents the grazing intensity of the surrounding area and intensity before the experiment started. The grazing intensity (4.2 TLUM/ha) has been calculated from the actual number of different types and breeds of livestock grazing in the watershed. However, the VHG treatment is not a control treatment, as faecal droppings remained on the grazed experimental plots whereas the farmers formerly collected the manure for fuel. Barbed wire fences were established in 1996 around the NOG and MDG plots. The flexible fence around the MDG plots is kept open for three days per week to allow free grazing. The NOG plots remain permanently shut to keep animals out. Grazing takes place during all days of the week in VHG plots and animals have free access at any time. By adopting a scoring index that was calibrated to avoid human bias, data on species composition and per cent ground cover of dominant species (5% or more) are being collected at 30-day intervals, from the second week of September to the second week of December each year from 1996 (only data up to 1999 are reported here), and the plots continue to be subjected to the same grazing pressure since the beginning of the study.

The method used to estimate plant cover permits total cover to exceed 100% due to overlapping of plants, hence cover may be considered a surrogate for leaf area index. All annual plants have one growing peak, which is September to December. Perennial plants have a second, smaller growing peak between mid February and end of March. The second peak is negligible as compared to the first one, and therefore, for all practical purposes, September to December represents the whole year in our study. The total set of plant species was grouped into non-graminoids and graminoids.

Each group was sub-divided into preferred (palatable) and avoided species.

RESULTS AND DISCUSSION

The species composition and the total ground cover in a given area differed with grazing regimes (Table 23.1). The total species number, as well as the number of grass species, initially decreased with increasing grazing pressure (first year) but tended to increase thereafter in heavily grazed plots compared to the non-grazed plots. There was a steady increase in plant cover per unit area in the MDG and VHG plots after the first year, with no change in the NOG plots up to the third year. Only in the fourth year were the species number and ground cover increased in the NOG plots. When plant cover exceeded 100%, plants overlapped, with some plants completely covering others. The data indicate that there is a significant influence of grazing regime on biodiversity.

Obviously, there will be more soil compaction and manure deposits in VHG areas compared to MDG (Mwendera et al., 1997). Since species richness is greater and ground cover is higher in the VHG plots compared to MDG plots, grazing and associated compaction of soil probably has not affected normal plant growth. Higher turnover of nutrients through greater manure deposition may have contributed to better plant establishment and recruitment in VHG compared to MDG plots (Zerihun and Mohamed Saleem, 2000). Differences in the species composition in the VHG and MDG compared to NOG plots may also indicate that animals transferred seeds to the grazing area from elsewhere. This might explain the increase in plant species richness in the third year of observation.

Some species tended to withstand very heavy grazing and persisted throughout the four years, while a few disappeared, new species also appeared and others oscillated over a period of differential grazing (Table 23.2). Non-expression of species in one year does not necessarily indicate permanent disappearance from the grazing area, nor does the appearance of species indicate invasion of uncommon

Table 23.1 1996–1999 annual species composition, accumulative species ground cover (%) and total species richness (in brackets) under three different grazing regimes (0, 1.8 and 4.2 Tropical Livestock Units per month and ha)

Year	Species present	NOG 0	MDG 1.8	VHG 4.2	TLUM Year	Species	NOG	MDG	VHG
1996	Andropogon abyssinicus	18	10	18	1998	Alchemilla gracilis	0	0	5
	Cyndon dactylon	15	13	13		Andropogon abyssinicus	18	10	22
	Cyperus rigidifolius	20	8	5		Chloris pychnotrix	0	0	30
	Eleusine floccifolia	0	5	5		Cyndon dactylon	5	10	12
	Hygrophilia auriculata	0	5	0		Cyperus rigidifolius	15	0	0
	Hyparrehenia arrhenobasis	20	0	0		Digitaria scalarum	10	0	8
	Pennisetum glabram	27	8	13		Eleusine floccifolia	0	10	12
	Pennisetum villosum	0	18	16		Eragrostis tenuifolia	0	5	5
	Scleria clathrata	30	0	0		Eriochloa nubica	0	0	10
	Scleria hispidior	0	8	0		Falkia oblonga	30	0	30
	Trifolium burchellianus	0	0	0		Fimbristiylis complanata	0	10	5
	Trifolium johnstonii	30	0	0		Hygrophila auriculata	20	0	20
	Trifolium semipilosum	0	3	10		Hyparrhenia arrehenobasis	10	15	15
	Verbena officinalis	10	0	0		Juncus sp.	31	12	28
	Total 1996	170 (8)	78 (9)	80 (7)		Pennisetum glabrum	18	18	16
1997	Alchemilla gracilis	0	0	3		Pennisetum villosum	5	15	15
	Andropogon abyssinicus	18	10	22		**Total 1998**	162 (10)	105 (9)	233 (15)
	Chloris pychnotrix	0	0	30	1999	Alchemilla gracilis	10	0	10
	Cyndon dactylon	10	5	8		Andropogon abyssinicus	10	7	8
	Eleusine floccifolia	15	0	0		Bidens pilosa	23	0	23
	Eragrostis tenuifolia	10	0	10		Bothrichloa insculpta	0	15	20
	Erichloa nubica	0	10	12		Chloris pychnotrix	0	0	5
	Falkia oblonga	0	5	10		Crotolaria spinosa	10	15	15
	Fimbristylis complanta	0	0	10		Cyndon dactylon	25	0	0
	Hygrophilia auriculata	30	0	30		Cyperus rigidifolius	10	10	10
	Hyparrehenia arrhenobasis	0	10	5		Digitaria sclarum	0	10	13
	Juncus sp.	20	0	20		Dischoriste radicans	10	0	10
	Pennisetum glabrum	10	15	15		Eleusine floccifolia	0	13	16
	Pennisetum villosum	41	6	21		Eragrostis tenuifolia	10	5	8
	Trifolium semipilosum	18	18	16		Eriochloa nubica	0	5	0
	Total 1997	172 (9)	79 (8)	212 (14)		Falkia oblonga	10	0	10
						Fimbristilis complanta	22	0	22
						Harpachne schimperi	15	5	8
						Hygrophilia auriculata	23	18	10
						Hyparrhenia arrhenobasis	18	0	15
						Juncus sp.	22	9	15
						Pennisetum glabrum	16	18	13
						Pennisetum villosum	5	4	9
						Polygonum nepalense	11	8	9
						Setaria sp.	0	23	28
						Trifolium semipilosum	0	18	18
						Trifolium sp.	10	15	13
						Trifolium tembense	0	15	10
						Total 1999	260 (18)	213 (18)	318 (24)

Table 23.2 Species dynamics after 3 years of different grazing regimes

	NOG (grazing stopped)	HVG (very heavy grazing)
Persistent	*Andropogon abyssinicus* *Cynodon dactylon* *Pennisetum glabrum* *Hyparrhenia arrehenobasis* *Cyperus rigidifolius*	*Andropogon abyssinicus* *Alchemilla gracilis* *Cynodon dactylon* *Pennisetum glabrum* *Chloris pychnotrix*
Disappeared	*Scleria clathrata* *Verbena officinalis* *Trifolium johnstonii* *Trifolim semipilosum*	*Erichloa nubica*
Emerging	*Achemilla gracilis* *Bidens pilosa* *Crotalaria spinaosa* *Dischoriste radicans* *Fimbristylis complanata* *Harpachne shcimperi* *Polygonum nepalense* *Trifolium tembense*	*Bothriochloa insculpta* *Digitaria scalarum* *Bidens pilosa* *Dichoriste radicans* *Polygonum nepalense* *Trifolium tembense* *Setaria* sp.

species. Disappearance of species from the grazed plots may be due to preferential removal of those species by animals or due to their lower ability to re-establish. On the other hand, species from the NOG plots may have disappeared due to competitive exclusion by species of the overgrowing canopy (Grime, 1979). Possibly, this triggered the recruitment of species which had remained dormant in the soil at the time when the experiment started. It is also possible that new propagules may have been introduced into the NOG plots from the adjacent grazed plots over time. However, the partly periodic appearance and disappearance of species in all three treatments makes interpretation very difficult. For example, there was a massive addition of species in all three treatments in 1999. However, it seems safe to conclude that grazing at high stocking rates was at least not negative for both cover and species richness over the four years of the experiment, so far.

Implications

Mountainous parts of east Africa are under severe economic and social pressure resulting from rapidly increasing population, land fragmentation and degradation, abject poverty and repeated cycles of drought and human tragedies. Under these circumstances, safeguarding biodiversity and attempting to restore pristine vegetation by protecting land areas from human interference are neither practicable nor appropriate. Increasing the usefulness of what is available will be a better strategy for guaranteeing sustainable livelihoods, preventing further biodiversity erosion and attempting to revert some species losses. Overstocking and overgrazing are considered among the major factors causing land degradation and loss of biodiversity of grazing lands. In most of the mountainous areas, grasslands are not the pristine vegetation but have been derived under prolonged human pressure. Livestock is a major component of the household economy in the mountains, and several land-use systems (cropping, forests, etc.), including the grazing lands, have to be managed in an integrated and compatible way to serve the multiple human needs.

The experiment reported here is the first of a series of studies, all planned to assist in developing strategies for managing biodiversity at different levels of grazing pressures, while serving both soil protection and feed needs of livestock in mountain watersheds. Contrary to the expected results, intensive grazing did not cause an immediate loss of biodiversity, nor did excluding animals cause significant gains. Certainly, we need longer observation periods to confirm longer term trends. Information on species persistence, oscillation of their presence and absence in different years and with treatments indicate that management can be tailored to favour certain species compositions and to optimise species diversity and land-use needs. How to operationalize such management practices should become a major

concern for development agencies. The rapid regeneration of vegetation after temporary withdrawal of livestock from steep sloping lands in the Tigray region of Ethiopia shows that livestock do not cause irreversible damage to biodiversity and, therefore, with careful planning, livestock can be integrated into sustainable land use (E. Feoli, L.G. Vuerich and Z. Woldu, 2001 unpublished results).

References

Anfray F (1967). Matara. *Annales d' Ethiopie*, 7:33–53

Fattovich R (1988). Remarks on the late prehistory and early history of northern Ethiopia. In Beyene T (ed.), *Proceedings of the Eighth International Conference of Ethiopian Studies*, Addis Ababa, 1984: Institute of Ethiopia Studies, pp. 85–104

Feoli E, Fermetti M and Woldu Z (1995). Vegetation of heavily anthropized areas in Tigray, northern Ethiopia. *Rivista di Agrocoltura Subtropicale Tropicale*, 89(2):223–41

Grime, JP (1979). *Plant Strategies and Vegetation Processes*. John Wiley & Sons, Chichester

Grove AT and Goudie AS (1971). Late quaternary lake levels and climatic changes in the Rift Valley of southern Ethiopia and elsewhere in tropical Africa. *Nature*, **234**:403–05

Hancock G (1985). *Ethiopia: The Challenge of Hunger*. Victor Gollancz Ltd., London

Kathryn AB (1997). *The environmental history and human ecology of northern Ethiopia in the late Holocene*. Istituto Universtitario Orientale, Dipartimento di Studi e Ricerche su Africa e Paesi Arabi, studi Africanisistici, serie 5

Messerli B, Hurni H, Kienholz H and Winiger M (1977). *Bale mountains: largest pleistocene mountain glacier system of Ethiopia*. XINQUA congress, Birmingham (abstract)

Messerli B and Winiger M (1992). Climate, environmental change, and resources of the African mountains from the Mediterranean to the equator. *Mountain Research and Development*, 12(4):315–36

Mohammed MU and Bonnefille R (1991). *The recent history of vegetation and climate around Lake Langano (Ethiopia)*. Laboratoire de Gélogie du Quaternarie, CNRS, Université de Luminy, Marseilles, France

Mwendera EJ, Mohamed Saleem MA and Woldu Z (1997). Vegetation response to cattle grazing in the Ethiopian highlands. Agriculture, Ecosystem and Environment, 64:43-51.

National Conservation Strategy Secretariat (1997). *Conservation Strategy of Ethiopia*. Environmental Protection Authority, Addis Ababa.

Pankrhust R (1961). *An Introduction to the Economic History of Ethiopia (from early times to 2800)*. Addis Ababa University, Addis Ababa

Pankrhust R (1992). The History of Deforestation and Afforestation in Ethiopia prior to World War I. In *Proceedings of the Michigan State University Conference on Northern Africa*. Michigan State University Press, East Lansing, pp 275–86

Pankrhust R and Ingrams L (1988). *Ethiopia Engraved*. Kengal Paul, London

Tamrat B (1993). Vegetation and ecology of remnant Afromontane forests on the central plateau of Shewa, Ethiopia. *Acta Phytogeogr. Suec.*, 79:1–64

Woldu Z and Backéus I (1991). The shrubland vegetation in western Shewa, Ethiopia, and its possible recovery. *Vegetation Science*, 2:173–80

Woldu Z and Mohammed Saleem MA (2000). Grazing-induced biodiversity in the highland ecozones of East Africa. *Agriculture, Ecosystem and Environment*, 79:43–52

Woldu Z (1986). Grassland communities on the central plateau of Ethiopia. *Vegetatio*, 67:3–16

Yemane K, Bonnefille R and Faure H (1985). Paleoclimatic and tectonic implications of Neogene microflora from the north-western Ethiopian highlands. Nature, vol. 318. No. 6047. pp 653–6

Balancing Conservation of Biodiversity and Economic Profit in the High Venezuelan Andes: Is Fallow Agriculture an Alternative?

24

Lina Sarmiento, Julia K. Smith and Maximina Monasterio

INTRODUCTION

Páramo: a biodiverse ecosystem

In the upper belt of the Northern Andes (3000–4800 m) the characteristic ecosystem is the paramo, a humid tropical ecosystem dominated by giant caulescent rosettes, shrubs and bunch grasses. The paramo flora is among the richest found in the high mountains of the world (van der Hammen and Cleef, 1986). Half of the estimated 3000 to 4000 species of paramo vascular plants are endemic (Luteyn *et al.*, 1992). This high biodiversity is related to the geographical distribution of the paramo which appears as a chain of islands, separated by lower altitude ecosystems. These islands repeatedly suffered processes of expansion and contraction during the Pleistocene and Pliocene that favoured alternatively the colonisation and speciation of the flora (Cleef, 1978, 1981). Also, the unique climatic conditions of the high tropical environment (drastic daily temperature fluctuations) have led to the evolution of a flora with very particular adaptations (Vuilleumier and Monasterio, 1986; Monasterio and Sarmiento, 1991).

Due to the paramo's high biodiversity, the originality of plant adaptations, the numerous medicinal plants, its importance for water availability in the lowlands, and the great potential for recreational and touristic activities, paramo qualifies as a high priority area for conservation. However, it has been subject to an accelerated process of degradation and transformation. Each year, the upper agricultural frontier rises and the pristine hill slopes are absorbed by agriculture at an alarming rate due to the pressure of increasing population. This agricultural expansion is reported for Colombia (Ferwerda, 1987; Verweij, 1995; Hofstede, 1995), Ecuador (Hess, 1990) as well as for Venezuela (Drost *et al.*, 1999; Sarmiento, 2000).

Man's use of the paramo ecosystem

The agricultural use of the paramo ecosystem is relatively recent (Ellenberg, 1979; Monasterio, 1980). In pre-Columbian times, the Venezuelan paramos were utilised exclusively for hunting and gathering (Wagner, 1978). It was only during the colonial period when the paramos began to be used for extensive grazing and for wheat growing (Monasterio, 1980). Later, wheat cultivation decreased, and more recently potato cropping in rotation with garlic and carrots has become an important economic activity. Initially, potatoes were cultivated with long fallow systems, as in many areas of the high Andes in Bolivia, Peru and Colombia (Brush, 1976; Sarmiento *et al.*, 1990; Hervé *et al.*, 1994; Pestalozzi, 2000). Recently, fallow is being eliminated by the utilisation of large amounts of mineral and organic fertilisers. This intensification is related to the geographical accessibility. In isolated areas more traditional systems persist, while easily reached areas are intensively cultivated. The unequal accessibility causes the coexistence of a variety of agricultural systems, including intensive, transitional and extensive systems, providing a good opportunity to assess the sustainability, environmental impact, conservation value and economic profitability of different management alternatives.

To evaluate the possibilities and challenges for the conservation of the high regional and local biodiversity in the paramo regions, it is essential to understand the social and economic importance of the human activities. In contrast to the agricultural marginality of most mountain regions (Rieder and Wyder, 1997), in tropical countries like Venezuela, many crops can only be cultivated in the cool mountain climate. This production is sold to the domestic market, providing food for an increasing national population and at the same time is the base of subsistence for a numerous and growing rural Andean population. Between 1984 and 1995, potato production rose five times, garlic four times and carrots nine times in the Venezuelan Andes (Gutierrez, 1996). This increase was accomplished by intensification as well as by expanding the agricultural frontier, frequently by an advance in altitude, incorporating fragile paramo areas that often lie inside the national parks.

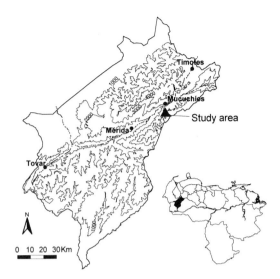

Figure 24.1 Localisation of the study area, Páramo de Gavidia, in the Venezuelan Andes

Long fallow agriculture and the maintenance of biodiversity

In the tropics and subtropics long fallow agriculture is not only widespread in low altitude but also in mountain areas (Grigg, 1974; Ferweda, 1987; Kellman and Tackaberry, 1997; Sarmiento et al., 1990, 1993; Knapp, 1991; Ramakrishnan, 1992; Hervé et al., 1994; Pestalozzi, 2000). An old polemic exists about the sustainability of this type of agriculture. Several authors agree that it can be sustainable providing the population or economic pressures are low, but others see long fallow agriculture as a wasteful form of land use, consuming large areas in support of few people (Ingram and Swift, 1989; Kleinman et al., 1995). In principle, a biodiversity comparable to that of the natural ecosystem can be maintained with a long fallow system, using the spatial coexistence of several successional stages, forming a mosaic landscape (Swift and Anderson, 1994). During the fallow period,

the typical behaviour of plant diversity is to increase in the early stages, as a result of the gradual colonisation of the area; to attain a maximum in intermediates stages, when competitors coexist; and to decrease in late stages, as the system approaches its competitive equilibrium and exclusion occurs (Huston, 1994). If the successional diversity is highest at intermediate stages, a landscape managed with a fallow system can be more diverse than the natural vegetation. Nevertheless, there are many exceptions to this general trend and a wide variety of successional patterns have been reported (Huston, 1994).

THE STUDY AREA: PÁRAMO DE GAVIDIA

A long fallow system in transformation

The study area, Páramo de Gavidia, is located in the Sierra Nevada National Park, in the state of Mérida, between 3200 and 3800 m asl (Figure 24.1). The area is a narrow glacial valley where agriculture is practised on steep slopes and small colluvial and alluvial deposits (Figure 24.2). The mean temperature ranges between 5° and 9 °C and the mean annual precipitation is 1300 mm. The present population

Figure 24.2   Panoramic view of the study area, the glacial valley of the Páramo de Gavidia (3200–3800 m asl)

(400 inhabitants) settled in the valley at the end of the nineteenth century, giving it a relatively short land-use history (Smith, 1995). The management system is long fallow agriculture, where potatoes are grown for 1 to 3 years and then the fields enter the succession–restoration phase (Sarmiento *et al.*, 1993). In some cases, wheat or oats can follow the potato crop. The majority of the potatoes are grown for the local market and the cereals for home consumption. The current average fallow period is 4.6 years, but a large variability, from 2 to more than 15 years exists. This management system generates a landscape mosaic where cultivated and fallow fields coexist with areas of natural vegetation. At present, the surface under fallow agriculture is 192 ha and the total surface of the study area is 464 ha. The 59% of the area still under natural vegetation is progressively being incorporated into the agricultural cycle. Between 1992 and 2000, 22 ha of natural vegetation were ploughed, corresponding to an increase of 11.5% of the area in the cultivation–restoration cycle.

In 1972, a road to the valley was built, facilitating the commercialisation and the introduction of mineral fertilisers. Since then, the fallow periods have been reduced but not eliminated. In the last three decades, the system has passed from a traditional fallow system, where potatoes and cereals were cultivated mainly for subsistence, to a semi-traditional system where mineral fertiliser allows a surplus of production for sale. Currently a new transformation towards an intensive agricultural system is beginning.

The ecological succession during the fallow period

Plant colonisation and replacement is a continuous process, but for practical reasons we differentiate four periods: early, intermediate and late succession, and restored paramo. During the early period (1 to 3 years) typical pioneer herbaceous species colonise, including *Rumex acetosella*, an introduced forb, which is the dominant species. Other species during this phase are *Vulpia muyrus*, *Lachemilla moritziana*, *Senecio formosus*, *Lupinus meridanus* and *Poa annua*. During the intermediate phase (4 to 6 years), *R. acetosella* and *L. moritziana* reduce their cover, while *L. meridanus* and *V. myurus* become more abundant. Newly arrived species include *Gamocheta americana*, *Geranium* sp. *Trisetum irazuense*, *Acaena elongata*, etc. Also, in this phase, some dominant paramo species increase their abundance, such as *Espeletia schultzii* and *Hypericum laricifolium*. In the late phase (more than 6 years) other species become dominant, including *Baccharis prunifolia*, *Noticastrum marginatum*, *Stevia lucida*, *Pernettya prostrata*, *Bromus carinatum*, etc. *E. schultzii* and *H. laricifolium* continue to increase their cover and the physiognomy changes from herbaceous to the typical rosette–shrub paramo. Finally, the restored paramo is dominated by *E. schultzii*, *H. laricifolium*, *P. prostrata*, *Calamagrostis effusa*, *Agrostis tolucensis*, *B. prunifolia*, *Nassella mexicana* and *Arcytophyllum nitidum* among many others (Figure 24.3).

RESEARCH QUESTIONS

In this study, biodiversity is addressed at two different scales, the field and the whole valley. At field level, we focus on the dynamics of restoration. If the fallow period is too short, the natural ecosystem will not be restored and consequently the agricultural practice will lead to ecological degradation. The research question

Figure 24.3 The paramo vegetation, with the characteristic rosettes of *Espeletia schultzii*

METHODOLOGY

Part 1: Biodiversity and richness along the succession gradient

For the estimation of species richness and diversity along the succession gradient, 150 fields with different fallow lengths (1 to 12 years) and eight areas with natural, never ploughed, vegetation were selected using a spatial database with information on the fallow lengths of 1200 fields.

The vegetation was sampled using the point-quadrat method (Greig-Smith, 1983). A pin was placed 100 times at random in each field and the touching species were recorded. Richness was estimated as the total number of species recorded in each plot. Species abundance was calculated as the number of contacts. Alpha diversity was calculated from the species abundance using the Shannon index. Beta diversity along the successional gradient was calculated using the equation of Shmida and Wilson (1985).

is: how much diversity can be attained with the present fallow length in relation to the natural paramo?

At local or valley scale, we examine the spatial coexistence of fields in several successional stages. This coexistence could theoretically generate greater biodiversity than that of the natural ecosystem, by the juxtaposition at local level (in neighbouring fields) of species characteristic of the different successional phases (pioneers, intermediates and climax species). The central question is whether this long fallow system increases or decreases biodiversity at the local scale compared to the valley with only natural vegetation.

The current changes that the agroecosystem is undergoing, such as fallow time reduction and ploughing of natural areas, are also analysed. The problems addressed are how important is maintaining areas of natural vegetation for local biodiversity, and how does shortening of the present fallow lengths affect local biodiversity?

Finally we analyse which agricultural land-use system could optimise the relationship between biodiversity and economic profit and thus answer the questions: Are long fallow systems the best alternative to maintain high plant diversity or are there other alternatives that could optimise the relationship between conservation of plant diversity and the needs of the human population?

Part 2: Biodiversity at local scale and possible future scenarios

Local biodiversity was estimated by extrapolating the data obtained in the individual plots to the entire valley. All the fields with the same fallow time and natural vegetation areas were considered as landscape units. The species abundance of each landscape unit was calculated as the average abundance of all the studied plots with the corresponding fallow time. Then the species abundance in the valley was calculated by weighting the abundance of each landscape unit by its surface:

$$\mu_i = \sum_{k=1}^{n} a_{ik} s_k \bigg/ \sum_{k=1}^{n} a_{ik}$$

where μ_i is the abundance of the ith species in the valley, a_{ik} is the abundance of the ith species in the kth landscape unit and s_k is the surface occupied by the kth landscape unit, obtained from the spatial database. The values of μ_i were utilised to calculate the biodiversity using the Shannon index (H'):

$$H' = -\sum_{i=1}^{n} pi \ln pi \quad \text{and} \quad pi = \mu i \bigg/ \sum_{i=1}^{n} \mu i$$

where p_i is the proportional abundance of the ith species.

Using this methodology, the local biodiversity corresponding to the current management system was calculated. In order to evaluate the effect of fallow shortening, the same methodology was utilised to calculate local biodiversity for scenarios with fallow lengths from 1 to 10 years. In this case, the surface occupied by each landscape unit (s_k) was calculated by dividing the total surface of the study area (464 ha) by the fallow length plus 2, considering that the fields are cultivated for 2 years before entering fallow. To examine the consequences of the yearly incorporation of never ploughed areas into the agricultural cycle, different relationships between areas under agriculture and natural vegetation were considered.

Definition of the scenarios

Three groups of scenarios were defined, considering the land-use systems currently practised in the high Venezuelan Andes. In each group the relationship between agricultural and natural area is modified until 100% of the valley is occupied by agricultural land use. The three groups are:

- Fallow lengths between 1 and 10 years, with 2 years of potato cropping.
- Continuous cropping, where organic manure replaces fallow (intensive system).
- Spatial combination of intensive and 10-year fallow system

Calculation of the economic profit of the different scenarios

The economic profit of each modelled scenario was calculated as the gross income minus the cost of production. The costs considered were labour, transport, mineral fertiliser and organic manure, where applied. The transport costs include the carrying of fertilisers, seeds and production between the fields and the road on horseback. For the calculation, a function of the distance to the agricultural fields, obtained from the spatial database, and the carrying capacity of mules was established. The calculation of all inputs was based on the prices of 1999 and the average amounts applied in the area. The workforce was calculated considering the preparation of the field, planting, hilling and harvesting, which varies depending on the production system.

For the yield calculations of the fallow system the restoration of soil fertility during the fallow and the fertility loss during cropping were considered. A model developed by Sarmiento (1995), using yield data from the same area, was applied. This model considers that soil fertility, defined as the capacity of the soil to produce potatoes without mineral fertiliser, increases as an exponential function of the fallow time:

$$F_t = 14 \, (1\text{-}e^{-0.03t})$$

where F_t is the soil fertility level in t ha^{-1} after a fallow time of t years.
Yield was calculate as:

$$Y = 12 \, F_t \, 0.5^n$$

where Y is the crop yield in t ha^{-1}, 0.5 is a factor of fertility reduction after each year of cropping, n is the number of years under cultivation and 12 is a factor that considers the yield increase by the application of the average dose of mineral fertiliser for the area (1.8 t ha^{-1} of NPK 16–16–08). In the intensive system, a constant yield of 18 t ha^{-1} was used, which represents the regional average when using organic manure. In the intensive and the fallow system the same dose of mineral fertiliser was considered.

In order to model the intensive–extensive agricultural system, a 150-m buffer zone was drawn around the existing road in the valley

287

bottom. The area close to the road was modelled using the intensive system and the area outside the buffer zone was modelled as a 10-year fallow system.

RESULTS

Part 1: Biodiversity during the fallow period

Species richness doubles during the first 4 years of the fallow, passing from an average of 10 to 20 species (Figure 24.4). Hereafter, the number of species stabilises, and after 12 years of fallow, the richness still remains significantly lower than in the natural vegetation, where an average of 35 species per plot was found. Alpha diversity presents a similar tendency, with an important increase during the first 4 years and a posterior stabilisation. These results show that neither the species richness nor the diversity of the original ecosystem are restored after 12 years of succession.

Beta diversity, that quantifies species turnover during succession, decreases exponentially and after 9 years, 60% of the species still have to be replaced to obtain the community structure of the natural vegetation (Figure 24.5). A successional deceleration in species turnover is evident, and the time necessary for a complete restoration seems to be much longer than the studied interval and the current fallow lengths.

Part 2: Local biodiversity

Effect of the fallow length and the proportion of natural vegetation on biodiversity

Different fallow lengths and proportions of the valley under agricultural use lead to changes in local diversity. In Figure 24.6, it can be observed that local biodiversity depends more on the remaining natural vegetation than on the fallow time, which is only important when almost all the area is incorporated into the agricultural cycle. Without natural vegetation, local biodiversity is very sensitive to fallow duration. A system with only 1 year of fallow generates very low local biodiversity, which

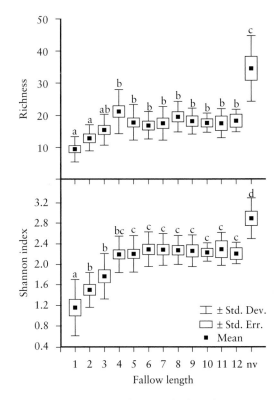

Figure 24.4 Species richness and plant diversity estimated, using the Shannon Index, along the succession and in natural vegetation (nv). Same letters indicated not significantly different ($p < 0.05$) between successional stages (Duncan test after one-way analysis of variance)

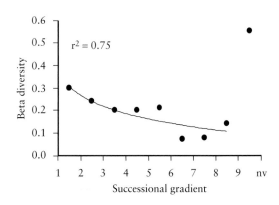

Figure 24.5 Plant beta diversity along the successional gradient, fitted to a negative exponential model (nv = natural vegetation)

quickly rises until a 4-year fallow system is reached. After that, the increase in fallow length has little effect on biodiversity. This tendency reflects earlier results showing that

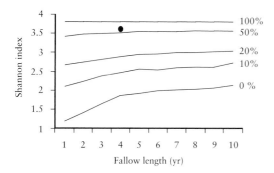

Figure 24.6 Simulated effect of the fallow length and of the percentage of natural vegetation on local plant diversity. The different fallow lengths refer to the duration of the fallow in the valley, e.g. a system with a fallow length of 4 years has 33% of its surface under crop and the remaining 66% is divided between 1 and 4 years of fallow. The point indicates the current agricultural system

the successional rate becomes very slow after 4 years. On the other hand, we can see that biodiversity rises quickly with the increase in natural vegetation area. For example, with 50% of the area under natural vegetation, a biodiversity comparable to that of the entire valley under paramo is attained and the fallow length with which the other 50% is managed has little importance. In the current agricultural system, local biodiversity is not very different from that of the natural vegetation. This is not due to the current fallow lengths but to the high proportion of the natural area remaining in the valley.

Effect of different management systems on biodiversity and net economic profit

The quantification of local biodiversity and net economic profit of the scenarios are presented in Figure 24.7. The response of the net economic profit is very dependent on the management system. The intensive system obtains much higher profits than the combination between intensive and extensive or the fallow systems. The profit difference between the 5- and 10-year fallow system is not very large and is due to the fallow effect on soil fertility. In all scenarios, biodiversity rapidly decreases when less natural vegetation remains in the area.

The current management system, while maintaining high local biodiversity, has a very low economic profit. To achieve higher profits more natural areas would have to be incorporated into the agricultural cycle, but this would lead to a rapid reduction in biodiversity and only a small increase in profit. A better alternative seems to be changing the management to an intensive system, but conserving large areas of natural vegetation. For example, a biodiversity of 3 can be reached by a 10-year fallow system with 20% of natural vegetation or by an intensive system with 40% of natural vegetation. In this case, the economic profit is three times greater in the intensive system. The same profit can be obtained using a significantly smaller area intensively and conserving the rest of the area under natural vegetation.

DISCUSSION AND CONCLUSIONS

In the paramo environment, succession proceeds too slow to restore plant diversity in a time interval compatible with an agricultural system. However, after 12 years of succession a considerable number of paramo taxa have colonised, forming a semi-natural vegetation, physiognomically comparable to the natural paramo. Even if part of the diversity is lost, long fallow agriculture allows the maintenance of a semi-diverse system. Contrasting with the most common tendency in secondary succession, in the paramo, the diversity at intermediate stages is lower than in the climax community. This is due to the fact that just a small number of species are exclusive to the succession. Only *Rumex acetosella*, *Poa annua* and a few others, most of which are introduced species, act as real colonists. The rest of the species abundant in early and intermediate stages are paramo species with better dispersal mechanisms and other characteristics that permit higher performance during these stages. As in other extreme environments (MacMahon, 1981), there is not a real succession in terms of species replacement, but only variations in the relative abundance and the progressive arrival of the paramo species. The time necessary for a complete restoration of

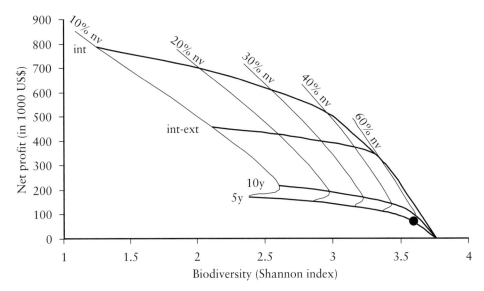

Figure 24.7 Relation between plant local diversity and economic profit for the studied scenarios. 5y = 5 year fallow system, 10y = 10 year fallow system, int-ext = combination of intensive–extensive systems (outside a 150 m buffer zone around the road a 10 year fallow system is practised) and int = intensive agricultural system. The point indicates the current agricultural system. The percentages indicate the proportion of the natural vegetation remaining in the area

the paramo vegetation cannot be extrapolated using our data, as both diversity and richness stabilise after 4 years of succession. The fast decrease in beta diversity indicates that a total restoration of the natural paramo would take many years. Ferwerda (1987) calculated 70 years to restore the natural vegetation under a similar management system in a Colombian paramo. In our area, the figure is probably of the same order.

The dynamics of plant diversity throughout the succession period suggest that 4 or 5 years of fallow are enough to reach a stable level, and few changes will take place with fallow prolongation. Nevertheless, during the first years of succession, the dominant species are non-native and consequently these stages are less interesting for the conservation of local biodiversity and can be seen as degraded systems – less diverse and dominated by introduced species. Only in the intermediate and late phases does the vegetation begin to be dominated by indigenous taxa.

The analysis of local biodiversity of the entire valley did not confirm the initial idea that the spatial coexistence of different successional stages would enhance plant biodiversity compared to the natural vegetation. Therefore, the highest local biodiversity would be achieved with the whole valley under paramo vegetation. Nevertheless, due to the large areas of natural vegetation remaining in the valley, the current agricultural system is very close to the biodiversity measured in the paramo. Apart from the positive effect of natural vegetation on diversity, it is also essential as a reservoir from which paramo plants can spread and colonise the fallow fields, as recolonisation depends on the surrounding mosaic and its rate is higher when patches are closer together (Forman, 1997).

Although the present fallow system is very diverse, it has a low productivity compared to an intensive system. In order to increase net earnings without changing the management system, more natural areas would have to be ploughed, resulting in a large negative effect on biodiversity. A reduction in the fallow time is also possible, but in this case the augmentation in the cultivated area is counterbalanced by the decrease in productivity due to incomplete restoration of soil fertility. The effect of these tendencies, the incorporation of new areas, as well as the reduction of fallow time currently taking place in the study area, will be a progressive reduction of plant diversity.

Analysing the different scenarios, the best relationship between economic profit and biodiversity can be achieved by combining intensive land use with the preservation of large natural vegetation areas, avoiding the existence of areas of disturbed or incompletely restored paramo. Even if this is the best theoretical system, its practical implementation confronts serious difficulties. The main problem is controlling agricultural expansion over natural areas when the intensive system brings so much higher economic gains. In contrast to intensive agriculture, fallow systems *per se* regulate the use pressure and oblige the farmer to maintain a high proportion of land at rest. The intensive system, on the other hand, has no internal limitations with respect to the area under cultivation, apart from the limits imposed by capital or workforce availability. In this case, regulations must come from outside, as local or national policies, which are much more difficult to implement and control. The fallow system has the advantage that it is self-regulated. The intensive system, on the other hand, requires external regulation, which is subject to the power of the economic interests.

Apart from difficulties in controlling the extension of agricultural areas, further aspects related to intensive agriculture need to be explored. The first issue is the dependency on external factors which replace the ecological functions of the fallow period. Intensive agriculture depends on large amounts of inputs (fertilisers, pesticides, seeds, irrigation, mechanisation, etc.) and is capital-intensive and highly market-oriented. Large investments are necessary that can only be compensated if the prices obtained for the end products are high enough. This fact makes the system very sensitive to market oscillations. Consequently, it is fragile, unstable and at the same time increases social inequality. The second issue is the environmental impact. The excessive and unbalanced use of artificial inputs can have serious ecological, economic and socio-political

repercussions (Reijntjes *et al.*, 1992) and pesticides may not only be a hazard to the water and soil, but also to the population's health. The overall sustainability of the intensive land-use system needs to be assessed. Fallow systems are less dependent on external inputs and, through crop field spreading, different risks, such as the impact of crop diseases and the loss of the entire harvest in the event of night frost, can be reduced. All in all, even though an intensive system in restricted areas is the best alternative from the biodiversity point of view, the negative aspects cannot be ignored and other alternatives need to be explored.

In conclusion, long fallow agriculture is not the ideal alternative to conserve the paramo plant biodiversity, but is a more secure alternative than an intensive system, where uncontrollable economic pressures can cause the reduction or elimination of natural and seminatural areas. Nevertheless, fallow systems are progressively being transformed as a consequence of their low economic profit and following the building of new roads. Consequently, the future conservation of the paramo ecosystems will depend on effective implementation of local and national regulation policies or, if these policies are lacking, will be subject to unpredictable economic forces and cyclic variations in market prices.

ACKNOWLEDGEMENTS

This research was supported by the European Union, within the TROPANDES project (N° IC18-CT98-0263) and by the IFS (grant No. C/2668-1). We wish to thank Tarsy Carballas, the international coordinator of TROPANDES, for her constant support. Thanks also to Nelson Marquez, Auxiliadora Olivo and Jimmy Morales for their participation in the fieldwork and to Alexander Berg and Ana Escalona for the botanical identifications. Luis Daniel Llambí helped in collecting and processing the vegetation data. Special thanks to the community of Gavidia and particularly to Señora Rosa, who always welcomed us in her house with delicious coffee and arepas.

References

Brush SB (1976). Man's use of an Andean ecosystem. *Human Ecology*, 4(2):147–65

Cleef AM (1978). Characteristics of neotropical páramo vegetation and its subantarctic relation. In Troll C and Lauer W (eds), Geoecological relations between the Southern temperate zone and the tropical mountains. *Erdwiss Forsch*, 11:365–90

Cleef AM (1981). *The vegetation of the paramos of the Colombian Cordillera Oriental*. Dissertationes Botanicae 61

Drost H, Mahaney W, Bezada M and Kalm V (1999). Measuring the impact of land degradation on agricultural production: a multi-disciplinary research approach. *Mountain Research and Development*, 19(1):68–70

Ellenberg H (1979). Man's influence on tropical mountain ecosystems in South America. *Journal of Ecology*, 67:401–16

Ferwerda W (1987). *The influence of potato cultivation on the natural bunchgrass páramo in the Colombian Cordillera Oriental*. Internal report no. 220, Hugo de Vries-Laboratory, University of Amsterdam

Forman RTT (1997). *Land mosaics: the ecology of landscapes and regions*. Cambridge University Press, Cambridge

Greig-Smith P (1983). *Quantitative Plant Ecology*. University of California Press, Berkeley

Grigg DB (1974). *The Agricultural Systems of the World. An Evolutionary Approach*. Cambridge University Press, London

Gutiérrez A (1996). *Plan estratégico de desarrollo agrícola del Estado Mérida (PEDEM). Documento sobre el sector agrícola*. Centro de Investigaciones Agroalimentarias, Universidad de los Andes, Mérida, Venezuela

Hervé D, Genin D and Riviere, G (1994). *Dinámicas del descanso de la tierra en los Andes*. IBTA-ORSTOM, La Paz

Hess C (1990). 'Moving up – moving down': Agropastoral land-use patterns in the Equatorial Páramos. *Mountain Research and Development*, 10(4):333–42

Hofstede R (1995). *Effects of burning and grazing on a Colombian páramo ecosystem*. PhD thesis, University of Amsterdam, Amsterdam

Huston MA (1994). *Biological diversity: the coexistence of species on changing landscapes*. Cambridge University Press, Cambridge

Ingram J and Swift M (1989). *Tropical Soil Biology and Fertility (TSBF) Program*. Report of the Fourth TSBF Interregional Workshop, Harare, Zimbabwe, May 31–June 8, 1988. Biology International, Special Issue 20

Kellman M and Tackaberry R (1997). *Tropical Environments: the functioning and management of tropical ecosystems*. Routledge, London

Kleinman PJA, Pimentel D and Bryant RB (1995). The ecological sustainability of slash and burn agriculture. *Agricultural Ecosystems and Environment*, 52:235–49

Knapp G (1991). *Andean Ecology: Adaptive Dynamics in Ecuador*. Westview Press, Boulder

Luteyn JL, Cleef A and Rangel O (1992). Plant diversity in paramo: toward a checklist of paramo plants and a generic flora. In Balslev H and Luteyn J (eds), *Páramo: an Andean ecosystem under human influence*. Academic Press, London, pp 71–84

MacMahon J (1981). Successional processes: Comparisons among biomes. In West D, Shugart H and Botkin D (eds), *Forest Succession: Concepts and Application*. Springer-Verlag, New York

Monasterio M (1980). Poblamiento humano y uso de la tierra en los altos Andes de Venezuela. In Monasterio M (ed.), *Estudios Ecológicos en los Páramos Andinos*. Ediciones de la Universidad de los Andes, Mérida, pp 170–98

Monasterio M and Sarmiento L (1991). Adaptive radiation of *Espeletia* in the cold Andean tropics. *Trends in Ecology and Evolution*, 6(12): 387–91

Pestalozzi H (2000). Sectoral fallow systems and the management of soil fertility: the rationality of indigenous knowledge in the high Andes of Bolivia. *Mountain Research and Development*, 20(1):64–71

Raintree JB and Warner K (1986). Agroforestry pathways for the intensification of shifting cultivation. *Agroforestry Systems*, 4:39–54

Ramakrishnan PS (1992). *Shifting Agriculture and Sustainable Development: An Interdisciplinary Study from North-Eastern India*. Man and Biosphere Series 10. UNESCO, Paris and The Parthenon Publishing Group, New York

Reijntjes C, Haverkort B and Waters-Bayer A (1992). *Farming for the future: An introduction to low-external-input and sustainable agriculture*. Macmillan, Leusden

Rieder P and Wider J (1997). Economic and political framework for sustainability of mountain areas. In Messerli B and Ives J (eds), *Mountains of the World: a Global Priority*. The Parthenon Publishing Group, New York, pp 85–102

Sarmiento L (1995). *Restauration de la fertilité dans un système agricole à jachère longue des hautes Andes du Venezuela*. PhD thesis, University of Paris XI, Orsay

Sarmiento L (2000). Water balance and soil loss under long fallow agriculture in the Venezuelan Andes. *Mountain Research and Development*, **20**(3):246–53

Sarmiento L, Monasterio M and Montilla M (1990). Succession, regeneration and stability in high Andean ecosystems and agroecosystems: The rest-fallow strategy in the Páramo de Gavidia, Mérida, Venezuela. In Winiger M, Wiesmann U and Rheker J (eds), *Mount Kenya Area: Differentiation and Dynamics of a Tropical Mountain Ecosystem*. Geographica Bernesia, African Studies Series A8, pp 151–7

Sarmiento L, Monasterio M and Montilla M (1993). Ecological bases, sustainability, and current trends in traditional agriculture in the Venezuelan high Andes. *Mountain Research and Development*, **13**(2):167–76

Shmida A and Wilson MV (1985). Biological determinants of species-diversity. *Journal of Biogeography*, **12**(1):1–20

Smith, JK (1995). *Die Auswirkungen der Intensivierung des Ackerbaus im Páramo de Gavidia – Landnutzungswandel an der oberen Anbaugrenze in den venezolanischen Anden*. Diplomarbeit, University of Bonn, Germany

Swift MJ and Andenson JM (1994). Biodiversity and Ecosystem Function in Agricultural Systems. In Schulze E and Mooney H (eds), *Biodiversity and Ecosystem Function*. Springer Verlag, Berlin, pp 15–41

van der Hammen T and Cleef A (1986). Development of the high Andean paramo flora and vegetation. In Vuilleumier F and Monasterio M (eds), *High Altitude Tropical Biogeography*. Oxford University Press, Oxford, pp 53–201

Verweij P (1995). *Spatial and temporal modelling of vegetation patterns: Burning and grazing in the paramo of Los Nevados National Park, Colombia*. PhD thesis. ITC Publication Number 30. Enschede

Vuilleumier F and Monasterio M (1986). *High Altitude Tropical Biogeography*. Oxford University Press, New York

Wagner E (1978). Los Andes Venezolanos, arqueología y ecología cultural. *Ibero-Americanisches Archiv, Neue Folge*, **4**(1):81–91

Conserving Mountain Biodiversity in Protected Areas

Lawrence S. Hamilton

25

INTRODUCTION

Many scientists have made a strong case for the high biological diversity in mountain areas, particularly on their humid slopes. At least seven of the expanded 18 Global Biodiversity Hotspots of Myers (1990) are largely mountainous. A global mapping of species numbers of vascular plants (Figure 25.1) shows an amazing congruence with mountain ranges (Barthlott *et al.*, 1996). Major centres of vascular plant species diversity are in or include mountains: Costa Rica, tropical eastern Andes, Atlantic Forest of Brazil's coastal mountains, eastern Himalaya–Yunnan region, Caucasus, northern Borneo, Papua New Guinea, Western Ghats, Madagascar, Eastern Arc of Tanzania, Albertine Rift, Cape Peninsula mountains, and Adamawa (Mt. Cameroon range). In the long-used and abused Mediterranean Region, it is the Mediterranean mountains that are the last refuges of native plant and animal biodiversity. The neotropical mountain humid slopes have greater mammal biodiversity than the much publicized Amazon lowland rainforest (Mares, 1992). In North America also, the Rocky Mountains, the Mexican Sierras, and the Appalachians are recognized as being important reservoirs of native biodiversity. In Great Smoky Mountain National Park an All Taxa Biodiversity Survey initiated in 1999 is revealing amazing numbers of new species, and high numbers of known species, and is strongly suggesting that the Appalachians is possibly the richest biodiversity ecoregion in the temperate zone. We have simply not done our biodiversity homework in mountains for they are, after all, known for their inaccessibility, but as we investigate them, they are proving to be rich treasure houses of biodiversity heritage. The mountains and highlands harbour the wild ancestors for instance, of the five most important food staples of the world – maize, potato, barley, wheat and rice (Rhoades, 1985).

This rich biodiversity is due to the altitudinal zonation of life forms, the exposure to different orientations, the soil variability and the abundance of microhabitats associated with mountain topography and features (Hamilton, 1992). Moreover, the high level of endemism in mountains because of isolation factors has been well documented. This seems to be particularly true for my favorite ecosystems, the tropical montane cloud forests (Wuethrich, 1993; Long, 1995; Leo, 1995; Hamilton *et al.*, 1995). This Chapter will focus on how this biodiversity wealth can best be conserved in mountain areas in view of the changes in human use of mountains, in natural catastrophes that characterize mountains, and the impact of exogenous threats such as climate change and air pollution. To me there are three principal answers:

(1) Have a representative and adequate system of mountain protected areas (both public and private), particularly of IUCN Categories I, II, III and IV[1];

---

[1] IUCN Protected Area Categories: I Strict Nature Reserve/Wilderness (managed mainly for science or wilderness); II National Park (managed mainly for ecosystem protection and recreation); III National Monument (for conservation of specific natural features); IV Habitat/Species Management Areas (for conservation through management intervention); V Protected Landscape (landscape conservation with traditional uses and recreation); VI Managed Resource Protected Area (sustainable use of natural resources)

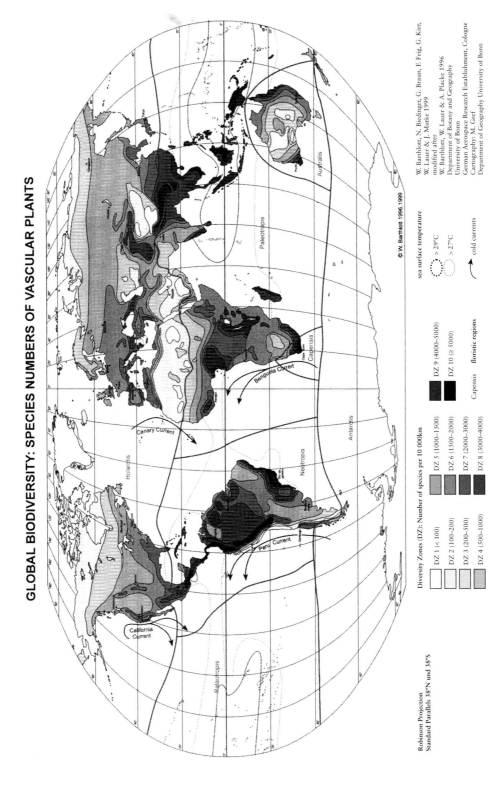

Figure 25.1   A global map of species numbers of vascular plants

(2) Obtain land-use management regimes principally in agriculture and forestry that are nature (biodiversity)- friendly (these could be for areas actually designated equivalent to IUCN Categories V and VI);

(3) Reduce the anthropogenic adverse air quality and climate change impacts by getting at the causes of these changes. I shall briefly discuss only items 1 and 2.

MOUNTAIN PROTECTED AREAS

Even using strict criteria as to what constitutes a mountain protected area (MtPA), in terms of a definition of a mountain and a minimum size, there are at least 473 sites totaling 264 million hectares of IUCN Categories I–IV scattered in 63 countries of the world (Thorsell, 1997). This inventory, compiled by the UNEP-World Conservation Monitoring Centre, included only those larger than 10,000 ha and with a minimum relative relief of 1,500 m. In addition there are a few private MtPAs (such as the 263,000 ha area extending across Chile which will become Pumalin Park) and many famous mountain protected areas (such as the Cairngorms in Scotland) which do not meet the 1,500 m criterion. If the relief criterion was relaxed to 1,000 m, another 200 sites would be included.

The substantial number of these, and the total considerable size, are no grounds for complacency. One single unit, Greenland National Park, is 97 million ha, 37% of the total area, and many mountain ranges are unrepresented, e.g. the Atlas Mountains, Papua New Guinea, mountains of Myanmar. Moreover, far too many of them protect mainly the high summits rather than the entire slopes; they are like islands in a sea of altered landscape; they are unconnected to the next protected area in the same mountain range; and they are far too small to protect the full suite of biodiversity. Most were originally selected as national parks or national monuments to give protection to particularly scenic or geologically unusual peaks, of appeal to visitors, and not on the basis of conserving

biodiversity. To conserve biodiversity we need to base our selection and design on different criteria. The global network of MtPAs is in need of additions and redesign in order to achieve the goals of the Convention on Biological Diversity.

Noss (1983, 1987, 1991, 1992) has laid out an approach and guidelines for maintaining biodiversity in a regional landscape based on the science of conservation biology. He suggests that there are four fundamental objectives to maintain the native biodiversity of a region in perpetuity:

- Represent all ecosystem types and successional stages across their natural range;
- Secure viable populations of all native species in their natural patterns of abundance and distribution;
- Maintain ecological and evolutionary processes, such as disturbance, nutrient and water cycling and predation; and
- Ensure that the biological diversity of each region can respond naturally to short-term and long-term change.

No single strategy alone can achieve all of these goals everywhere, and in some countries or regions all of these goals may not be possible. The need, however, is to move toward them as best and as rapidly as we can.

The science of conservation biology has given the basis for the selection and design of protected areas in order to conserve species and genetic flows that keep populations viable. The testing of the theories of island biogeography have shown for instance, the relationship between the size of an island, its distance from the nearest source of replenishment, and the rate at which species are lost from the island (MacArthur and Wilson, 1967; Diamond, 1975). It has become evident that nature protection areas surrounded by an 'unfriendly' landscape from which recruitment (species migration and gene flow) cannot occur, are indeed similar to islands. In essence, the smaller the 'island' the more rapid and greater is species loss; i.e. the larger the better. Moreover, if the ecosystem contains low-density, wide-ranging species (such as wolves, bears or mountain lions), protected areas need to be **very** large in order to maintain minimally viable populations of these species. And these species are essential to the health of the whole system – umbrella species that can carry with them a large spectrum of the other biodiversity of the area. If we conserve the habitat for these generally large carnivores, we include most of the area's other species of plants and animals, and in a healthier state. Data on size of area to maintain viable populations of montane large carnivores are largely lacking, but some figures from lowland Africa point in the general direction. Conserved habitats to maintain a viable population (at least 500 individuals) of lion may have to be on the order of 2,413 miles^2, and for the endangered wild dog may be 100,000 km^2 (Newmark, 1992). We know that Yellowstone National Park, large as it is at 800,000 ha, is not large enough to maintain its population of elk or bison, both of whom must range outside the park (to their peril) into what is being called the Greater Yellowstone Ecosystem. Some estimates are beginning to emerge from North America as to minimum size for a viable population. The Vermont Nature Conservancy is using for home range of a female: bobcat – 2,000 ha, lynx – 1,300 ha, fisher – 1,200 ha and marten – 485 ha (TNC, personal communication, 2000). If minimum viable population is 200–300 females, these areas are very large.

Most of our formally protected areas are too small to fulfill the function of maintaining all of the native biodiversity of an area. Small areas are more susceptible to massive invasion by aggressive alien species due to the large perimeter to area ratio. They are also more vulnerable to being eliminated or irreparably damaged by natural or human caused disturbance. Mountain areas will be especially impacted by upward shifts in altitudinal habitat belts in a global warming scenario, and small sky Islands will fare badly, as highest elevation species will have no place to go. Alpine vegetation and associated wildlife on summits

(e.g. Australian Alps) are already in trouble from diminished snow cover (Green, 1998).

What then of the small protected areas – do they not have value? They may certainly play an important role in biodiversity conservation especially for plants. Moreover, due to past history of landscape alteration those wide-ranging, low-density animal species may have been extirpated and the job is to maintain other 'keystone' species of more limited range, or to simply slow the rate of species loss, and here 'small' protected areas have a role to play in biodiversity conservation. They may also serve as 'stepping stones' for migratory species, especially birds. The point is that they need to be as large as is possible. In addition, these small areas, even 200 ha, may play other valuable roles in protecting scenic beauty, water supplies, natural monuments, or provide opportunities for education, nature-based tourism, or mountaineering.

The concept of having conservation buffer zones or transition areas around the protected cores is also critical. This is well expressed in the idealized design of a UNESCO Biosphere Reserve. In these transition areas more nature-friendly human uses can extend the viability of the protected core for many species. The shape of a protected area is also important, for minimizing perimeter. And, if the protected areas are small and isolated, their geographic distribution in the landscape matrix is a factor. Some of these size, shape, and distribution criteria are shown in Figure 25.2 from Diamond (1975).

Finally comes one of the most important criteria for MtPAs – connectivity. Connecting fragmented 'Islands' is of paramount importance to maintain pathways for species migration and gene flow. These are especially needed to accommodate climate change:

- Connectivity altitudinally up and down mountains, for seasonal migration and for longer-term shifts in species and communities to higher elevations;
- Connectivity along the ranges, longitudinally to permit poleward shifts in response

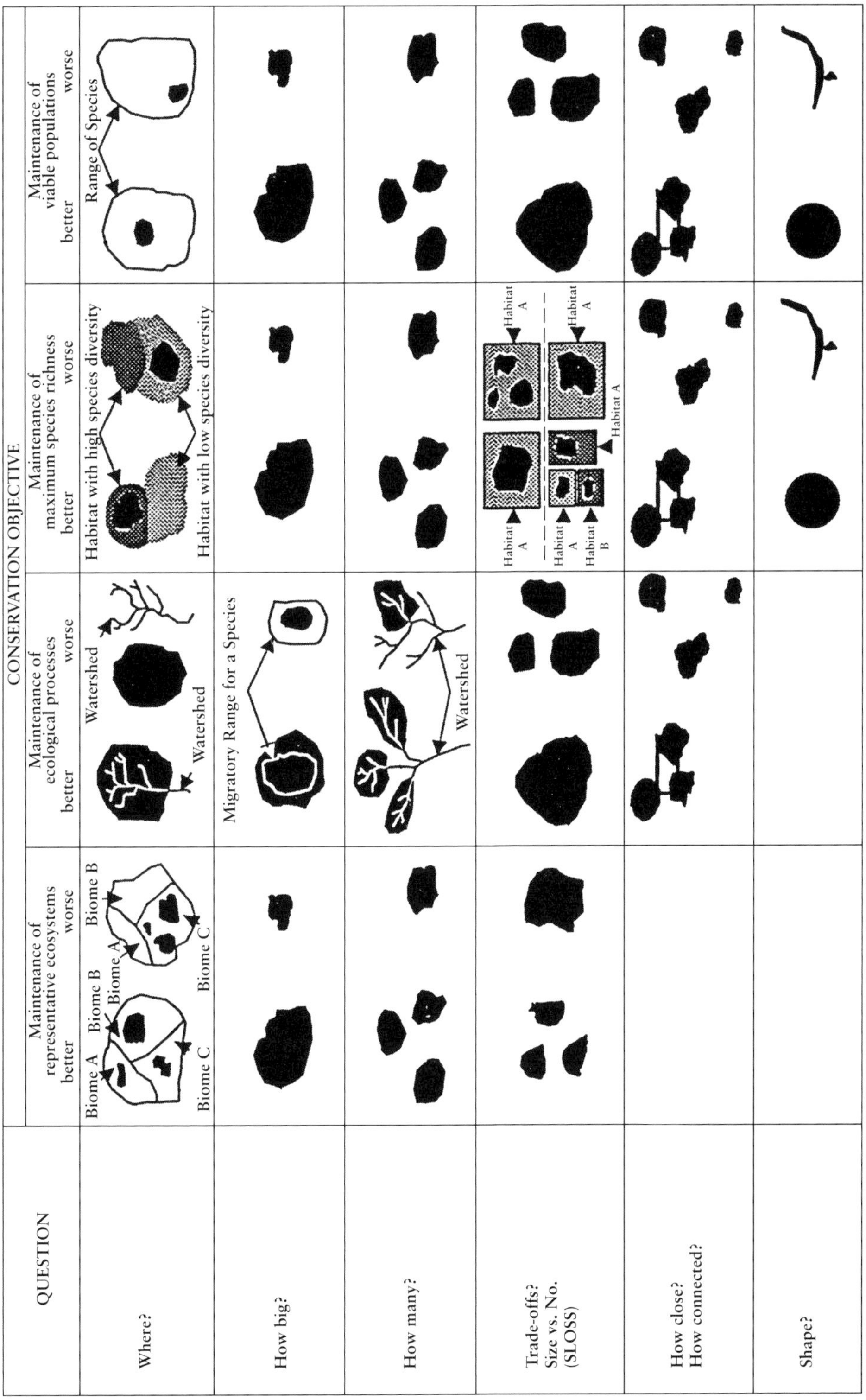

Figure 25.2 Guidelines for selection and design of protected areas in relationship to four objectives for conserving living resources. The preferable guideline for the selection and design of protected areas in relation to conservation objectives and design are denoted as 'better', while the less preferable option is denoted 'worse' (After Diamond, 1975)

to temperature change (e.g. the Andes or the Appenines) and latitudinally for response to changing precipitation patterns (e.g. the Himalayas).

- Connectivity across national or state boundaries in common, cooperative management regimes for biodiversity conservation is a challenging and important enterprise. Native species may seasonally migrate across irrelevant political boundaries, as do the ibex in the southern European Alps. Mercantour (France) and Alpi Maritime (Italy) Parks have cooperative policies and management for these animals (Rossi, 1996). Programmes of cooperation characterize Krkonoše (Czech Republic) and Karkonosze (Poland) National Parks' biodiversity research under a Transboundary Biosphere Reserve umbrella, including jointly meeting challenges of invasive exotic species which do not respect political boundaries in the Giant Mountains (Flousek, 1995).

The achieving of mountain conservation corridors of linked areas to protect biological diversity (and provide many other ecosystem services at the same time) is not easy. Except in quite remote, unpopulated mountains and ranges, the opportunity to establish such large areas of protection (Categories I–IV) has passed. The recently declared Kamchatka Volcanoes World Heritage area perhaps represented one of the last opportunities, or Antarctica which is now essentially protected by treaty. Mainly, the task will be to connect formal MtPAs by 'nature-friendly' corridors of human-altered and human-used landscapes. One might call these 'Stewardship lands'. For a connection between two or more protected core areas to function effectively it cannot be too narrow, since other conservation biology studies have shown the several disadvantages of such narrow links (Simberloff and Cox, 1987). Consequently, it is desirable to make these connections as wide as possible so that the idea of a narrow corridor, such as a hallway, needs to be dispelled. These intervening lands will have to be large conservation landscapes. Miller (1996) calls them 'bioregional management areas'.

STEWARDSHIP LANDS

Private or communal lands between the core areas of protection will be extremely important components of the large conservation corridors or bioregions. These lands could well become designated as Category V, Protected Landscape, or Category VI, Managed Resource Protected Area. Some, such as forest reserves that are managed for sustained multiple use, might also be government lands. And, many areas officially designated by governments as Regional Nature Parks (especially in Europe) are classified as Protected Landscapes in the IUCN system. Even many National parks, e.g. Cevennes (France) or Hohe Tauern (Austria) are classified as Category V, because though they have core areas of strictly protected lands owned by the state, much of the land is privately owned but used with certain restrictions. If done well, such lands, even privately owned, should be managed in a sustainable 'nature-friendly' fashion that conserves biological diversity (e.g. no hunting of certain species, restrictions on pesticides, etc.). Brown and Mitchell (1998) presented a good discussion of the myriad aspects of incentives, policies, philosophies, requirements for success, and some case examples from around the world for these Stewardship Lands. The George Wright Society FORUM journal (Vol 17, No 1, 2000) has the theme 'Landscape Stewardship' and has several pertinent articles, including one on an Andean mountain corridor from Colombia to Bolivia, La Ruta Cóndor/ Wiracucha (Sarmiento et al., 2000). While Europe has substantial experience and success with protected landscapes, it has been slower to develop elsewhere, but it is an urgent need if biodiversity is to be conserved. A good example from the Himalaya is the Annapurna Conservation Area (Sherpa et al., 1986). The challenge of converting the heretofore, often nature-hostile land management practices, to a regime of stewardship is formidable, and demands creativity, patience and hard work with local landowners and community groups. In this work, indigenous knowledge is a key element, and it often resides mainly with women and the 'elders' (Gurung, 1994). The use of conservation easements, held by land trust organizations, can preclude biodiversity-destroying land development and promote good stewardship. The Nature Conservancy has recently (with a local partner, Pronatura A.C.) implemented the first conservation ease-ment in Mexico – a cloud forest in Veracruz (TNC, personal communication, 2000). On government lands, administered by a commo-dity production agency, such as many tradi-tional forestry departments, the task is almost as difficult, though progress is occurring in Australia, in Canada and the USA, as examples.

MOUNTAIN CONSERVATION CORRIDORS ON THE GROUND

Is this practicable? Some progress is being achieved by organizations (mostly NGO-initiated, but some government agencies also) to develop these large landscape level, ecore-gional or bioregional corridors, clusters or constellations. I am currently working with Dr. Kenton Miller of the World Resources Institute on a joint WRI/WCPA project to iden-tify and map the various proposed scenarios of this kind around the world. There are at least 31 mountain corridors or clusters. The earliest one, and most formally developed is the ambi-tious proposal for a Yellowstone to Yukon Conservation Initiative (Figure 25.3). This grouping of conservation NGOs on both sides of the USA–Canada border now has a fulltime executive officer and staff of two others. One very early initiative developed in the Appenines, where the Superintendent of Abruzzo National Park conceived linking his park and seven other MtPAs to form a Central Appenines Green Region (Figure 25.4). An altitudinal corridor is being worked on in Ecuador, from the high peaks to the coast in a Choco-Andean Corridor involving eight existing protected areas and three potential biocorridors (Figure 25.5). Many of the most advanced of these initiatives and the concept behind them have been described by Hamilton (1997). A methodology for assessing the 'biopermeability' of these intervening areas between formally protected

reserves has been developed by Romano (in press) using the Central Appenines as a test area. The Mountain Theme of the World Commission on Protected Areas has as its chief programme element the fostering of these corridor initiatives and promoting interaction between them. There are many potential areas where very long and wide corridors might be achieved. A map of the protected areas in the European Alps invites the capturing of this vision (Figure 25.4). Locally, where I live in Vermont, we are working to make our interstate Great Northern Forest part of the Greater Laurentian Region, and link the mountainous Adirondack Park to Algonquin Park in Ontario. And for a super-vision, one can dream of a 'Corridor of Conservation of the Americas', from Tierra del Fuego to the Bering Sea, a dream first articulated by IUCN's Jim Thorsell. The Meso-American Biological Corridor agreed to by the seven Central American countries in 1997 gives this dream an encouraging touch of reality when added to the other initiatives in the Andes, and the Rocky Mountains and the Southwestern Sky Islands of the USA. And indeed, in mid-1999 the Wildlife Conservation Society, out of San Pedro, Costa Rica has adopted this vision and launched a programme, 'A Biological Corridor of the Americas' (Figure 25.5). Key action in North America is being undertaken by The Wildlands Project which currently has 25 project initia-tives, 13 of them in mountains.

The IUCN/WCPA Mountain Theme is attempting to support and to provide inter-change between various mountain corridor initiatives around the world. Among these are: Albertine Rift (Uganda/Rwanda/Burundi/ Democratic Republic of Congo and Zambia); Drakensberg/Maloti (South Africa and Lesotho); Indian Plains to Bhutan Alpine Area; Southeastern Great Escarpment (NSW, Australia); Australian Alps; Andean Bear Ecological Corridor (Venezuela); Serra do Mar (Brazil); Naya Watershed Conservation Corridor (Colombia); Ruta Cóndor/Wiracucha (Colombia to Bolivia); Condor Bioreserve in Ecuador; and Western Ghats (India). Several other notable bioregional mountain corridors are in concept stage. As these corridors

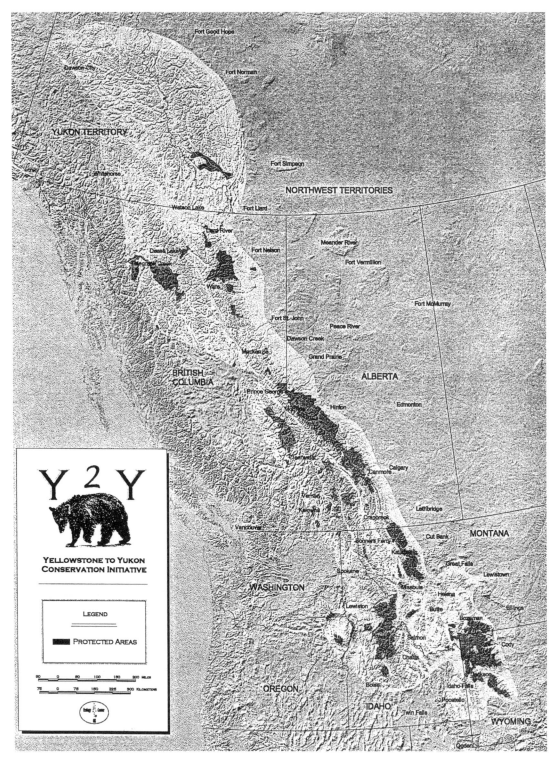

Figure 25.3 The Yellowstone to Yukon Conservation Initiative

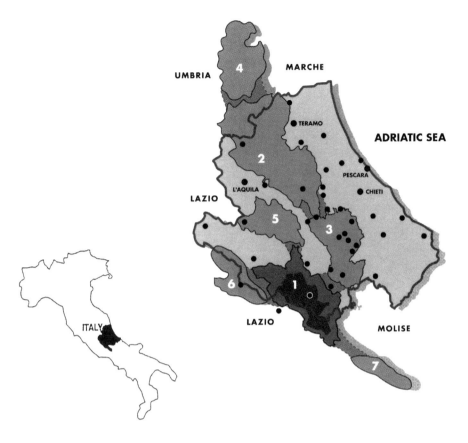

LEGEND:

☐ Abruzzo National Park and its peripheral protected zone

▨ New protected areas or in process of establishment

● Nature reserves, refuges and oases

▨ Connectivity corridors needed under conservation regime

1   Abruzzo National Park and peripheral protected zone
2   Gran Sasso–Laga National Park
3   Majella National Park
4   Monti Sibillini National Park
5   Sirente–Velino Regional Park
6   Proposed Monti Ernici–Simbruini Inter-regional Park
7   Proposed Matese Inter-regional Park

Figure 25.4   The Central Appeniness Conservation Corridor. The map was compiled by the World Commission on Protected Areas/IUCN and the World Resources Institute, based on a map supplied by Dr Franco Tassi, Parks Centre of Italy. It illustrates one of a selection of cases of various corridors or bioregional approaches presented at the WCPA Symposium (1997) From Islands to Networks

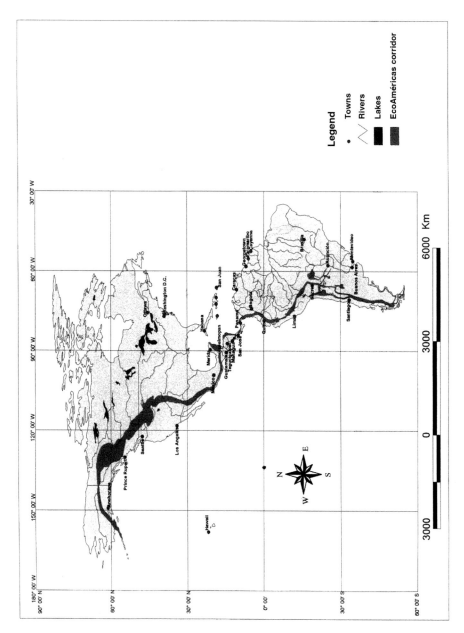

Figure 25.5 The proposed Biological Corridor of the Americas

materialize, conditions improve for the processes of recovery of missing elements of the flora or fauna, either through natural recolonization or careful reintroduction. Some extirpated biodiversity can be regained, and in mountain regions this can often mean the return of the large predators such as the lynx, bear, wolf, bearded vulture and condor. Mountain Theme support of re-wilding activities is another thrust of the World Commission on Protected Areas programme, but is another topic, for another time.

References

Barthlott W, Lauer W and Placke A (1996). Global distribution of species diversity in vascular plants: Towards a world map of phytodiversity. *Erdkunde*, 50:317–27

Brown J and Mitchell B (1998). The stewardship approach and its relevance for Andean landscapes. In Sarmiento F and Hidalgo J (eds), *Entendiendo las Interfaces Ecologicas para la Gestion de los Paisajes Cultural en los Andes*. Asociacion de Montanas Andinas, Quito, Ecuador, pp 281–6

Diamond JM (1975). The Island dilemma: lessons of modern biogeographic studies for the design of natural resources. *Biological Conservation*, 7:129–46

Flousek J (1995). Conserving biodiversity in the Czech and Polish Krkonoše National Parks and Biosphere Reserve. In Arends J, Čeřovský J and Picková G (eds), *Transboundary Biodiversity Conservation: Selected Case Studies from Central Europe*. Ecopoint, Prague, pp 23–9

Green K (1998). Problems in Australia's alpine zone. *Mountain Protected Areas UPDATE* (newsletter of IUCN/WCPA Mountain Theme) June 1, pp 8–9

Gurung J (ed.) (1994). *Indigenous Knowledge Systems and Biodiversity Management*. Proceedings of a MacArthur Foundation ICIMOD Seminar, Kathmandu

Hamilton LS (1992). Biodiversity: future wealth. *In An Appeal for the Mountains*. The Mountain Agenda, Bern, pp 8–9

Hamilton LS (1997). Maintaining ecoregions in mountain conservation corridors. *Wild Earth*, 7(3):63–6

Hamilton LS , Juvik JO and Scatena FN (eds) (1995). *Tropical Montane Cloud Forests*. Springer-Verlag, New York

Leo M (1995). The importance of tropical montane cloud forest for preserving vertebrate endemism in Perú: the Río Abiseo National Park as a case study, In Hamilton LS, Juvik JO and Scatena FN (eds), *Tropical Montane Cloud Forests*. Springer-Verlag, New York, pp 198–211

Long A (1995). The importance of tropical montane cloud forests for endemic and threatened birds. In Hamilton LS, Juvik JO and Scatena FN (eds), *Tropical Montane Cloud Forests*. Springer-Verlag, New York, pp 79–106

MacArthur RH and Wilson EO (1967). *The Theory of Island Biogeography*. Princeton University Press, Princeton

Mares M (1992). Neotropical mammals and the myth of Amazonian biodiversity. *Science*, 225:967–79

Miller KR (1996). *Balancing the Scales. Guidelines for Increasing Biodiversity's Chances Through Bioregional Management*. World Resources Institute, Washington

Myers N (1990). The biodiversity challenge: expanded hot-spots analysis. *The Environmentalist*, 10(4):243–56

Newmark WD (1992). The selection and design of nature reserves for the conservation of living resources. in Lusigi W (ed.), *Managing Protected Areas in Africa*. UNESCO World Heritage Fund, Paris, pp 87–99

Noss RF (1983). A regional landscape approach to maintain diversity. *BioScience*, 33:700–06

Noss RF (1987). Protecting natural areas in fragmented landscapes. *Natural Areas Journal*, 7:2–13

Noss RF (1991). *Protecting Habitats and Biological Diversity Part I: Guidelines for Regional Reserve Systems*. National Audubon Society, New York

Noss RF (1992). The Wildlands Project, land conservation strategy. *Wild Earth*, Special Issue (undated) on The Wildlands Project

Rhoades R (1985). Farming on high. In Tobias M, (ed.), *Mountain Peoples*. University of Oklahoma Press, Norway, pp 39–44

Romano B (In press). Biopermeability, a key concept in the establishment of biological corridors. Proceedings of the *Workshop on Mediterranean Protected Areas*. World Commission on Protected Areas/IUCN, Cilento, Italy, 4–7 November,1999

Rossi P (1996). Maritime Alps/Mercantour Parks – nature without frontiers. In Hamilton LS,

Mackay JC, Worboys GL, Jones RA and Manson GB (eds), *Transborder Protected Area Cooperation*. Australian Alps National Parks and IUCN, Canberra, pp 53–63

Sarmiento FO, Rodríguez G, Torres M, Argumedo A, Muñoz M and Rodríguez J (2000). Andean stewardship: tradition linking nature and culture in Protected Landscapes of the Andes. *The George Wright FORUM*, 17(1):55–69

Sherpa MN, Coburn B and Gurung CP (1986). *Annapurna Conservation Area, Nepal, Operational Plan*. King Mahendra Trust for Nature Conservation, Kathmandu

Simberloff D and Cox J (1987). Consequences and costs of conservation corridors. *Conservation Biology*, 1(1):63–71

Thorsell J (1997). Protection of nature in mountain regions. In Messerli B and Ives JD (eds), *Mountains of the World: A Global Priority*. Parthenon Publishing, New York and London, pp 237–48

Wuethrich B (1993). Forests in the clouds face stormy future. *Science News*, 144(2):23

How Effective is Protected Area Management in Mountains?

26

Douglas Williamson

INTRODUCTION

How much can protected areas contribute to the conservation of mountain biodiversity? The answer to this question depends on whether a theoretical or a practical approach is adopted. In theory, protected areas should have a crucial role to play in conserving biodiversity, but in practice it is uncertain to what extent they actually do so because, until recently, surprisingly little attention has been paid to the question of how effectively protected areas are conserving biodiversity, whether in mountains or in other ecosystems.

Where this question has been addressed, the answers have not always been encouraging. For instance, Canada has recently acknowledged that most of its National Parks are suffering from serious environmental deterioration as a result of mismanagement, overuse and encroaching development (Parks Canada, 2000). The conclusion of a two-volume report by a task force which spent a year reviewing the status of Canadian protected areas was that: 'We have come to recognise that Canada's wild places are not endless and that even our protected areas are not safe from undesirable change.'

A preliminary survey of management status and threats in the protected areas of ten key forest countries concluded that only 1% of protected areas in these countries were secure and that 25% were suffering degradation and loss (Stolton and Dudley, 1999). Direct external threats to protected areas in these countries included: encroachment by human settlements, agriculture and over grazing, forestry operations, mining and fossil fuel extraction, bushmeat hunting, collection of commercially important plants (e.g. orchids) and animals (e.g. parrots) for sale, fire, pollution and climate change, invasive species, war, tourism and recreational pressure.

These threats have two key underlying causes: high levels of consumption by the richest quarter of the world's population, and the poverty of the poorest segment of the global population. In poor countries they are often exacerbated by lack of money and technical capacity of protected area authorities.

Uncertainty about the effectiveness of protected area management is worrying, both in terms of operational considerations and in terms of the issue of accountability. In operational terms, it creates doubts about the success of a crucial component of global efforts to conserve biodiversity. The issue of accountability arises because, in a world of ever increasing competition for land, conservationists are accountable not only to their own constituency but also to the wider world for their stewardship of the large areas of land that have been set aside as protected areas. Ineffectual management of this land results in a dangerous reduction of the credibility of biodiversity conservation as a land use.

This Chapter examines the reasons for uncertainty about the effectiveness of protected area management and outlines one possible approach to addressing the issue. In recent years there has, in fact, been a growing interest in management effectiveness in protected areas, but it is difficult to be sure that achievement of management objectives will result in effective biodiversity conservation because there no agreement on exactly what biodiversity is or how to conserve it (Brandon *et al.*, 1998; Freese, 1998). For the sake of brevity,

307

the definition of biodiversity in the Convention on Biological Diversity (CBD) will be adopted and discussion will be confined to the question of how to conserve biodiversity in protected areas.

WHAT ARE THE GOALS OF BIODIVERSITY CONSERVATION?

What is at issue here is not *why* we should conserve biodiversity but, having decided that it should be conserved, *how* we should do so. This is a question which has to be addressed because if we do not know what we are trying to accomplish, it is obviously impossible to decide whether or not we are succeeding. From conceptual and reported practical difficulties in knowing or deciding whether or not biodiversity conservation programmes have been successful (Salafsky and Margolouis, 1999; Salafsky *et al.*, 1999), it can be inferred that people implementing conservation activities in the field are in need of a clear conceptual basis for deciding whether or not they are working effectively, but no such basis currently exists. What does exist is a diversity of approaches (e.g. CBD, 1994; IUCN, 1994; Noss, 1996: p 574; Mangel *et al.*, 1996; Janzen, 1998; Angemeier, 2000; Parks Canada, 2000; Myers *et al.*, 2000).

From the point of view of people implementing conservation activities on the ground, e.g. protected area managers, there are at least two major difficulties with the existing diversity of approaches and goals. One is that it is unclear how much congruence there is between them. The other is that practical ways of applying these approaches or achieving these goals have yet to be worked out.

Consider, for example, the notion of maintaining biodiversity at genetic, species and ecosystem level in a large mountain protected area containing a diversity of ecosystems and landscapes and thousands of species in a range of taxa. The problem of knowing whether or not biodiversity is actually being maintained is awesome. It involves spatial scales covering a range of at least 10 to 12 orders of magnitude and temporal scales from the instantaneous to decades, generations or however far one cares

to look ahead. There will, of course, be obvious visible changes which are easy to detect but many ongoing changes will be impossible to detect. This could either be because the entity that is changing is simply not known to the protected area manager, e.g. a species population that has not been detected and may not even be know to science, or because there is simply too much going on, e.g. changes in the size, sex and age structure, and rate of growth in the hundreds or thousands of species populations that are known to be present.

At the level of ecosystems, there is also the problem that even the best available understanding of their functioning and the role biodiversity plays in this, is inadequate. A common response to this situation is to advocate the use of adaptive management, but this in itself calls for a level of technical expertise that is not currently available in many financially-poor countries that are richly endowed with biodiversity. The difficulty of applying the ecosystem approach has evoked from the Director of the US Fish and Wildlife Service a call for guidance from scientists in identifying the biological goals and objectives that the Service should be striving for (Clark, 1999). If it is difficult for the US Fish and Wildlife Service to apply the ecosystem approach, what is it like for the conservation agencies of poor, developing countries?

The conceptual and practical difficulties of applying the approaches to biodiversity conservation that are now being advocated are an unwelcome addition to the already daunting problems faced by protected area managers. An approach to protected area management that addresses this additional difficulty is outlined below.

MANAGEMENT OF MOUNTAIN PROTECTED AREAS

An approach to effective protected area management [PAM] in mountains and other ecosystems could have the following elements:

- A set of strategic objectives;
- An assessment of the conservation value of each protected area;

- A participatory management planning process which produces a management plan with achievable objectives;
- Implementation of management plans;
- Monitoring of plan implementation; and
- Periodic review of management effectiveness and revision of management objectives as required.

Each of these elements is elaborated below.

A set of strategic objectives

The existence of a World Conservation Strategy and numerous National Conservation Strategies establishes that strategic thinking is nothing new in conservation, but what is advocated here is something specific to protected area management (PAM). A set of strategic objectives for PAM would certainly be valuable at national level to accommodate unique national circumstances and to facilitate consistent values and approaches in different protected areas, but whether it would be helpful at a global level is less certain, although there are indications that it could well be. For instance, in the context of assessing management effectiveness, it has been suggested that although a diversity of approaches will be needed because of the diversity of circumstances in which PAM is implemented, these should be derived from a single, broad conceptual framework (Hocking and Phillips, 1999). In the context of sustainable forest management, a set of generic principles, criteria, indicators and verifiers of sustainable forest management has been developed (CIFOR, 1999) which is, in effect, a global template which can be adapted to accommodate local circumstances. A set of global strategic objectives for PAM could serve a similar purpose and also provide the broad conceptual framework proposed by Hocking and Phillips (1999).

For the purpose of articulating strategic objectives, the approach developed by Keeney (1994) seems particularly useful because it is explicitly based on values, which are a core component of conservation thinking (Roebuck and Phifer, 1999). We use values to make judgements about the actual or potential consequences of action and inaction, of proposed alternatives, and of decisions. Questions of value arise in relation to ethics, desired traits, characteristics of consequences that matter, guidelines for action, priorities, value tradeoffs, and attitude to risk.

The function of strategic objectives is to propagate strategic values, so a logical starting point for the development of strategic objectives is the clear identification and articulation of strategic values. In the context of PAM, strategic values would of course include the biodiversity values that will be the focus of conservation efforts. Time and effort expended on faithfully capturing strategic values in clearly stated strategic objectives is a good investment. Well thought out and articulated strategic objectives should be stable for years and can provide stable points of reference to guide management planning for a long time. They also provide a sound place to start thinking in situations where it is difficult to know where to begin (Keeney, 1994). At the level of strategic objectives, it would be possible to reflect national concerns and commitments on issues such as poverty alleviation, which is a major issue wherever there is heavy dependence on natural resources, as in Africa (Ntiamoa-Baidu et al., 2000).

An assessment of conservation values

The reason why a given area receives protected status is that it has particular conservation value or values. A diversity of values have motivated the establishment of mountain protected areas, including aesthetic, cultural, ecological, economic, educational, recreational, scientific, spiritual and world heritage value through the possession of attributes such as spectacular or unique landscape features. Protection of mountain ecosystems may also have strong practical motives, such as maintaining water supplies for large cities, or preventing downstream damage from flooding or siltation.

Spiritual and cultural values have often been the primary motives for the protection of mountains by traditional societies, and it is believed that: 'the reverence that mountains tend to awaken in cultures around the world provides a powerful source of motivation for preserving biodiversity' (Bernbaum, 1999). This point has wide implications for conservation in general, because 'as a modern scientific concept, biodiversity may have little or no meaning in traditional societies.' Traditional relationships with the natural world 'based on what other species, including plants, have in common with people, may actually provide a more meaningful and sustainable basis for biodiversity conservation than the abstract rationales given by science' (Bernbaum, 1999). The maintenance of spiritual and cultural values requires explicit sensitivity to them and to the interests and concerns of those who cherish them (Hamilton, 2000; Bernbaum, 2000).

In mountains, and all other environments, a detailed assessment and documentation of values and existing or potential threats to them provides a necessary basis for management planning and can also provide benchmarks for future monitoring. Such assessments often include detailed descriptions of the biophysical attributes of the area and an account of how it has been impacted by human activities. Copious documentation of the status of the area can be a valuable resource for future reference.

Participatory management planning

In addition to their staff, protected areas often have local people living in or around them, and they and other groups, such as traditional and local authorities and local and international NGOs, have legitimate interests in how the area is managed. It is now generally accepted that all such interest groups should be involved in the process of management planning. Where appropriate, one or more of the interest groups may also be involved in the implementation of management plans. One of the circumstances in which involvement of local people in implementing management activities may be appropriate, or even essential, is when they provide the only means for overcoming resource constraints on the management of the area.

Management objectives for an individual protected area will seldom, if ever, be identical to those in the set of national strategic objectives for PAM, but they should be consistent with the strategic objectives and should contribute to their fulfilment. The linkage of management and strategic objectives is a mechanism for propagating strategic values.

In mountain protected areas, management objectives must make provision for the physical fragility of mountains, for the slow regeneration of vegetation, and for the interests of distant downstream populations whose welfare depends on rivers originating in the mountains. An important role that management objectives can play in any environment is to provide measures of conservation success. For instance, to give a very simple example, assume that one management objective is to maintain a given number of hectares of undisturbed primary forest. Then the measured number of hectares of undisturbed primary forest at a given time in the future is a quantitative measure of conservation success. If a management objective lacks obvious attributes that can be measured, then it is often possible to decompose it into subobjectives, or subsubobjects and so on, which may have measurable attributes.

As well as being consistent with strategic objectives, management objectives must be achievable. What is achievable depends on what financial and human resources are available for management. This means that, in poor countries, what can be achieved is very modest and that in these countries, for the foreseeable future, it will be impossible to have direct knowledge of whether or not biodiversity is being maintained at genetic, species and ecosystem levels. However, it will still be possible to implement relatively simple activities that contribute to the maintenance of biodiversity. For example, it does not necessarily take sophisticated science to identify and reduce threats to biodiversity, but this does contribute meaningfully to its conservation.

The question of management capacity in poor countries is strongly linked to the question of how protected area management should be financed in these countries. The idea that the conservation of globally significant biodiversity should be paid for by the global community is receiving increased attention and support. It is unlikely to have an immediate impact, but in the medium to long term it could result in a substantial enhancement of management capacity in poor countries.

Implementation of management plans

A management plan obviously has little or no meaning unless it is implemented, so this is an issue where management capacity is critical. It is also an area where there is a payoff for being realistic. Even if capacity is severely limited it is still possible to identify modest objectives which will make a contribution to biodiversity conservation. For example, simply having people on the ground at the site and patrolling the area on foot can discourage and reduce activities that would otherwise have a detrimental effect on the protected area.

Other important issues here are who is involved in implementation and who controls it. In practice, arrangements cover a spectrum from sole control and implementation by the protected area authority, to collaborative management involving a large number of interest groups, to full control and implementation by communities. What option is appropriate tends to be site-specific, but a relevant consideration is that there is a substantial body of opinion to the effect that ultimate control should always be in the hands of the protected area agency, even if there is substantial involvement by other interest groups in the implementation of management objectives.

Monitoring

The results of monitoring management activities provide the basis for the final stage of the management cycle – assessment of management effectiveness and review and revision of management objectives. To serve this purpose, monitoring needs to provide information on the extent to which objectives have been implemented, and the level of agreement between expected and actual results. The nature and timing of monitoring activities depend both on what is being monitored and what resources are available for the exercise. Monitoring should clearly be an important issue in the management planning phase because, with intelligent design, for instance by including reporting procedures in implementation activities, much can be done to enhance its cost effectiveness.

Assessment of management effectiveness and review and revision of management objectives

The assessment of management effectiveness has recently received considerable attention (Dudley et al., 1999) and a wide variety of approaches is emerging. These range from what is effectively an internal review by a protected area agency, such as that described by Singh (1999), to assessments implemented on a participatory basis, as described by Courrau (1999).

An approach that may be of particular interest in countries where resources are severely constrained has been developed in the United Kingdom (Alexander and Rowell, 1999). In this system, management and monitoring are tightly coupled. Its development was driven by the need to comply with reporting obligations for a very large number of sites. Two aspects of the system's design were that: only those features which were the reasons for the site being selected would be monitored, and that a strict definition of monitoring was used which involved making observations against a standard for each feature, and categorising the feature accordingly. The standard would be developed individually for each feature on each site, in the form of a conservation objective within the context of a management plan. This meant that any feature could be categorised as either meeting its conservation objective (favourable condition) or not (unfavourable). If the condition of a feature is categorised as unfavourable, this signals the need for recovery management. The system thus facilitates both reporting and management. It is a simple matter to translate this approach into the language of PAM. 'Features' become the conservation values of a protected area. For each conservation value in the protected area 'favourable condition' becomes 'conservation value being maintained', while 'unfavourable' becomes 'conservation value not being maintained', which signals the need for a management response.

In terms of the approach to effective PAM being proposed here, the assessment of management effectiveness would be based on the results of monitoring. The assessment would be done by the interest group(s) involved in the management planning process. On the basis of the assessment of management effectiveness, management objectives would be reviewed and then revised by amendment or augmentation as required.

CONCLUSIONS

One of the few things that is certain about PAM in the twenty-first century is that it will have to contend with a daunting array of problems. The efforts that are currently being made to develop approaches to assess and improve the effectiveness of management in protected area are thus timely and essential. They would be facilitated and enhanced if greater conceptual clarity could be achieved about the biological goals of conservation and the practical ways in which these goals could be achieved. What this means for protected areas is that 'there needs to be a substantial rethinking about parks and about what can realistically be expected of efforts to manage and protect them. This will require a tremendous amount of focus and conceptual clarity' (Brandon *et al.*, 1998).

References

Alexander M and Rowell TA (1999). Recent developments in management planning and monitoring on protected sites in the UK. *Parks*, 9(2):50–5

Angermeier PM (2000). The Natural Imperative for Biological Conservation. *Conservation Biology*, 14(2):373–81

Bernbaum E (1999). Mountains: The Heights of Biodiversity. In Posey DA (ed.), *Cultural and Spiritual Values of Biodiversity*. UNEP, Nairobi, pp 325–44

Bernbaum E (2000). The Cultural and spiritual significance of mountains as a basis for the development of interpretative and educational materials at national Parks. *Journal of Parks*, 10(2):30–4

Brandon K, Redford KH and Sanderson SE (1998). Introduction. In Brandon K, Redford KH and Sanderson SE (eds), *Parks in Peril*. The Nature Conservancy/Island Press, Washington DC; Covelo, California, pp 1–23

CBD (1994). Convention on Biological Diversity Secretariat. United Nations, Environment Programme, Montreal, Canada. http://www.biodiv.org

CIFOR (1999). *The CIFOR Criteria and Indicators Generic Template*. CIFOR, Jakarta

Clark JR (1999). The Ecosystem Approach from a Practical Point of View. *Conservation Biology*, 13(3):679–81

Courrau J (1999). Monitoring protected area management in Central America: a regional approach. *Parks*, 9(2):50–5

Dudley N, Hockings M and Stolton S (1999). Measuring the Effectiveness of Protected Area Management. In Stolton S and Dudley N (eds), *Partnerships for Protection*. Earthscan, London, pp 249–57

Freese CH (1998). *Wild Species as Commodities*. Island Press, Covelo, CA

Hamilton LS (2000). Some guidelines for managing mountain protected areas having spiritual or cultural significance. *Parks*, 10(2):26–9

Hocking M and Phillips A (1999). How well are we doing – some thoughts on the effectiveness of protected areas. *Parks*, 9(2):5–14

IUCN (1994). *Guidelines for Protected Area Management Categories*. IUCN, Gland, Switzerland

Janzen D (1998). Gardenification of Wildland Nature and the Human Footprint. *Science*, 279:1312–13

Keeney RL (1994). *Value-focused Thinking – A Path to Creative Decision making*. Harvard University Press, Cambridge, MA

Mangel M *et al.* (1996). Principles for the Conservation of Wild Living Resources. *Ecological Applications*, 6(2):338–62

Myers N, Mittermeier RA, Mittermeier CG, da Fonseca GAB and Kent J (2000). Biodiversity

Hotspots for conservation priorities. *Nature*, **403**:853–8

Noss RF (1999). Conservation of Biodiversity at the landscape scale. In Szaro RC and Johnson DW (eds), *Biodiversity in managed landscapes: Theory and practice*. Oxford University Press, New York, pp 574–89

Ntiamoa-Baidu Y, Zéba S, Gamassa D-GM and Bonnéhin L (2000). *Principles in practice: Staff observations of conservation projects in Africa.* Biodiversity Support Program, Washington DC

Parks Canada (2000) *Report of the Panel on the Ecological Integrity of Canada's National Parks.* *http://parkscanada.pch.gc.ca/EI-IE/index_e.htm*

Roebuck P and Phifer P (1999). The Persistence of Positivism in Conservation Biology. *Conservation Biology*, **13**(2):444–6

Salafsky N, Cordes B, Parks J and Hochman C (1999). *Evaluating Linkages Between Business, The Environment and Local Communities.* Biodiversity Support Program, Washington DC

Salafsky N and Margolouis R (1999). Threat Reduction Assessment: a Practical and Cost-Effective Approach to Evaluating Conservation and Development Projects. *Conservation Biology*, **13**(4):830–41

Singh S (1999). Assessing management effectiveness of wildlife protected areas in India. *Parks*, **9**(2):34–49

Stolton S and Dudley N (1999). A preliminary survey of management status and threats in forest protected areas. *Parks*, **9**(2):27–33

UNEP/CBD (1994). *Convention on Biological Diversity.* UNEP/CBD/94/1, Gland, Switzerland

313

National Action Plans for Mountain Biodiversity Conservation and Research

27

Manab Chakraborty

IMPORTANCE OF MOUNTAIN BIODIVERSITY

Mountains cover 24% of the Earth's land surface, are home to one tenth of the world's poorest people, and account for as much as 80% of humanity's fresh water. About a third of the 785 million hectares of designated protected areas worldwide are in mountain areas. Mountains shelter half of the 90 000 species of higher plants in the neotropics alone, yet their cultural heritage and rich resources are threatened by grazing, mining, poaching and ill-conceived infrastructure projects.

Mountain areas, due to their inaccessibility, are the last refuge of many endangered and rare birds, butterflies, plants, and animals. In the tropics, mountains contain a greater concentration of genetic resources than lowland forest ecosystems with typically higher levels of endemic species. Mountains are also often rich in endemics, in both tropical and temperate areas. Levels of plant endemism are low in high latitude montane areas which were once heavily glaciated (UNEP, 1995: p 181).

NATIONAL BIODIVERSITY STRATEGIES AND ACTION PLANS (NBSAP)

This Chapter discusses research priorities emerging from national biodiversity strategies and action plans which more than 117 countries around the world are undertaking to meet their obligations under the Convention on Biological Diversity (CBD). CBD is one of the most important international legal instruments for the conservation of biodiversity and sustainable use of its components. In the Convention, particular consideration is to be given to the special situation of developing countries, including those that are environmentally most vulnerable, such as those with mountainous areas. According to Decision II/17 of the second meeting of the Conference of the Parties of the CBD, all the parties agreed to prepare assessments of national biodiversity, set priorities for action, and report to the subsequent meeting of the COP. These overviews based on national scientific consensus, are contained in the NBSAP.

PRIORITISATION FOR CONSERVATION

A major problem faced at the national and international level is what elements of biodiversity should be conserved and what methodology should be used for ranking different priorities. Broadly speaking, priority setting process for NBSAP in developing countries consisted of three steps:

(1) The creation and updating of biodiversity inventories;
(2) The selection of specific components of biodiversity to be targeted for protection, based on inventory data; and
(3) The development of action plans for protecting the selected component of biodiversity.

As described by Perlman and Adelson (1997: p 143), the priority setting process embodies one or more of four distinct central goals:

(1) Protecting *all* biodiversity;
(2) Protecting *important species*;
(3) Protecting the *most diverse subset* of diversity;
(4) Protecting the *most valuable* biodiversity.

Since all of the world's current conservation resources, both human and financial, are insufficient to deal with the number of species that are seriously threatened, 'protect all' is not a feasible option. Protecting all 'important' species assumes that there is a way to measure biodiversity precisely and consistently. One way to determine priorities is to use species richness and allied measures of ecological diversity. Used in conjunction with other criteria, such as integrity, endemism, rarity, and taxonomic isolation, species richness yields rough and ready priorities. An extension of the species richness as a priority setting criterion is the 'megadiversity country' concept that considers degree of threat, endemism and presence of charismatic mammals (World Resources Institute, 1996–1997). A variation of this method is protecting representative samples of the biodiversity found in a region. This method emphasises protecting a representative sample of biodiversity available regionally and that succeeding conservation efforts should protect as much of the as-yet unprotected biodiversity as possible (Pressey *et al.*, 1993: p 125). The advocates of protecting the most diverse subset of biodiversity argue that, if protection of all of biodiversity is not feasible, then we should protect as diverse a subset as possible. Confronted with constraints allowing only protection for a small proportion of the biodiversity, these procedures drawn from plant systematics suggest methodologies for deciding which subset to protect. The main weakness of this method is its attempt to identify the 'most diverse subset of biodiversity' by using a single measure, which is rather limiting.

Mittermeier *et al.* (2000: p 27) have suggested that, though biodiversity of each and every nation is critically important to that nation's survival, to achieve maximum impact with limited resources, heavy concentration should be placed on those areas rich in diversity

and yet severely threatened, i.e. 'hotspots'. Their list of some of the high biodiversity areas include: tropical rain forests, Mediterranean-type ecosystems, and mountain chains in central and southeast Asia. According to the authors, biodiversity priorities must be based on actual data, first and foremost on species diversity and endemism, on phyletic diversity, on ecosystem diversity, and subsequently on degree of threat.

The NBSAP preparation process has largely relied on identifying 'most valuable' biodiversity, determined by key stakeholders and important decision makers. These consultations encompass a broad range of values and preferences articulated by groups manifesting their cultural, economic, social and political background and interests. The consultative process have three distinct advantages over the other three central goal-setting methods:

(a) It allows a trade-off among priorities for allocation of scarce human and financial resources;
(b) It benefits from the accumulated knowledge and information of a cross section of people and disciplines. No single individual or organisation can express the complete range of values that humans hold toward components of biodiversity or identify all types of worth found in the biodiversity in the region.
(c) It achieves a better fit between conservation goals and conservation actions when priority setting is carried out in a participatory process.

CURRENT RESEARCH AND NATIONAL BIODIVERSITY PLANNING

In order to appreciate what might be the best approach for research in the future, it is necessary to elucidate what kind of research would be useful for national biodiversity planning. At the outset, mountain research must be designed to address the utilitarian, social, cultural and environmental needs of mountain people. From this point of view, mountain research is seen as a goal-oriented process that supports the mountain communities by providing information, knowledge and tools that improve the quality and the effectiveness of the development, maintenance and implementation of biodiversity projects and programmes. The underlying principles for such research are:

- **Application-oriented**: Research priorities are set in consultation with mountain peoples and designed to meet the needs according their priorities.
- **Economically feasible**: Given the high costs and inherent uncertainties in research, research projects should be cost-effective, pragmatic and time-bound.
- **Scientifically sound**: Research activities should be based on scientifically sound approaches and should preferably be multidisciplinary in character.
- **Consistent with national interest and priorities.**
- Add to existing national and local research capacity and knowledge.

Current mountain research is plagued by four main deficits relating to information, technology, focus and finance. These deficits inhibit full realisation of research benefits to flow to mountain communities.

Information deficit

Information regarding mountains is highly concentrated in developed country institutions and capitals within developing countries. The cost of accessing such information is prohibitive to the mountain communities. Generation of knowledge, its dissemination, and practical application should be part of research design.

Some areas are grossly under researched. The major problem is that current research so far is based on existing data, obviously discriminating against regions, which have little or scanty data available (Ch Körner pers com., 2000). Information on globally important conservation sites is scarce and mostly contained in 'grey' literature. It is expected that GEF

emphasis on in situ protection of systems of conservation areas in Mesoamerican, Andean, East African, and Himalayan regions (including the Hindu Kush – Karakoram – Pamir – Tien Shan range), and the montane regions of the Indochinese peninsula, as well as mountain chains on tropical islands, would partially alleviate this deficit.

Technology deficit

In mountain areas, technology more often comes from the plains, rather than moving laterally from contiguous regions with comparable physical and social attributes. For example, agricultural implements for hoeing and ploughing in Garhwal (a mountain area) are the same as those used in the plains because alternative, more appropriate tools does not attract organised distribution. Crop exchange within mountain ecoregions is limited due to lack of institutions with strong rural extension services. Besides, there are technical and knowledge barriers which inhibit crop exchange. Grains such as quinoa (*Chenopodium quinoa*) and amaranth (*Amaranthus caudatus*); tubers such as oca (*Oxalis tuberosa*) and olluco (*Ullucus tuberosus*), and roots like arracha (*Arracacia xanthorriza*) and yacon (*Polymnia sonchifolia*) are some examples of introduced crops from the Andes to the Himalayas. It seems that the African highlands could also benefit from the introduction of Andean grains, especially to alleviate nutritional deficiencies.

Focus deficit

The primary focus of conservation has been protected areas. A preliminary inventory of Tropical Mountain Cloud Forest (TMCF) sites compiled by the World Conservation Monitoring Centre identified a total of 605 TMCF sites in 44 countries (WCMC, 1997). This draft inventory recorded 44% of TMCF sites as having some form of protection status consistent with IUCN management categories I–VI, although in reality many of these sites designated as protected areas are threatened and fragmented. For example, mining is damaging the Podocarpus National Park in Ecuador and the Jaua-Sarisarinama National Park. The protected forests of Mount Kenya, Amboselli and Nakuru are severely threatened by agricultural encroachment and logging. The issue is to identify carefully under what ecological, social and institutional conditions protected area systems might yield desired conservation and where it might not.

Protected areas also draw a major portion of conservation resources. Yet most biodiversity remain outside protected areas. There has been increasing demand for rewards to be given to local communities in recognition of their services in conservation. However, should such rewards be at the cost of reduction of investment in protected areas, then the economic and institutional implications need to be better understood.

Finance deficit

Undoubtedly, mountains deserve more attention and resources. As a result of intensive biodiversity planning processes, many countries such as India, Nepal, Peru, China and the Philippines have earmarked special funds for mountain areas. At the global level, the GEF allocated US$ 79 million in the year 2000. The financial resources required over the period of the first three years of the GEF Operational Strategy, adopted by the GEF Council in October 1995, are estimated to range from US$ 85 to 100 million. However, actual available finance has been much lower. Cumulative GEF allocation for projects in the mountain ecosystems operational programme (1992–1999) stood at US$ 64.1 million i.e. approximately 17% of the entire biodiversity portfolio (GEF, 1999, Annex B, para 10). Allocation of funding to mountain areas is the least of the four biodiversity focal areas. This is partly due to poor uptake, slow absorptive capacity, and lack of quality proposals.

PRIORITIES FOR RESEARCH

What should be the priorities for targeted research? From the perspective of national biodiversity planning, and in accordance with principles outlined above, a number of categories of activities may be identified.

Inventory, assessment and monitoring

The NBSAP included preparation of national biodiversity inventories. Biodiversity inventorying refers to surveying, sorting, cataloguing, quantifying and mapping of entities such as genes, individuals, populations, species, habitats, biotopes, ecosystems and landscapes or their components, and the synthesis of the resulting information for the analysis of patterns and processes. Monitoring consists of repeated inventorying over time and space and hence it measures change (UNEP, 1995: p 457). Though inventories are required at the genetic, species and ecosystem level, in reality rarely does genetic diversity appear in an inventory, and frequently ecosystem diversity is omitted as well. Considering diversity of species only, most taxonomic groups are filtered out of inventories, in particular invertebrates, microorganisms and fungi only rarely appear in inventories. Also, fewer that 5% of the protected sites have been thoroughly inventoried for even one major group of their biota, and this has restricted their value for the understanding and conservation of biodiversity (UNEP, 1995: p 549).

The value of long-term monitoring has been well-demonstrated (Pollard and Yates, 1993; Woiwod, 1991). It not only helps to track declines but also to 'see how we are doing' when undertaking successive environmental interventions. Yet, in many parts of the world, institutional weaknesses, financial constraints and lack of trained human resources, would not permit such repeated inventories. In 28 countries where UNEP has assisted developing countries to undertake national biodiversity inventories, the average cost has ranged between US$ 100 000 to about 150 000. Therefore, research on rapid assessment, rapid inventory preparation techniques and baseline calibration procedures is of utmost importance.

Capacity building for biodiversity planning

A great deal of the effort of international environmental organisations such as UNEP is directed towards building capacities of developing countries to manage their biological assets. Capacity building refers to augmenting the ability of individuals, institutions and society as a whole to make and implement decisions and perform functions in an effective, efficient and sustainable manner. At the individual level, capacity building encompasses the process of changing attitudes and practices, and developing skills, while increasing the benefits of participation, knowledge exchange and ownership. At the institutional level, it focuses on the overall organisational performance, functioning capabilities and the ability to deliver results as per changed expectations. Within the constraints of public policy frameworks and societal norms, individuals and organisations operate and develop formal and informal relationships which ultimately affect the degree to which capacities are utilised.

The NBSAP process has largely focused on identifying capacity building needs at the national and institutional level, and has suggested how these might be addressed through action plans. Wide differences exists among countries due to their stage of economic development, administrative traditions, and ecological characteristics. However, the major needs could be summarized as below:

- Assessment of overall policy, regulatory and legislative frameworks governing the conservation of biodiversity particularly in agriculture, forestry and fisheries sectors.
- Strengthening the competencies of key national institutions with the introduction of new resources (financial, human and information) and providing adequate infrastructure.
- Broadening and updating the skills and expertise of the individuals and institutions involved in biodiversity planning.
- Facilitating access to technical tools and scientific models for integrated assessments.
- Increasing access to data and indicators for use in local and national assessments.
- Developing and disseminating new approaches for linking local level expertise and assessments with national, regional and global expertise and assessments.
- Enhanced use of a clearing house mechanism for rapid dissemination of information and networking at the local, national and regional scale.
- Widening participation of all stakeholders in the distribution of economic benefits of biodiversity conservation.
- Higher international financial and technical support for long term projects.

Management actions within conservation areas

If protected areas are to become more effective in maintaining biodiversity, serious obstacles must be overcome. The obstacles include:

- Inadequate biogeographic distribution. Ninety-four percent of the 605 TMCF sites compiled by the WCMC were found in 27 countries, 46% of the total were found in 12 countries of Latin America.
- Tension and open conflicts with local populations, pastoralists and farm communities.
- Ineffective management and inadequate funding.
- Lack of preparedness of plans against natural hazards and risks.

319

- Limited understanding of potential roles in contributing to development of national economies.

Individual biodiversity conservation objectives need to be established for each protected area and, in most cases, they need to be better integrated into the social, environmental and economic welfare structures. Research priorities could include:

- Ways to identify site-specific root and proximate causes of biodiversity loss (e.g. deforestation, desertification, trade or the introduction of invasive species);
- Use of geographical information systems and participatory planning tools in designing and managing protected areas and their buffer zones;
- Ways to enhance the national revenue base for conservation areas; and
- Incentives for training and retention of necessary technical and institutional capacity at the site level.

Development of sustainable use alternatives

Wells *et al.* (1999) found very little substantive involvement in NBSAP preparation of the key agencies responsible for land use decision making in agriculture, forestry, mining, transportation, energy or other areas. The main reasons for non-involvement identified by the study are:

(1) Lack of biodiversity knowledge and awareness outside the traditional biodiversity constituency,
(2) Institutional arrangements which do not encourage biodiversity or other environmental concerns to be taken into consideration by decision makers,
(3) Lack of methodologies or guidelines for incorporating biodiversity into other sectors in ways that are meaningful to planners, and, most seriously,
(4) An unwillingness to identify and start to address the real and politically difficult

tradeoffs which will be involved within countries if current rates of biodiversity loss are to be reduced.

Key international issues, such as poverty eradication, structural adjustment policies, trade liberalisation, privatisation, modification of export and import policies of other countries, are generally ignored. Hence, the research priority could be to explore ways to integrate biodiversity considerations into sectoral developments in the buffer zones and influence zones of key habitats; as well as inventory analysis, documentation of indigenous knowledge, protection of collective property rights and dissemination of information about local sustainable use practices.

Applied policy research

There is recognition that policy should be devised in a context that provides understanding of the root and proximate causes driving biodiversity loss (Stedman-Edward, 1998). Identification of entry points of intervention strategies to reduce threats to biodiversity requires critical analysis that distinguishes symptoms from root causes. Additional research is required to provide methods and information that establishes a basis for long term assessment of implications of intervening in complex systems to counteract biodiversity loss. This will entail a long-term commitment to provide support for policy-relevant research, identification of interventions, and allowing evaluation of the impact of intervention strategies over time on both socio-economic and environmental conditions. The pros and cons of various new methodologies and their policy significance could be the subject for further research. These include:

- Use of economic and social incentives,
- Economic valuation of biological resources,
- Natural resource accounting,
- Technology transfer, adaptation and absorption, and
- Ecosystem approaches.

Conflict resolution

In recent years, mountain areas are increasingly the site of social conflict, ethnic strife and war, resulting from a combination of very complex issues. Conflict in mountain areas can be caused by diverse factors such as poverty, ethnic tension, competition for scarce resources and even the geographical isolation of mountain areas which makes them likely refuges for armed groups. Efforts to carry out conservation and sustainable development initiatives are often hindered by these conditions. The issue of security and conflict resolution is very relevant to efforts to achieve sustainable approaches to mountain development and conservation. One possible way to promote inter-state cooperation is through transboundary projects, viz. the Transcarpathian Biodiversity Protection project, in which Poland, Slovakia and Ukraine cooperate. Professional networks such as the Mountain Protected Area Network of the World Conservation Union (IUCN), the Mountain Forum, and DIVERSITAS could potentially play a role in promoting peace and dialogue among states in conflict.

CONCLUSION

In order to be effective, mountain research must address the needs of mountain communities as well as conservation. Research priorities could be designed in such a way that they reduce the threats to biodiversity at the national level and, at the same time, increase the contribution to economic development and welfare. From the national biodiversity planning point of view, it is suggested that the research priorities should assist in improved and cost-effective inventorisation and monitoring; strengthening of protected area networks; promotion of the sustainable use dimension in conservation; while striving to be policy-relevant and to reduce potential conflict at local and inter-state level by generating better options for resource management.

References

Bandyopadhay J (1999). *Development of U.P. Himalayas, (mimeo)*, United Nations Development Programme, New Delhi

Global Environment Facility (1999). *Draft Report of the GEF to the Fifth Meeting of the Conference of the Parties to the Convention on Biological Diversity*. Global Environment Facility, GEF/C.14/13

Ministry of Nature and the Environment (1998). *Biological Diversity in Mongolia – First National Report*. Ministry for Nature and the Environment of Mongolia

Ministry of Forests and Soil Conservation (1997). *National Report on Implementation of the Convention on Biological Diversity in Nepal*. Ministry of Forests and Soil Conservation, Nepal

Mittermeier RA, Myers N and Mittermeier CG (2000). *Hotspots: Earth's Biologically Richest and Most Endangered Terrestrial Ecoregions*, Conservation International, USA, 27 pp

Perlman DL and Adelson G (1997). *Exploring Values and Priorities in Conservation*. Blackwell Science, UK

Pollard E and Yates TJ (1993). *Monitoring Butterflies for Ecology and Conservation*. Chapman and Hall, London

Pressey RL, Humphries CJ, Margules CR, Vane-Wright RI and Williams PH (1993). Beyond Opportunisms: Key Principles for Systematic Reserve Selection. *Trends in Ecology and Evolution*, **8**:124–8

Stedman-Edward P (1998). *Root Causes of Biodiversity Loss – An Analytical Approach*. World Wide Fund for Nature, Gland

United Nations Environment Programme (1995). *Global Biodiversity Assessment*. Cambridge University Press, Cambridge

Wells M *et al.*(1999). *An Interim Assessment of Biodiversity Enabling Activities – National Biodiversity Strategies and Action Plans. A Study for the Global Environment Facility*. Global Environment Facility, Washington DC

Wiesmann U (1999). *Introduction to the Report on Mountains of the World Tourism and Sustainable Mountain Development,* presented at CSD-7, Side

Event on Mountains (available at http://www2.mtnforum.org/mtnforum)

Woiwod IP (1991). The ecological importance of long-term synoptic monitoring. In Firbank LG, Carter N, Darbyshire JF and Potts GR (eds), *The Ecology of Temperate Cereal Fields*. Blackwell Scientific Publications, Oxford, pp 275–304

World Conservation Monitoring Centre (1997). *Global Biodiversity* – Status of the Earth's Living Resources. Chapman & Hall, London

World Resources Institute (1997). *World Resources 1996–97*. Oxford University Press, New York

Part V

Synthesis

A Global Assessment of Mountain Biodiversity: Synthesis

Eva M. Spehn, Bruno Messerli and Christian Körner

INTRODUCTION

The compression of climatic zones along elevational gradients causes mountain biota to represent hot spots of biological richness. At very high elevations, biodiversity diminishes gradually but so does land area, leading to very high biodiversity/land area ratios which often exceed those of lower elevations (Körner, Chapter 1).

The causes of high biological diversity are manifold. Mountain terrain is commonly highly fragmented and topographically diverse and this high geodiversity is strongly related to biological diversity, as it reflects the multitude of life conditions in a given area (Körner, 1999). Patterns of snow distribution reinforce landscape diversity by influencing soils, length of growing season, and microclimate. Selective microenvironments, e.g. due to insufficient or presistent snow cover, constrain species richness in alpine plant communities. In the European Alps communities with moderate snow cover are therefore richer in species than very strongly exposed communities or snowbed sites (Virtanen *et al.*, Chapter 7). Other causes of high biological richness in mountains are the small size of organisms, geographical isolation combined with effective genetic systems ensuring genetically variable populations (e.g. by polyploidy and a high degree of self-incompatibility), and moderate disturbance (Chapters 1 and 2).

Intraspecific genetic diversity in alpine and montane plant species seems to be mainly determined by the reproductive system that a species employs, overriding any specific effect of the alpine environment, as shown in a review by Till-Bottraud and Gaudeul (Chapter 2). There is a high degree of ecotypic differentiation with elevation. High elevation populations within one species exhibit distinct genetic differences from low elevation populations (e.g. the well known elevational provenance effects in forest trees). A comparison by Beall (Chapter 16) of indigenous populations of Andean and Tibetan high-altitude natives suggests that such natural selection has acted even on humans with different traits relevant for offsetting high-altitude hypoxia selected in two, geographically distant populations adapting to the same environmental stress. This raises the possibility of similar diversity within other species that have colonized more than one mountain ecosystem.

HOW MUCH BIODIVERSITY IS THERE?: REGIONAL ASSESSMENTS

The Rocky Mountains

The Rocky Mountains have higher community and landscape diversity than surrounding lowlands. Differences in plant diversity in the Rocky Mountain alpine belt is the product of historical factors associated with climate change and biogeographic isolation, as well as probable influences of regional climatic differences and variation in parent material. Circumpolar species contribute more to the floras of the northern part of the Rockies in comparison to the southern part, possibly as a result of greater isolation from the Arctic flora and greater speciation in the south (Bowman and Damm, Chapter 3). Arctic species in general seem to be limited to more moist alpine sites. Species distribution in the alpine zone at six sites in the Rocky Mountains of Alberta (in the eastern ranges of the North American

28

Cordillera) along a 600-km transect suggest that a thorough mixing of Circum-Arctic, North American, Cordilleran and Beringian alpine species had taken place during cold events when the region was not glaciated. The fact that 73% of the species in the alpine zone only occur in North America suggests that considerable evolution of the species occurred on this continent when it was not connected to Eurasia (Harris, Chapter 4). Endemism is generally low throughout the Rockies relative to other alpine regions (Bowman and Damm, Chapter 3).

Alpine habitats in the Cascade and Rocky Mountains studied by Raphael *et al.* (Chapter 15) make a strong contribution to overall vertebrate diversity: Although the area covers less than 1% of a 58-million ha area of the northwestern United States, it supports a distinctive set of resident or migratory terrestrial vertebrate species which make up nearly 10% of the total vertebrate species of this area.

The Andes

Linked to large floristic databases, satellite and GIS data are used by Braun *et al.* (Chapter 6) to analyse altitudinal gradients of species and family richness. The Andean flora shows a very high beta diversity, resulting in exceptional high species numbers at a landscape level. Although the highest numbers of tree species can be found in the lowlands, shrubs show a peak at 2000–3000 m asl. In general, shrubs, herbs and epiphytes have a higher species per family ratio in the Peruvian montane flora than do trees (Chapter 6).

Kessler (Chapter 5) studied plant species richness and endemism in tropical montane vegetation in the Bolivian Andes. Along gradients of natural succession and of human disturbance, species richness peaks in mature, undisturbed forests within the montane zone. Above about 2000 m, vascular plant alpha diversity in the humid parts declines with elevation at a rate of about 10% per 100 m. At the treeline, open habitats are richer in species than adjacent forests and plant species richness varies regionally at least fourfold, mainly in relation to humidity. Natural timberline elevation in the Bolivian Andes is at 3800–4200 m,

but human activities have lowered it to 3300–3600 m, destroying over 95% of the original forest cover above 3500 m (Kessler, Chapter 5). In general, human activities caused a much higher impact on the forestline in the humid eastern Andes than in the subhumid and semiarid ranges (Braun *et al.*, Chapter 6).

Asian mountains

The Caucasus seems to be richer in species than the mountains of central Asia, when related to surface area (Agachanjanz and Breckle, Chapter 9). In mountains with a continuous forest belt, the maximum percentage of endemics is above the treeline, whereas in arid mountains which lack a forest vegetation, the maximum percentage of endemism occurs in the foothills. Wang *et al.* (Chapter 11) found in the northeastern Tibetan plateau that the reduction in montane plant species richness with latitude corresponds with a drop in annual average temperature and precipitation. The arid high mountains of the Karakorum, separating Central Asia and the Indian subcontinent, comprise considerable areas of less than 100 species per 10,000 km^2 (Dickoré and Miehe, Chapter 10). These distinct 'cold spots' of plant diversity align to sharp major phytogeographical boundaries. The Karakorum links the East Himalaya–Yunnan hot spot to other Holarctic mountain floras, and its ecological status is critical to potential genetic interchange between these floras in space and time.

European mountains

Alpine vascular plant species richness across 43 European mountain areas decreases from south to north (Virtanen *et al.*, Chapter 7). This has probably resulted from glacial extinctions which strongly depauperated the flora of the North European mountains. In contrast, bryophyte and macrolichen species richness in the alpine zone increases towards the north. Vascular plant species richness is greater on calcareous than on siliceous substrates, a difference more pronounced in the mountains of northern Europe. The richer calcareous flora may reflect a predominant evolution of alpine taxa in environments favouring calcicoly. It seems the alpine flora has better survived on calcareous substrates in the north (Virtanen *et al.*, Chapter 7). The type of substrate is also the predominant factor affecting species richness of vascular plants in the Swiss Alps (Wohlgemuth, Chapter 8), where continental regions are also richer in species than oceanic regions.

Australian and African mountains

Alpine vegetation in Australia occurs in more than 100 discrete habitat islands, from the Snowy Mountains in New South Wales down to southwestern Tasmania, that were all part of a few large habitat islands at the height of the Last Glacial, and where most species are Australian endemics. The mean geographic range for the obligate alpine floras is much larger for mountain islands on the mainland of Australia than for such mountain islands in Tasmania. The concentration of alpine local endemics on the largest of the relictual alpine areas suggests that local extinction processes are related to area, rather than to recent speciation and subspeciation (Kirkpatrick, Chapter 12).

In South Africa, the most significant mountains are the Drakensberg Range and the Cape Fold Mountains. The latter fall within one of the world's hot spots of plant diversity, the Cape Floristic Region. About 68% of the 8550 plant species in this region are endemic and a large proportion of the species is confined to the mountains (McDonald *et al.*, Chapter 21).

GLOBAL CHANGE IMPACTS

Climate change

Global warming will reduce available land area for cold-adapted organisms and therefore it will be a threat to mountain plant species richness, especially in front ranges where alpine plants are restricted to small and isolated summits (Wohlgemuth, Chapter 8). The uppermost zones or the uppermost ecotones often host unique species which are only exceptionally found in the zones below and are therefore distinguished as nival from the other alpine species (Gottfried *et al.*, Chapter 17). These nival species react quite differently to microclimatic factors such as night-time temperatures and snow-cover duration compared to species typical for alpine grasslands. As a consequence of earlier snowmelt and/or climatic warming, typical alpine species may fill the nival niche, exerting strong competitive pressure. The migration of alpine species into former nival habitats may lead to remarkable biodiversity losses (Chapter 17).

In the Central Andes, drought appears to be the most significant consequence of climatic change, as was exemplified for plant communities in the Cumbres Calchaquies of NE Argentina (Halloy, Chapter 18). Recent drought events can be related to significant losses of diversity and decreases in growth rates, using lake water levels, soil temperatures and long-lived plants as practical and inexpensive long-term recorders and indicators of the environment.

With predicted higher temperatures, longer summers with greater incidence of drought would also be expected in the Cape Fold Mountains of South Africa. This would lead to increased incidence of fires, threatening the fynbos (macchia-like) plant communities in many moist highland and marginal arid localities, and would probably allow C4 grasslands to continue dominating this region into the future. Substantial plant species replacements are possible (McDonald *et al.*, Chapter 21). Increasing summer temperatures, predicted for this century, are also likely to cause a loss of species in the Australian mountains, and a stochastic loss of species from surviving alpine islands related to increased isolation and decreased area are expected (Kirkpatrick, Chapter 12). The responses of mammals and birds to a 30% reduction in snow cover in the Snowy Mountains of Australia over the last 45 years has already caused a higher abundance of feral mammals in alpine/high subalpine areas, along with the prolonged winter presence of browsing macropods (Green and Pickering, Chapter 19). The predicted impacts of global warming on snow cover will result in a significant change in distribution of animal communities, both spatially and temporally. Large-scale changes in distributions of perma-frost, biomes and net primary production are also expected on the Tibetan Plateau due to global climate changes (Dickoré and Miehe, Chapter 10).

If regional climates continue to warm, invasions of subalpine fir (*Abies lasiocarpa*) into subalpine meadows are likely to continue, as was documented in the last century for three North American mountain ecosystems (Olympic, North Cascades and Glacier National Parks) by Fagre *et al.* (Chapter 20). In addition, these authors used models to examine future biodiversity trends under a climate scenarios of 0.5 °C warmer annual temperatures and a 30% increase in annual precipitation. Mesic tree species are predicted to invade valleys and cause a rise in treeline, a decrease in area and biomass of treeline species, and a tendency for growth of trees near the treeline to become more nitrogen-limited.

Levels of within- and among-population diversity found in alpine and montane plants are comparable to those found at lower elevations. Therefore, alpine species are not immediately at risk in the event of environmental change from the genetic viewpoint, assuming that the existent neutral variation correlates with adaptive potential. Yet, any decrease in population size might lead some populations into the extinction vortex (Till-Bottraud and Gaudeul, Chapter 2).

Glacier foreland succession is an example of a rapidly changing alpine environment (Kaufmann, Chapter 14). Invertebrates and vascular plants, during colonisation and primary succession, have shown a rapid establishment of diverse and well structured communities within 50 years.

Land-use changes

Although global climate changes can dramatically affect the distribution of plant species in the alpine zone, these changes will most likely be subsumed by heavy anthropogenic impacts, such as overgrazing and inappropriate land management in the short term. Of all global change impacts on mountain biodiversity, land use is the most important factor. It is encouraging that traditional upland grazing systems and land management have contributed to the establishment of rich biota in a sustainable way. One example is the Andringitra Massif in south-central Madagascar, which is known as a biodiversity hot spot, characterised by high local endemity (Bloesch *et al.*, Chapter 13). Traditional land use appears to be a key to the preservation of this hot spot of mountain biodiversity as it replaces former natural drivers of biodiversity, such as fire and (extinct) large herbivores. This traditional knowledge is currently either overtaken by population pressure and poverty or becomes lost as traditional land-use methods disappear (Körner, Chapter 1). However, different approaches to alleviate increasing human pressure and its consequences on biodiversity in mountain ecosystems are needed.

Sarmiento *et al.* (Chapter 24) demonstrated that the traditional long fallow system in the paramo, located in the upper belt of the Northern Andes, reduced local biodiversity and generated only a small economic income for the Andean farmers. Currently, high land-use pressure is pushing up the agricultural frontier into the pristine paramo, representing a risk to plant diversity conservation. The paramo is characterised by a very diverse flora, with many endemisms and particular adaptations to these cool tropical environments. The best theoretical alternative to retain or enhance both local biodiversity and economic profit is to conserve large areas of natural vegetation and to manage the rest with an intensive system.

In the East African mountains, the natural vegetation has almost vanished, except for a few patches. In regions such as these, human population growth and survival needs exceed land carrying capacity and biodiversity protection becomes a low priority. A four-year test with different grazing regimes, including exclosures, revealed that high stocking rate pasturing is not necessarily detrimental to species richness and ground cover (Saleem and Woldu, Chapter 23). There is also evidence that a certain level of forest use in tropical montane

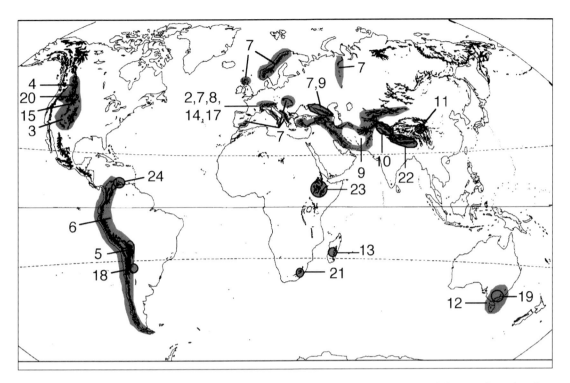

Figure 28.1 A first attempt at a global assessment of mountain biodiversity, its causes and threats. This is based on research located in the regions indicated on the map. The numbers indicate the relevant chapters in this book

forests in Bolivia is compatible with the conservation of endemic plant taxa (Kessler, Chapter 5), where endemism reaches a maximum in moderately anthropogenically disturbed forests at about 3500 m asl.

Globalization, economic liberalization and the introduction of intellectual property rights open up new opportunities as well as challenges to mountain societies in using mountain-specific natural resources. Medicinal plants are one of the most valuable resources at high altitudes. A survey of the available literature reveals that about 2500 species from the Indian subcontinent are in local medicinal use or commerce/trade, including the pharmaceutical industry (Purohit, Chapter 22). Of these, 1748 species are from the Indian Himalayan region and 44% of such plants are from subalpine and alpine zones. These are also the species with high economic returns. As this resource is more and more exploited, production and/or processing-based strategies to ensure the sustainable use of medicinal mountain plants need to be developed.

CONSERVATION POLICY

The current state of mountain biodiversity conservation in mountain regions and the future needs under global change scenarios is reviewed by Hamilton (Chapter 25). Clearly, there is a dual need for adequate systems of mountain protection as well as for sustainable land-use management regimes. Most of the formally protected areas are too small to fulfill the expected function of maintaining the native biodiversity of an area and Hamilton presents worldwide initiatives to create large mountain conservation corridors of connected areas in order to protect the biological richness of mountains. The effectiveness of mountain biodiversity conservation in protected areas may be enhanced by employing a set of strategic objectives, by assessing the conservation value of a given protected area, by following a participatory management planning and implementation process, and by monitoring the success and revising the management objectives as required (Williamson, Chapter 26).

A major problem faced at the national and international level is what elements of biodiversity should be conserved and in what order of priority. Chakraborty (Chapter 27) argues that mountain conservation research must be designed to jointly address the utilitarian, social, cultural and environmental needs of mountain people. Mountain research should provide information, knowledge and tools that improve the quality and the effectiveness of the development, maintenance and implementation of biodiversity projects and programmes.

OUTLOOK

Mountain biodiversity is perhaps the best indicator value of the integrity of mountain ecosystems (Chapin and Körner, 1995). Functional implications of high mountain biodiversity for ecosystem integrity are largely related by slope stability, the centrepiece of any mountain ecology (Figure 28.1, Körner, Chapter 1). Streams and lakes, and also water reservoirs, depend on the integrity of upslope systems, and mountain hydrology is strongly influenced by the type of vegetation and its stability. More than 50% of the world's population depends directly or indirectly on mountain resources and services, such as water supply, and thus on the integrity of these upland ecosystems (Messerli and Ives, 1997).

Major research needs

Biodiversity research is often seen as an inventory effort and we certainly need more and better (i.e. in terms of comparability) documents on the biological richness of regional mountain biota. However, it would be dangerous to wait for this to happen and to consider this as the sole task of such research. Science needs to underpin the functional significance of biodiversity with facts, on top of the general ethical, aesthetic and economic attributes. In other words:

1. Both available and future inventories of biological richness need to be analysed with quantitative methods in order to distill functional linkages to abiotic and biotic determinants.
2. We need empirical evidence to underpin the insurance hypothesis, the strongest scientific foothold supporting the need for diversity and the sustained integrity of ecosystems.
3. Ecosystem services, such as productivity of upland pastures or erosion control, need to be demonstrated and quantified, which requires experimental methods.
4. Research needs to explore management scenarios which serve both the sustained integrity of diverse mountain biota and human needs.

These four themes should become the focus of future assessments of mountain biodiversity and its ecological significance. The current scientific basis on which the benefits of diversity and the potential drawbacks of change can be judged is sparse. Much of the current debate rests on observational, plausibility-oriented and theory-based reasoning. We hope that this book, which attempts to bring together knowledge from around the world, will assist in developing a deeper and more functionally-oriented search for answers, one of the major tasks of the Global Mountain Biodiversity Assessment (GMBA) and its international networking activity.

References

Körner Ch (1999). *Alpine Plant Life*. Springer, Berlin
Chapin FS III and Körner Ch (eds) (1995). *Arctic and alpine biodiversity: Patterns, causes and ecosystem consequences*. Ecological Studies 113, Springer, Berlin, Heidelberg, New York

Messerli B and Ives JD (eds) (1997). *Mountains of the World: A global priority*. The Parthenon Publishing Group, New York, London

Index

Abruzzo National Park, Italy 301, 303
acute mountain sickness (AMS) 204–205
Africa *see* Ethiopia; Lesotho; southern Africa
Alberta, Canada 49–52
    vascular timberline flora biodiversity 49–55
        geographical ranges of species 53
        species distribution within study region 53
        species present at all sites 53–55
Alps
    alpine versus nival plant zones 213–214, 218
        climate change scenarios 219–220
        generalised niche schemes 218–219
        snow cover and 214, 217–218
        temperature and 214, 217
    see also Rotmoos glacier foreland; Swiss Alps
altitude
    genetic diversity relationship 27–29
        among-population differentiation 29–30, 31
        environmental heterogeneity 27–29
        post-glacial recolonisation 29
        selection 27
    human adaptations to hypoxia 199
        haemoglobin concentration 201–202
        oxygen saturation of haemoglobin 200–201
        physical work capacity 202–203
    human reproduction and 203
    metabolic adjustments 7
    snow cover gradients 217
    species richness patterns 5–7
        Andes 63–65, 79, 82–86, 87
        floristic drainage 120
        Karakorum Mountains 135–137, 140–142
    temperature gradients 217
    see also mountain habitats
Andes 75
    biodiversity 78, 326
        altitudinal trends 79, 82–86, 87
    forestline/treeline 79–82, 86–86
    geodiversity quantification 78–79
    GUS and remote sensing data 76–78
    see also Bolivian Andes; Cumbres Calchaquíes
        mountains, Argentina; paramo ecosystem, Venezuela
Andohariana plateau *see* Andringitra Mountains
Andringitra Mountains, Madagascar 165–173
    biota 166–167
        comparisons 170–171
    causes of forest absence on plateau 171–172
    climate 165–166
    fire effects 167, 172–173
    grazing effects 172–173
    invasion of *Philippia* 168–170, 173
        threat to biodiversity 171

land use history 167–168
    management recommendations 173
area size
    species richness and 8–9, 93–94
        Swiss Alps 105–106, 111–112
    viable carnivore populations and 298
Argentina *see* Andes; Cumbres Calchaquíes mountains
Asian mountains 117–118
    biodiversity 326
    chorology 123–125
    endemism 123–125
    life forms 122–123
    species numbers 118–122
    see also specific mountains and ranges
Australia
    alpine plant species distributions 155–163
    endemism 155, 158–159
    see also Snowy Mountains, Australia
Azorella adaptive strategy, Cumbres Callaquíes
        mountains 234

bedrock type, species richness and 94–96
    Swiss Alps 112–113
biodiversity 23, 117, 295
    African mountains 261–264, 327
        Andringitra Mountains 171–173
        Cape Fold Mountains 261–262
        Drakensberg–Maloti Mountains 262–263
        potential effects of climate change 263–264
    Andes 326
        Cumbres Calchaquíes mountains 226, 228–231
        paramo ecosystem, Venezuela 283
    Asian mountains 326
    Australian mountains 327
    conservation of *see* conservation
    current research and national planning 316–318
        finance deficit 318
        focus deficit 317–318
        information deficit 317
        technology deficit 317
    European mountains 326–327
        Rotmoos glacier foreland 180–184
    functional groups assessment 178, 181–184, 186,
        187–188
    human populations *see* human population biodiversity
    importance of 315
    insurance hypothesis 3–5, 23
    invertebrates 177–178
    medicinal plants 267–272
        Himalayan region 269–270
        potential impact of climate change 270–271
    mountain biota 5–7